THERAPEUTIC APPLICATIONS OF NITRIC OXIDE IN CANCER AND INFLAMMATORY DISORDERS

THERAPEUTIC APPLICATIONS OF NITRIC OXIDE IN CANCER AND INFLAMMATORY DISORDERS

THIRD INTERNATIONAL CONFERENCE ON THERAPEUTIC APPLICATION OF NITRIC OXIDE IN CANCER AND INFLAMMATORY DISORDERS

Edited by

VALENTINA RAPOZZI
University of Udine, Italy

LUIGI XODO
University of Udine, Italy

BENJAMIN BONAVIDA
Department of Microbiology, Immunology and Molecular Genetics, David Geffen School of Medicine, University of California, Los Angeles, United States

ELSEVIER

ACADEMIC PRESS
An imprint of Elsevier

Academic Press is an imprint of Elsevier
125 London Wall, London EC2Y 5AS, United Kingdom
525 B Street, Suite 1650, San Diego, CA 92101, United States
50 Hampshire Street, 5th Floor, Cambridge, MA 02139, United States

ISBN: 978-0-443-22354-9

For information on all Academic Press publications visit our website at https://www.elsevier.com/books-and-journals

Publisher: Stacy Masucci
Acquisitions Editor: Linda Buschman
Editorial Project Manager: Michaela Realiza
Production Project Manager: Jayadivya Saiprasad
Cover Designer: Miles Hitchen

Typeset by TNQ Technologies

Contents

Part IV

Nitric oxide and neurodegeneration

5. Nitric oxide-mediated neuroinflammatory pathways as treatment targets in neurodegeneration

Jennifer Cale, Tracy D. Farr, Sébastien Serres, and Joern R. Steinert

Part V

Nitric oxide and therapeutics

6. The role of nitric oxide in gemcitabine resistance of pancreatic cancer cells

Mariachiara Gani, Eros Di Giorgio, Raffaella Picco, Luigi E. Xodo, and Valentina Rapozzi

7. Exploring the dual role of nitric oxide in glioblastoma

Inesa Navasardyan and Benjamin Bonavida

8. Nitric oxide donating systems and their potential in shaping tumor immunoregulation

Greta Varchi

9. Control of nitric oxide synthase 2: Role of NRF2-regulated distal enhancer

Eros Di Giorgio, Valentina Rapozzi, and Luigi E. Xodo

10. Nitric oxide in tumor biology: From stemness to metabolic reprogramming

Antonio Rodríguez-Ariza

11. Impact of nitric oxide on hemeprotein maturation and its relevance to cancer and pulmonary diseases

Chaitali Ghosh, Mamta Sumi, and Arnab Ghosh

Part VI

Educational session

12. Using artificial intelligence to discover new cancer therapies

Apostolos Zaravinos

Contributors

Lorenzo Berra Department of Anesthesia, Critical Care and Pain Medicine, Massachusetts General Hospital and Harvard Medical School, Boston, MA, United States; Respiratory Care Department, Massachusetts General Hospital and Harvard Medical School, Boston, MA, United States

Benjamin Bonavida Department of Microbiology, Immunology and Molecular Genetics, David Geffen School of Medicine, Jonsson Comprehensive Cancer Center, University of California, Los Angeles, CA, United States

Jennifer Cale Division of Physiology, Pharmacology and Neuroscience, School of Life Sciences, University of Nottingham, Nottingham, United Kingdom

Mattias Carlström Department of Physiology and Pharmacology, Karolinska Institute, Stockholm, Sweden

Valerio Ciccone Department of Life Sciences, University of Siena, Siena, Italy; Fondazione Veronesi, Milano, Italy

Eros Di Giorgio Department of Medicine, Laboratory of Biochemistry, University of Udine, Udine, Italy

Tracy D. Farr Division of Physiology, Pharmacology and Neuroscience, School of Life Sciences, University of Nottingham, Nottingham, United Kingdom

Giuseppe Filomeni Department of Biology, University of Rome "Tor Vergata", Rome, Italy; Redox Biology, Danish Cancer Institute, Copenhagen, Denmark

Mariachiara Gani Department of Medicine, University of Udine, Udine, Italy

Chaitali Ghosh Department of Biomedical Engineering, Lerner Research Institute, The Cleveland Clinic, Cleveland, OH, United States

Arnab Ghosh Department of Inflammation and Immunity, Lerner Research Institute, The Cleveland Clinic, Cleveland, OH, United States

Natalia Jarzebska Department of Internal Medicine III, Technische Universität Dresden, Dresden, Germany

Arduino A. Mangoni Discipline of Clinical Pharmacology, College of Medicine and Public Health, Flinders University, Adelaide, SA, Australia; Department of Clinical Pharmacology, Flinders Medical Centre, Southern Adelaide Local Health Network, Adelaide, SA, Australia

Gianmarco Matrullo Department of Biology, University of Rome "Tor Vergata", Rome, Italy

Lucia Morbidelli Department of Life Sciences, University of Siena, Siena, Italy

Inesa Navasardyan Department of Microbiology, Immunology and Molecular Genetics, David Geffen School of Medicine, Jonsson Comprehensive Cancer Center, University of California, Los Angeles, CA, United States; College of Osteopathic Medicine of the Pacific, Western University of Health Sciences, Pomona, CA, United States

Chiara Pecorari Redox Biology, Danish Cancer Institute, Copenhagen, Denmark

Raffaella Picco Department of Medicine, University of Udine, Udine, Italy

Vinitha N. Ragavan Discipline of Clinical Pharmacology, College of Medicine and Public Health, Flinders University, Adelaide, SA, Australia; Department of Internal Medicine III, Technische Universität Dresden, Dresden, Germany

Valentina Rapozzi Department of Medicine, Laboratory of Biochemistry, University of Udine, Udine, Italy

Emanuele Rezoagli School of Medicine and Surgery, University of Milano-Bicocca, Monza, Italy; Department of Anesthesia and Intensive Care Medicine, Fondazione IRCCS San Gerardo dei Tintori, Monza, Italy

Roman N. Rodionov Department of Internal Medicine III, Technische Universität Dresden, Dresden, Germany

Antonio Rodríguez-Ariza Instituto Maimónides de Investigación Biomédica de Córdoba (IMIBIC), Córdoba, Spain

Sébastien Serres Division of Physiology, Pharmacology and Neuroscience, School of Life Sciences, University of Nottingham, Nottingham, United Kingdom

Davide Signori School of Medicine and Surgery, University of Milano-Bicocca, Monza, Italy; Department of Anesthesia and Intensive Care Medicine, Seriate, Italy

Joern R. Steinert Division of Physiology, Pharmacology and Neuroscience, School of Life Sciences, University of Nottingham, Nottingham, United Kingdom

Mamta Sumi Department of Inflammation and Immunity, Lerner Research Institute, The Cleveland Clinic, Cleveland, OH, United States

Sara Tommasi Discipline of Clinical Pharmacology, College of Medicine and Public Health, Flinders University, Adelaide, SA, Australia; Department of Clinical Pharmacology, Flinders Medical Centre, Southern Adelaide Local Health Network, Adelaide, SA, Australia

Greta Varchi Italian National Research Council, Institute of Organic Synthesis and Photoreactivity, Bologna, Italy

Lashika Weerakoon Discipline of Clinical Pharmacology, College of Medicine and Public Health, Flinders University, Adelaide, SA, Australia

Luigi E. Xodo Department of Medicine, Laboratory of Biochemistry, University of Udine, Udine, Italy

Apostolos Zaravinos Department of Life Sciences, School of Sciences, European University Cyprus, Nicosia, Cyprus; Cancer Genetics, Genomics and Systems Biology Laboratory, Basic and Translational Cancer Research Center (BTCRC), Nicosia, Cyprus

About the editors

Benjamin Bonavida, PhD

Dr. Benjamin Bonavida is currently a Distinguished Research Professor at the University of California, Los Angeles (UCLA). He is affiliated with the Department of Microbiology, Immunology and Molecular Genetics, UCLA David Geffen School of Medicine, and the Jonsson Comprehensive Cancer Center at UCLA. His research career, thus far, has focused on investigations in the fields of basic immunochemistry and cancer immunobiology. His research investigations have ranged from the biochemical, molecular, and genetic mechanisms of cell-mediated killing and tumor cell resistance to chemo-immuno cytotoxic drugs. The reversal of tumor cell resistance was investigated by the use of various selected sensitizing agents based on molecular mechanisms of resistance. In these investigations, there was the newly characterized dysregulated NF-κB/Snail/YY1/RKIP/PTEN loop in many cancers that was reported to regulate cell survival, proliferation, invasion, metastasis, and resistance. Emphasis was focused on the roles of the tumor suppressor Raf kinase inhibitor protein (RKIP) and the tumor promoter Yin Yang 1 (YY1) and the role of nitric oxide (NO) as a chemo-immuno-sensitizing factor. Many of the aforementioned studies are centered on the clinical challenging features of cancer patients' failure to respond to both conventional and targeted therapies.

Dr. Bonavida has been active in the organization of regular sequential international miniconferences that are highly focused on the roles of YY1, RKIP, and nitric oxide in cancer and their potential therapeutic applications. Several books edited or coedited have been published. In addition, he is the Series Editor of books (over 20) published by Springer/Nature: *Resistance to Anti-Cancer Targeted Therapeutics*. In addition, he is presently the Series Editor of *Three Series* published by Elsevier/Academic Press on "*Cancer Sensitizing Agents for Chemotherapy*," "*Sensitizing Agents for Cancer Resistance to Cell Mediated Immunotherapy*," and "*Breaking Tolerance to Anti-Cancer Antibody Immunotherapy*." He is the Editor-in-Chief of the Journal *Critical Reviews in Oncogenesis*, Editor-in-Chief of *Onco Therapeutics*, and Associate Editor of "*Critical Reviews in Immunology*." Dr. Bonavida has published over 500 research publications and reviews in various scientific journals of high impact.

Valentina Rapozzi

Valentina Rapozzi is an Associate Professor in Biochemistry at the Department of Medicine in the University of Udine, President of the Italian Society of Photobiology, Founder and Member of the International Society for Nitric Oxide and Cancer, and Member of the board of the PhD in Biomedical Science and Biotechnology at the University of Udine.

Valentina Rapozzi has a strong background in cell biology, molecular biology, and *in vivo* animal tumor.

Her research interests can be divided into three fields: (1) Application of new photosensitizers in tumor cells culture and in vivo system and study of molecular pathways involved in host's response to the photodynamic therapy; (2) Molecular strategies (antigene and antisense) to study neoplastic proliferation; (3) Influence of psychological stress on metastasis progression in animal model.

Her current research interest is the study of molecular mechanisms in the tumor response, particularly in the field of redox signaling and photooxidative stress. She is focusing on some aspects of cell response to photodynamic therapy and its relevance in molecular oncology and cancer.

She has contributed to the understanding of the molecular mechanisms involving reactive oxygen and nitric species (ROS/RNS) in coordinating cellular responses to photooxidative stress in different cell/animal models.

She is the coauthor of about 80 international publications, 10 book chapters, and a participant in 138 congresses, Guest Editor for Forum on Immunopathological Diseases and Therapeutics in 2011 with the Volume *Molecular Pathways in the Response of Tumors to Photodynamic Therapy*, and Editor together with Prof. Giulio Jori from the University of Padova, Italy, for the Volume Resistance to Photodynamic Therapy in Cancer 2015, Vol 5. Resistance to Targeted Anti-Cancer Therapeutics, Series Editor Benjamin Bonavida, Springer. Guest Editor together Prof. Benjamin Bonavida for "Circadian Rhythms and Oncogenesis" for Critical Reviews in Oncogenesis, 2021.

Luigi Xodo

Luigi Xodo is a Full Professor of Biochemistry at the Department of Medicine of the University of Udine, Italy. His work focuses on unraveling the oncogenes involved in carcinogenesis and on the development of new cancer therapies. In particular, Dr. Xodo's work led to the discovery that the KRAS oncogene, a key player in cellular transformation toward malignancy, is intricately regulated by guanine-rich sequences that adopt unique G-quadruplex structures. This discovery paved the way for targeted interventions against KRAS-related cancers. Dr. Xodo developed effective anticancer drugs that could inhibit the KRAS oncogene. One notable approach has been the use of cationic porphyrins that specifically target G-quadruplexes in KRAS mRNA and trigger transcript degradation upon photoactivation. In addition, Dr. Xodo's studies contribute to the discovery of a metabolic reprogramming phenomenon in KRAS-related cancers by elucidating how arginine metabolism shifts to promote the production of polyamines, phosphocreatine, and nitric oxide. He is the author of 120 publications in high-quality peer-reviewed journals. His academic contributions are evidenced by 120 publications in peer-reviewed journals with high-impact factor, underlining his profound impact on the field of biochemistry and oncology.

Preface

Chapter 1 by Davide et al, titled Pathogenesis and therapeutic management of cardiac surgery associated acute kidney injury: Role of nitric acid biologic pathways, reviews the deadly diseases of patients undergoing cardiac surgery with an acute kidney injury (AKI) leading to increased morbidity and mortality. As nitric oxide (NO) plays a role in cardiovascular pathogenesis, the authors have discussed the pathophysiology of cardiac-associated acute kidney injury (CSA-AKI) and its dysregulation by NO and the various approaches used to treat this disease through NO-mediated biological signaling pathways. The authors present several lines of evidence on the dysregulated role of NO in CSA-AKI and, unfortunately, there are no current investigations preclinically both in animal models and in humans. They suggest potential therapies that include inhaled NO or dietary supplementations with nitrate that are useful for research investigations in the hope of providing means to treat and prevent CSA-AKI.

Chapter 2 by Arduino et al, titled The emerging role of dimethylarginine dimethylaminohydrolase 2 in regulating vascular mimicry in cancer, reviews the role of NO in tumor neovascularization and its vasculogenic mimicry. The authors discuss the pathogenesis and clinical significance of vasculogenic mimicry and the implication of dimethylarginine dimethylaminohydrolase 1 (DDAH1) and DDAH2. These findings are important to unravel the underlying mechanisms of vasculogenic mimicry and cancer dissemination. Evidence was presented on the potential inhibition of DDAH1 through the inhibition of NO as a possible novel therapeutic; however, the role of DDAH2 is not clear in vasculogenic mimicry and requires further research investigations.

Chapter 3 by Matrullo et al, titled Decoding S-nitrosylation: Theoretical aspects and analytical approaches, reviews the various physiological roles mediated by NO and its enzymatically regulated expression and chemical reactions with cellular molecules. Not only NO is involved in the nitrosylation of proteins but also it is denitrosylation by denitrosylases. In this review, the authors present the important role of NO in both physiology and pathology such as its roles in vasodilation and muscle cell contraction, neurotransmission, metabolism and the immune response. The role of NO in cancer is discussed on tumorigenesis (impact on mutations, cell death, angiogenesis, and EMT). The authors also discuss the various approaches and challenges re the assays used for the detection of NO and nitrosylated proteins. The authors conclude with the evolving discovery of new S-nitrosylase and denitrosylase proteins and their intrinsic activities in health and disease.

Chapter 4 by Morbidelli and Ciccone, titled Preclinical models to assess the pharmacological properties of NO derivatives, reviews the importance of the model choices to be used to investigate NO-mediated effects in various diseases and the use of synthetic NO derivatives for in vivo applications. The authors review the transition from preclinical to clinical trials of new NO and drugs and the various steps that need to be undertaken for the ultimate use in the clinics. These include the cell-based in vitro pharmacological screening, cell source and maintenance, survival assay and proliferation of the cell culture, the matrigel assay to assess endothelial network formation, assay for platelet aggregation, clonogenic assays, tumorsphere formation, and usage of both chemotaxis and invasion. They also discuss ex vivo experiments to evaluate the efficacy of novel molecules on the vascular system and in vivo experiments to determine the efficacy and safety of the NO derivatives. In addition, the effect of the new derivatives on tumor-bearing animals was investigated. Overall, this chapter is very useful for the guidelines that need to be followed to investigate new NO-derived molecules for their potential therapeutic applications.

Chapter 5 by Steinert, titled Nitric oxide-mediated neuroinflammatory signaling as treatment targets in neurodegeneration, reviews the topic of vascular insufficiency and its association with astrocytic responses to endothelial malfunction. This association results in the release of NO that is involved in bidirectional communication with the vascular tissues, astrocytes, and neurons. The author discusses the redox signaling of the neurovascular unit (NVU) and the brain−blood barrier (BBB), oxygen- and NO-mediated signaling, redox stress in the vascular, neuronal and astrocytic compartments and therapeutic targets of NO signaling.

Chapter 6 by Gani et al, titled Role of nitric oxide in gemcitabine resistance in pancreatic cancer cells, reviews the underlying mechanisms by which NO regulates the resistance to chemotherapeutic drugs used to treat human pancreatic cancer. The authors' discussions center primarily on preclinical studies with human pancreatic cancer cell lines and their resistance to gemcitabine. Gemcitabine induces the release of ROS that generate NO and RNS. They discuss the relationship and commonality between the signaling pathways mediated by gemcitabine and NO. They review the role of NO in chemoresistance, properties of pancreatic cancer cells, the development of gemcitabine-resistance cancer cells, and the metabolic changes that took place in the gemcitabine resistant cell line. They also propose several approaches they will undertake to target NO signaling for the identification of therapeutic targets.

Chapter 7 by Navasardyan and Bonavida, titled Navigating the dual role of nitric oxide in glioblastoma, reviews the deadly disease glioblastoma and the lack of any effective therapies to treat glioblastoma (GBM). In this review, they discuss the hypothesis whether NO or derivatives might be a novel therapeutic strategy in the treatment of GBM. In this review, the authors briefly provide an overview on the current treatments of GBM, the reported literature on the roles of iNOS and NO in GBM, and several mechanisms pertaining to NO as a tumor suppressor (e.g., nitrosative DNA damage, inhibitor of survival pathways, inhibitor of angiogenesis and inhibitor of EMT). Furthermore, the authors postulate several therapeutic strategies targeting NO for the treatment of GBM and they also present various perspectives and future directions for considerations.

Chapter 8 by Varchi, titled Nitric oxide-donating systems and their potential in shaping tumor immunoregulation, reviews the contrasting roles NO plays in the regulation of cancer cells and the immune response. She presents an important overview of newly synthesized NO donors for the treatment of cancer, used alone or in combinations. Several classes of NO donors are presented with their chemical structures and functions and their effects on the immune response and reviews the current literature. Of interest, Doctor Varchi introduces innovative approaches overriding the many limitations of administering NO donors in vivo in humans who lack stability, precise targeting, and retention. She, nevertheless, is optimistic on the development of effective and nontoxic-mediated agents in the treatment of various human cancers.

Chapter 9 by Di Giorgio et al, titled Control of nitric oxide synthase 2: Role of NRF-2 regulated distal enhancer, reviews recent investigations in the authors laboratory of the discovery of nuclear factor erythroid 2-related factor 2 (NRF2) as a regulator of redox homeostasis and NOS2 expression. They review the structure of NOS2, NOS2 and cancer, transcriptional regulation of NOS2, and NOS2 expression controlled by NRF2-binding to its distant enhancer. In their model of pancreatic cancer cell lines, they identified the molecular interactions driven by NRF2, which is stimulated by the KRAS oncogene via the MAPK and AKT pathways. The authors propose the potential usage of novel combination therapies such as KRAS and NOS2 inhibitors to treat pancreatic cancer cells in vitro and in animal models for their potential translation into human cancer patients.

Chapter 10 by Rodriguez Ariza, titled Nitric oxide in tumour biology: from stemness to metabolic reprogramming, reports his own findings on the imbalance between nitrosylation and denitrosylation mediating the expansion of cancer stem cells (CSCs) in ER+ breast cancer cells. This imbalance is mediated by NO. This chapter reviews the role of NO in the acquisition of both stem properties and therapy resistance in breast cancer, the metabolic shift that underlies tumor progression and immune evasion in S-nitrosoglutathione reductase deficient cancer cells. It is proposed to investigate various NO-mediated therapies to exploit the metabolic vulnerabilities that are linked to abnormal NO signaling.

Chapter 11 by Ghosh C et al, titled Impact of nitric oxide on hemeprotein maturation and its relevance to cancer and pulmonary diseases, reviews a different aspect of NO on heme functions such as the impact of NO on hemeprotein maturation, the significance of NO-sGC-cGMP pathway in globin maturation, and its relevance to cancer progression and the role of NO in asthma and pulmonary diseases. This extensive review with various analyses of underlying biochemical and molecular mechanisms is well illustrated by various schematic diagrams. In addition, the authors present various suggestions for future directions in this important and evolving field.

Chapter 12 by Zaravinos, tilted Artificial intelligence (AI), the ability of machines to learn from new input, is a broad term for a range of computing methods: AI is changing drug discovery, was an educational session that was presented in a field that is rapidly evolving with many applications in science and drug development. This chapter is primarily presented by slides each of which presents a specific information geared to the untrained professional for their adaptations. Details are shown on drug development, a table describing targets, selected descriptors and statistics for selected exemplary models, in silico prediction of drug—target interaction, the available drug target databases, the framework of AI techniques for drug discoveries, AI models used to enhance patient cohort selection, and advantages of the application of AI technologies. In addition, Dr. Zaravinos discusses the various challenges that need to be recognized and addressed on all studies using AI.

Overall, it is clear that the above contributors offer a variety of recent investigations on the complex role of NO and its applications in various diseases and also address the contrasting roles NO plays under various conditions and diseases. These contributions provided opportunities for the near future to solve the various challenges introduced as to develop the proper means and agents that can be targeted selectively with minimal toxicities and maximal effectiveness. The contents of this volume are also educational for new investigators and also opportunities for biotech companies.

The Editors sincerely thank the various authors who contributed to this volume for their efforts and significant valuable information and we are indebted to them. Furthermore, the Editors acknowledge Ms. Michaela Realiza, Editorial Project Manager, for her consistent efforts in assisting and working closely with the Editors and the contributors and facilitating the execution of this volume in a timely manner.

Benjamin Bonavida, PhD
Editor

UDINE 2023
Third International Conference

Therapeutic Applications of Nitric Oxide in Cancer and Inflammatory-Related Disorders

4–5 September 2023

Organizers: Valentina Rapozzi, Luigi Xodo, & Benjamin Bonavida

Co-Organizer: Massimo Libra

UNIVERSITÀ DEGLI STUDI DI UDINE

Università di Catania

UDINE 2023

Dear Distinguished Participants and Guests,

On behalf of the organizing committee, it is our honor and privilege to welcome you all to this esteemed conference *"**Therapeutic Applications of Nitric Oxide in Cancer and Inflammatory-Related Disorders.**"* We gather here as a diverse group of professionals, scholars, researchers, and experts from various fields, united by our shared pursuit of knowledge, collaboration, and innovation.

This conference serves as a platform for us to come together, exchange, ideas, and engage in meaningful discussions. The depth and breadth of expertise represented in this conference are truly remarkable, and we are confident that the interactions and insights we will gain over the course of this event will be invaluable.

The theme of this conference, *"**Therapeutic Applications of NO**"* reflects our commitment to exploring cutting-edge research, emerging trends, and transformative ideas within our respective disciplines. It is a call to collectively push the boundaries of knowledge, challenge conventional thinking, and seek innovative solutions to the complex challenges that lie before us.

Throughout the course of this conference, we have developed a diverse program, comprising keynote addresses, panel discussions, and educational workshops. We have invited distinguished speakers who are leading authorities in their fields, ensuring a rich and stimulating intellectual experience for all participants.

Moreover, this conference is not only an opportunity for professional growth and networking but also a chance for us to forge new connections and foster interdisciplinary collaborations. The diversity of backgrounds, perspectives, and expertise in this conference presents an unparalleled opportunity for us to learn from one another, establish lasting partnerships, and create a collective impact that transcends individual disciplines.

We would also wish to express our sincere gratitude to the University of Udine, our sponsors, staff, and volunteers, who have dedicated countless hours to ensure the success of this conference. Your unwavering support and commitment are the foundation upon which this event stands.

Finally, we encourage each and every one of you to actively participate, contribute your unique insights, and engage in lively discussions throughout the conference. Remember that the true value of such an event lies not only in what we take away individually but also in what we collectively contribute to the body of knowledge and the advancement of our respective fields.

Once again, our warm welcome to this conference that will be filled with enlightening conversations, fruitful collaborations, and lasting connections. Let us embrace the opportunities that lie ahead and embark on this intellectual journey together.

Thank you from the organizers

Valentina Rapozzi

Luigi E. Xodo

Benjamin Bonavida

Table of contents

Sponsors

UNIVERSITÀ DEGLI STUDI DI UDINE

Rafael Teixeira, Elsevier/Academic Press
Content Acquisition Lead-Biochemistry Journals

Sponsorship of a Book ***"Therapeutic Applications of Nitric Oxide in Cancer and Inflammatory Diseases"***

Massimo Libra Director of **Bioinformatics Master - University of Catania.** https://www.unict.it/it/didattica/master-universitari/2022-2023/bioinformatica

Università di Catania

Biometec
Dipartimento di Scienze Biomediche e Biotecnologiche
Università di Catania

Le Fondazioni Casali
Fondazione Benefica Kathleen Foreman Casali

Fondazione Benefica Alberto e Kathleen Casali **awarded 5 prizes for the participation of young researchers**

MDPI Academic Open Access Publishing since 1996

MDPI awarded two prizes for Poster session

International Society for Nitric Oxide and Cancer ISNOC

BRUKER

Aim and scope

The main objective and focus of this conference will highlight the major recent advances in the therapeutic applications of nitric oxide-based drugs, nitric oxide-generating systems, nitric oxide nanodelivery systems, and nitric oxide derivatives in various human cancers and inflammatory disorders.

Emphasis will be placed on the underlying biochemical and molecular mechanisms of action, preclinical model systems, and clinical studies.

This conference will also feature special educational sessions for young and new investigators interested in exploring and pursuing nitric oxide applications. In addition, the current state of the art in the development and design of new nitric oxide drugs will be presented.

Useful information

Venue
Palazzo Garzolini di Toppo Wassermann
Via Gemona, 92
Udine

Workshop Organizers
Valentina Rapozzi
Luigi Xodo
Benjamin Bonavida
Massimo Libra

Contact
Valentina Rapozzi
University of Udine
Email: valentina.rapozzi@uniud.it

Workshop website
https://valentinarapozzi.wixsite.com/no-udine

Wi-Fi
Eduroam

Program

Monday, 4th September 2023

08:00–8:30 Registration

08:45–9:15 Greetings and Introduction

Session 1: Nitric Oxide-Mediated Signaling Pathways
Chairs: *Giuseppe Filomeni & Stavroula Baritaki*

9:20–9:35	**Philip Eaton** William Harvey Research Institute, Queen Mary University of London, UK *Redox sensing and signaling-from basic mechanisms to therapeutic opportunities*
9:40–9:55	**Anna Grochot-Przeczek** Jagiellonian University, Krakòw, Poland *Orchestrate S-nitrosation: A novel function of the KEAP1 redox sensor*
10:00–10:15	**Gianmichele Massimo** Queen Mary University of London, UK *Natural mutations of human XDH promote the nitrite (NO_2-)-reductase capacity of xanthine oxidoreductase: a novel mechanism to promote redox health?*
10:20–11:00	**Educational Session 1:** Methods for measuring nitric oxide content in biological systems. *Nitric oxide sensing by a blue fluorescent protein* **Thomas Gensch** Institute of Biological Information Processing, Forschungszentrum Jülich, Germany *RNS and ROS- EPR detection and quantitative analysis* **Manuela Liberi** *Bruker Company*
11:00–11:20	**Coffee break**

Session 2: Nitric Oxide and Cancer
Chairs: *Jordi Muntanè & Valentina Rapozzi*

11:20–11:35	**Arduino Mangoni** Flinders University, Adelaide, Australia *The emerging role of dimethylarginine dimethylaminohydrolase 2 in regulating vasculogenic mimicry in cancer (Online)*
11:40–11:55	**Costanza Montagna** University of Rome Tor Vergata, Rome, Italy *GSNOR safeguards S-phase entry and genomic stability in rhabdomyosarcoma cells: unraveling new molecular players*
12:00–12:15	**Salvatore Rizza** Danish Cancer Society Research Center, Copenhagen, Denmark *Protein denitrosylation in tumor progression and therapy*
12:20–12:35	**Annie Beuve** New Jersey Medical School, Rutgers, The State University of New Jersey, Newark, USA *Blocking the soluble guanylyl cyclase/Thioredoxin 1 complex increases caspase-3 activity in Jurkat T cells*
12:40–13:00	**FC AWARDS**

13:00–14:15	**Lunch with Poster Session**
14:15–14:45	**Educational Session 2:** Nitric oxide-modulated signaling: canonical and noncanonical signaling pathways **Sharon Glynn,** University of Galway, Ireland

Session: 3: Nitric Oxide and Metabolism
Chairs: *Luigi Xodo & Mattias Carlstrom*

14:50–15:05	**Antonio Rodríguez-Ariza** Instituto Maimónides de Investigación Biomédica de Córdoba (IMIBIC), Spain *Nitric oxide in tumor biology: from stemness to metabolic reprogramming*
15:10–15:25	**Jean-Luc Balligand** Université catholique de Louvain, Belgium *Nitric Oxide and cardiac remodeling: a question of metabolism*
15:30–15:45	**Eros Di Giorgio** University of Udine, Italy *NRF2 interacts with distal enhancer and inhibits nitric oxide synthase 2 expression in KRAS-driven pancreatic cancer cells.*
15:50–16:05	**Emanuele Buratti** University of Trieste, Italy *hnRNP modifiers of TDP-43 pathology in Amyotrophic Lateral Sclerosis can act through the control of NOS1AP expression in neurons*
16:10–16:30	**Coffee break**

Session 4: Nitric Oxide and Health
Chairs: *Bruno Grassi & Antonio Rodríguez-Ariza*

16:30–16:45	**Andrew Jones** University of Exeter, UK *Dietary nitrate supplementation: Effects on skeletal muscle and exercise performance in healthy humans*
16:50–17:05	**Simone Porcelli** University of Pavia, Italy *Effects of increased nitric oxide bioavailability on skeletal muscle function in aging*
17:10–17:25	**Vincenzo Brancaleone** University of Basilicata, Italy *Novel molecular mechanisms regulating the interplay between NO and H_2S in hyperglycemia*
17:30–17:45	**Jenny Chang** Houston Methodist Hospital, Texas USA *Dissecting the synergistic effects of nitric oxide synthase and PI3K inhibitors in metaplastic breast cancer*
17:55–18:25	**Educational Session 3**: Decoding S-nitrosylation: Theoretical aspects and analytical approaches **Giuseppe Filomeni** Danish Cancer Society Research Center, Copenhagen, Denmark.
19:45	**Gala dinner at *Casa Della Contadinanza***

Tuesday, 5th September 2023

Session 5: Nitric Oxide and Inflammation
Chairs: *Khosrow Kashfi & Stephanie Plenchette-Colas*

9:00–9:15	*ISNOC Presentation* **Khosrow Kashfi** University of New York, NY
9:20–9:35	**Moran Benhar** Technion-Israel Institute of Technology, Haifa, Israel *Widespread cysteine persulfidation in activated macrophages as a protective mechanism against oxidative-inflammatory stress*

9:40–9:55 **Mattias Carlstrom**
Karolinska Institutet, Stockholm, Sweden
Modulation of oxidative stress and inflammation by the nitrate-nitrite-NO pathway in cardiometabolic and renal disorders

10:00–10:15 **Mark Crabtree**
University of Surrey, Guildford, UK
Unravelling a novel Immunoregulatory role for iNOS in macrophages

10:20–10:35 **Thaissa Horne**
Institute of Biomedicine of Seville (IBiS), Hospital University Virgen del Rocío/CSIC/ University of Seville, Seville, Spain
Nitric oxide plays a relevant role during cell resistance to Sorafenib treatment in liver cancer cells

10:40–10:50 **MDPI Awards Poster**

10:50–11:10 **Coffee Break**

11:10–11:50 **Educational Session 4:** Using Artificial Intelligence to discover new cancer therapies
Using AI to discover new cancer therapies
Apostolos Zaravinos
European University Cyprus, Nicosia, Cyprus
Applications of Artificial Intelligence techniques for therapeutic designs targeting modulation of gene expression
Alexandre Ferreira Ramos
Universidade de São Paulo, São Paulo, Brazil

Session 6: Nitric Oxide and Vascular/neurodegenerative diseases
Chairs: Moran Benhar & Sharon Glynn

11:55–12:10 **Adam Straub**
University of Pittsburgh, Pittsburgh, USA
CYB5R3 and Nitric Oxide Signaling

12:15–12:30 **Michael Heneka**
University of Luxembourg, Belvaux, Luxembourg
Nitric oxide-mediated posttranslational protein modifications in neurodegenerative disease

12:35–12:50 **Melanie Madhani**
University of Birmingham, Birmingham, UK
Interactions between persulfides, nitrite and nitric oxide in cardiovascular health and disease

12:55–14:20 **Lunch**

14:20–14:35 **Arnab Ghosh**
Lerner Research Institute, Cleveland, Ohio-44196, USA
Regions of hypoventilation as determined by the 4Dx scanner suggests that regional and not global inflammation impacts lung function in asthma (Online)

14:40–14:55 **Brant Isakson**
University of Virginia health System, Virginia, USA
Alpha hemoglobin in endothelium

15:00–15:15 **Joern Steinert**
School of Life Sciences, University of Nottingham, Nottingham, UK
NO-mediated neuroinflammatory pathways as treatment targets in neurodegeneration

Session 7: Therapeutic Applications of Nitric Oxide (Part 1)
Chairs: *Lucia Morbidelli & Greta Varchi*

15:20–15:35	**Gregory Giles** University of Otago, Dunedin, New Zealand *The design of photoactivated therapeutics for controlled nitric oxide release*
15:40–15:55	**Cristina Maccallini** University "G. d'Annunzio" of Chieti-Pescara, Chieti, Italy *Recent developments of selective iNOS inhibitors as anticancer agents*
16:00–16:20	Coffee Break
16:20–17:00	**Educational Session 5:** Novel nitric oxide delivery systems: from chemical synthesis to preclinical development *Nitric oxide donating molecules and nanosystems: advances and perspectives on the current chemical tools* **Greta Varchi,** ISOF, CNR, Bologna, Italy *Preclinical models to assess the pharmacological properties of NO derivatives* **Lucia Morbidelli,** University of Siena, Siena, Italy

Session 7: Therapeutic Applications of Nitric Oxide (Part 2)
Chairs: *Benjamin Bonavida & Massimo Libra*

17:05–17:20	**Moises Pinto** University of Lisbon, Instituto Superior Técnico, Lisbon, Portugal *Hybrid and composite porous materials as vehicles to deliver therapeutic nitric oxide*
17:25–17:40	**Emanuele Rezoagli** University of Milano-Bicocca, Milano, Italy *Nitric oxide pathophysiology and applications in patients undergoing major cardiac surgery*
17:45–18:00	**Lucas Carvalho** Karolinska Institute, Stockholm, Sweden *Effects of NO-supplementation Flexovital on a novel reno-cardio-metabolic disease model to study diabetic nephropathy*
18:05–18:20	**Claudio Laquatra** University of Padova, Italy *Oxidative stress induces Cysteines-mediated tetramerization of the mitochondrial chaperone TRAP1*
18:25–18:40	**Benjamin Bonavida, Luigi Xodo & Valentina Rapozzi** *General Discussions and Concluding Remarks*
	End of Meeting

Speakers' abstracts

The abstracts of the speakers are in chronological order according to the program

Redox sensing and signaling—from basic mechanisms to therapeutic opportunities

Philip Eaton

William Harvey Research Institute, Queen Mary University of London
Contact: p.eaton@qmul.ac.uk

Oxidants, nitro-oxidants, and electrophiles, which can be produced in cells and tissues, have been causatively implicated in the pathogenesis of most major diseases. While this paradigm is still prevalent and such molecules may indeed play important roles in the disease etiology, it is now understood that they can also serve as regulatory entities that are important for the maintenance of homeostasis during health and disease. Oxidants can be sensed and transduced into a biological event via their reaction with select cysteine thiol side chains on some proteins. Such reactions can result in oxidative posttranslational modifications of proteins that in some cases alter their function to enable adaptation and homeostasis. By identifying thiol-based redox sensor proteins susceptible to oxidant-dependent alterations in structure and function, we hope to gain a better understanding of the biological significance of these modifications during cardiovascular health and disease. Indeed, in some cases, we have developed new electrophilic drugs that modify precise redox regulated cysteines for therapeutic purposes to combat cardiovascular disease. The redox control of protein kinase G Iα, soluble epoxide hydrolase, and p90 ribosomal S6 kinases will be discussed together with our studies trying to develop new electrophilic molecules with therapeutic potential that covalently modify regulatory cystines in these proteins.

To orchestrate S-nitrosation: A Novel Function of the KEAP1 Redox Sensor

Anna Grochot-Przęczek

Department of Medical Biotechnology, Faculty of Biochemistry, Biophysics and Biotechnology, Jagiellonian University, Krakow, Poland

Contact: *anna.grochot-przeczek@uj.edu.pl*

A hallmark function of KEAP1, a redox sensor, is the repression of the NRF2 transcription factor. The NRF2/KEAP1 axis acts as a crucial modulator of cell homeostasis, orchestrating a response to oxidative stress. The biological significance of NRF2 is well established. The deregulation of NRF2 transcriptional activity has been described in the pathogenesis of multiple diseases and is associated with ageing. The focus on KEAP1, however, is hardly affirmed beyond NRF2 inhibition.

KEAP1 is an important mediator of NO biology. KEAP1 cysteine 151 is a cellular sensor of nitric oxide (NO). Furthermore, recently we have identified KEAP1 as a crucial modulator of S-nitrosation (SNO) in endothelial cells and blood vessels. SNO, previously thought to be a nonenzymatic posttranslational reaction, is governed in mammalian cells by a complex of three proteins: NO synthase (NOS), transnitrosating protein GAPDH, and KEAP1. The deletion of any protein of this complex inhibits SNO. Similar SNO machinery has been found in bacteria. We showed that KEAP1-mediated SNO, which is observed primarily in NRF2-deficient cells, determines endothelial cell fate, balancing premature senescence and cell death. Moreover, S-nitrosation of NOX4 has a fundamental significance in endothelium, protecting it from oxidative damage and apoptosis in the absence of NRF2. However, KEAP1-dependent hyper-S-nitrosation of proteins is associated with a loss of proteostasis. It can be found in ageing endothelial cells, which show strong protein aggregation. The non-canonical functions of KEAP1 may have clinical relevance in the context of age-related dysfunctions, when an imbalance between NRF2 and KEAP1 occurs.

References

1. Kopacz et al., Redox Biol, 2020, 28:101,304
2. Kopacz et al., Redox Biol, 2020, 34:101,572
3. Seth et al., Mol Cell, 2018, 69: 451–464

Natural mutations of human XDH promote the nitrite (NO_2-)-reductase capacity of xanthine oxidoreductase: a novel mechanism to promote redox health

Gianmichele, Massimo BSc PhD[1], R. S. Khambata BSc PhD[1], T. Chapman PhD[2], K. Birchall PhD[2], Claudio Rimondi BSc PhD[1], A. Shabbir MD[1], Nicki Dyson MReS[1], Krishna Rathod MD, PhD[1], C. Borghi MD[3], and A. Ahluwalia BSc PhD[1]

[1]*Center For Cardiovascular Medicine and Devices, William Harvey Research Institute, Barts and The London Medical School, Queen Mary University of London, EC1M 6BQ, London (UK);* [2]*LifeArc, Accelerator Building Open Innovation Campus, Stevenage, SG1 2FX, (UK) 3 Department of Medical and Surgical Sciences, Faculty of Medicine, Alma Mater Studiorum, University of Bologna, 40,138, Bologna (IT)*

Contact: g.massimo@qmul.ac.uk

Hypothesis

We hypothesize that naturally occurring mutations of the *XDH* gene specifically alter enzymatic activity of xanthine oxidoreductase (XOR) and thus the products of its activity: uric acid (UA), superoxide anion (O_2-), and nitric oxide (NO).

Introduction

XOR produces UA and reactive oxygen species (ROS) during purine metabolism, and NO as a consequence of its capacity to reduce inorganic nitrite (NO_2-) [1]. Elevation of its expression and activity is associated with rises in UA and ROS accumulation: both of which are pathogenic. Conversely, NO production in the setting of raised levels of inorganic NO_2- exerts a range of protective and positive effects on vascular health [2]. A few studies have identified that nonsynonymous single nucleotide polymorphisms (SNPs) in the *XDH* gene result in alterations in UA generation [3]. Whether these mutations might affect nitrite-reductase activity, and how this might relate to ROS and UA generation is unknown and has been explored in this study.

Aims

To test whether genetic mutations alter the broader XOR activity profile, particularly whether targeting the nitrite-reductase profile in those carrying such mutations may provide novel therapeutic options.

Methods

We investigated the effect of human-isolated mutations (Asn909Lys, His1221Arg, and Ile703Val) and an artificial double mutation (Trp336Ala/Phe337Leu) stably expressed in HEK-293 cells. The effect of the mutations upon gene and protein expression as well as the XDH/XO proportion was assessed. The impact of the mutations upon biochemical function was determined by measuring O_2-, UA, and NO generation in the presence of XOR reducing agents and pharmacological inhibitors.

Results

qPCR analysis revealed significant differences in *XDH* mRNA levels across the cell lines, with WT showing a 1.9 ± 0.68, 2.5 ± 0.65, 3.3 ± 1.2, and 4.9 ± 1.01-fold higher expression than cells bearing Trp336Ala/Phe337Leu, Ile703Val, His1221Arg, and Asn909Lys mutations, respectively. This was in line with immunoblotting ($p<0.0001$). XDH + XO activity assay revealed that His1221Arg and Ile703Val had a significantly greater proportion of XO versus XDH, while no differences in proportions were evident in WT. His1221Arg also showed a greater production of O_2- versus WT in the presence of xanthine ($p < .0001$), which was accompanied by higher UA ($p < .0001$). This same mutation also expressed significantly elevated NO production from nitrite (0.001) versus WT.

Conclusions

These results suggest that mutations of the *XDH* gene not only influence the well-described UA and O_2- -generating activity of XOR but also modifies nitrite reductase activity. Therefore, this should be considered when assessing the impact of these mutations in the general population, and this aspect may offer opportunities for a personalized medicine approach.

References

1. Zhang, Z., et al., *Generation of nitric oxide by a nitrite reductase activity of xanthine oxidase: a potential pathway for nitric oxide formation in the absence of nitric oxide synthase activity.* Biochem Biophys Res Commun, 1998. **249**(3): p. 767–72.
2. Gee, L.C., et al., *Inorganic nitrate attenuates cardiac dysfunction: role for xanthine oxidoreductase and nitric oxide.* Br J Pharmacol, 2021.
3. Kudo, M., et al., *Functional characterization of human xanthine oxidase allelic variants.* Pharmacogenet Genomics, 2008. **18**(3): p. 243–51.

Nitric Oxide Sensing by a Blue Fluorescent Protein

T. Gensch[1], C. Montali[2], K. Morreale[1,2,] S. Abbruzzetti[2], A. Franzen[1], G. Casini[1], V. Svensson[3], T. Drepper[3], S. Bruno[4], P. Delcanale[2], S. Burgstaller[5], J. Ramadani-Muja[5], S. Santiago-Schübler[6], R. Malli[5], and C. Viappiani[2]

[1] *Institute of Biological Information Processing (IBI-1: Molecular and Cellular Physiology), Forschungszentrum Jülich, Leo-Brandt-Straße, D-52,428 Jülich, Germany;* [2] *Dipartimento di Scienze Matematiche, Fisiche e Informatiche, Università di Parma, Parco Area delle Scienze 7A, 43,124 Parma, Italy;*[3] *Institute of Molecular Enzyme Technology (IMET), Forschungszentrum Jülich, Wilhelm-Jonen-Straße, D-52,428 Jülich, Germany;* [4] *Dipartimento di Scienze degli Alimenti e del Farmaco, Università di Parma, Parco Area delle Scienze 23/A, 43,124 Parma, Italy;* [5] *Molecular Biology and Biochemistry, Gottfried Schatz Research Center, Medical University of Graz, Neue Stiftingtalstraße 6/6, 8010 Graz, Austria;* [6] *Central Institute of Engineering, Electronics and Analytics (ZEA-3; Analytics), Forschungszentrum Jülich, Leo-Brandt-Straße, D-52,428 Jülich, Germany*
Contact: t.gensch@fz-juelich.de

S-Nitrosylation of cysteine residues is an important molecular mechanism for dynamic, posttranslational regulation of several proteins, providing a ubiquitous redox regulation. Cysteine residues are present in several fluorescent proteins (FPs), including members of the family of *Aequorea victoria* Green Fluorescent Protein (GFP)-derived FPs, where two highly conserved cysteine residues contribute to a favorable environment for the autocatalytic chromophore formation reaction. The effect of nitric oxide on the fluorescence properties of FPs has not been investigated thus far, despite the tremendous role FPs have played for 25 years as tools in cell biology.

We have examined the response to nitric oxide of fluorescence emission by the blue-emitting fluorescent protein mTagBFP2. To our surprise, upon exposure to micromolar concentrations of nitric oxide, we observed a roughly 30% reduction in fluorescence quantum yield and lifetime. Recovery of fluorescence emission is observed after treatment with Na-dithionite. Experiments on related fluorescent proteins from different families show similar nitric oxide sensitivity of their fluorescence. We correlate the effect with S-nitrosylation of Cys residues. Mutation of cysteine residues in mTagBFP2 removes its nitric oxide sensitivity.

Similarly, fluorescent proteins devoid of cysteine residues are insensitive to nitric oxide. In another set of experiments, we investigated the NO-transfer to FPs mediated by GSNO. We also show that mTagBFP2 can sense exogenously generated nitric oxide when expressed in a living mammalian cell. We propose mTagBFP2 as the starting point for a new class of genetically encoded nitric oxide sensors based on fluorescence lifetime imaging.

C. Montali et al. *Antioxidants* 2022, 11, 2229. https://doi.org/10.3390/antiox11112229

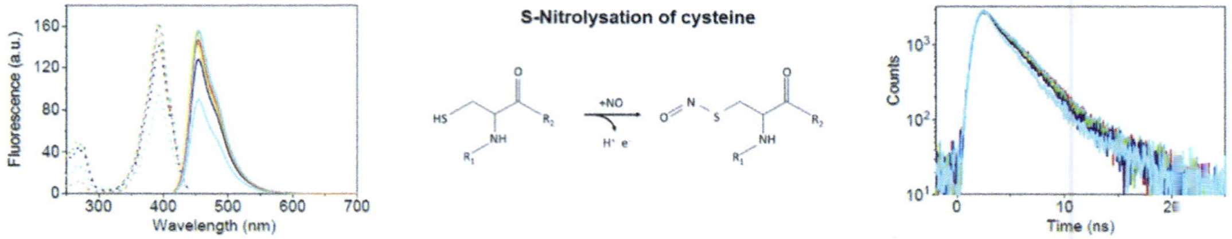

RNS and ROS-EPR detection and quantitative analysis

Manuela Liberi
Bruker Company

Contact: manuela.liberi@bruker.com

Electron paramagnetic resonance (EPR) spectroscopy is the only method for the direct detection of species that have unpaired electrons.

The EPR spectrum reflects the local and global environment of the unpaired electron: nearby nuclei, other unpaired electrons, solvent viscosity, molecular dynamics, and concentration of species.

Now, quantifying EPR species with high precision and reliability without additional user calibrations is also possible.

The emerging role of dimethylarginine dimethylaminohydrolase 2 in regulating vasculogenic mimicry in cancer

Arduino A Mangoni[1,2], Lashika Weerakoon[1], Vinitha N Ragavan[1,3], Natalia Jarzebska[3], Roman N Rodionov[3], Sara Tommasi[1,2]

[1]*Discipline of Clinical Pharmacology, College of Medicine and Public Health, Flinders University, Adelaide, Australia;* [2]*Department of Clinical Pharmacology, Flinders Medical Centre, Southern Adelaide Local Health Network, Adelaide, Australia;* [3]*Department of Internal Medicine III, Technische Universität Dresden, Dresden, Germany.*
Contact: arduino.mangoni@flinders.edu.au

Introduction: Both isoform 1 and isoform 2 of dimethylarginine dimethylaminohydrolase (DDAH1 and DDAH2) have been traditionally thought to metabolize the endogenous nitric oxide (NO) synthase inhibitors asymmetric dimethylarginine (ADMA) and monomethyl arginine (L-NMMA) to L-arginine and dimethylamine. More recently, DDAH1 has been shown to play a critical role in stimulating cancer cells, particularly triple negative breast cancer (TBNC) cell lines, to form blood vessel-like structures, a phenomenon called vasculogenic mimicry (VM) which increases cancer aggressiveness, metastasis, and adverse outcomes. Unlike DDAH1, however, the role of DDAH2 in regulating vasculogenic mimicry in cancer is unknown.

Aims: The capacity to undertake VM, assessed by measuring the number of branches, tubes, and loops, was investigated *in vitro* in the TNBC cell line MDA-MB-231, specifically in:

- Wild-type (WT) cells;
- *DDAH1* knock-out (CRISPR/Cas9 system) cells;
- *DDAH2* knock-out cells; and
- *DDAH1* + *DDAH2* knock-out cells.

Results: The loss of *DDAH1*, *DDAH2*, and both *DDAH1* and *DDAH2* function all significantly reduced the formation of branches, tubes, and loops by MDA-MB-231 cells on Matrigel matrix (Fig. 1) compared to the WT cell line. The reduction in tubes observed in *DDAH1* knock-out cell lines (80%) was comparable to what previously observed by our group using *DDAH1* silencing by siRNA (Hulin et al, Sci Rep 2017; 7(1):13,996). While the effect of *DDAH1* and *DDAH2* knock-out on tubes and loops formation was comparable, the effect size of *DDAH2* and *DDAH1* + *DDAH2* knock-out on the formation of branches was significantly greater than that of *DDAH1* knock-out (mean difference 30%, 95% CI 10.8%−49.2%, $p = 0.0003$, and 25%, 95% CI 5.8%−44.2%, $p = 0.0041$, respectively).

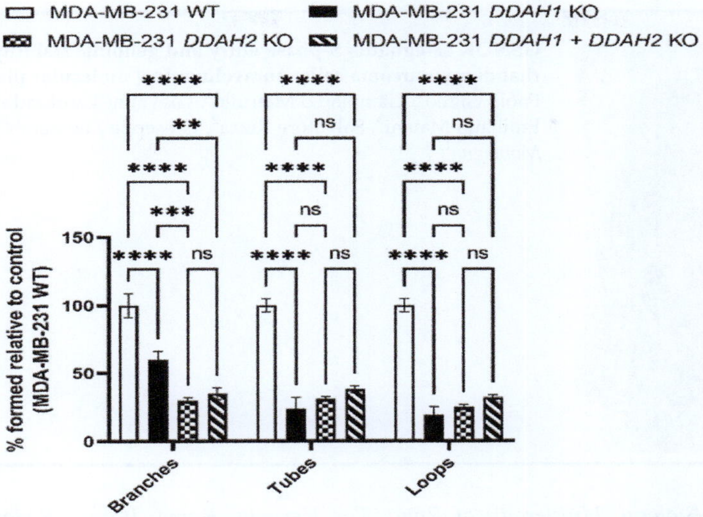

FIGURE 1 DDAH1 and DDAH2 loss of function significantly attenuates the formation of capillary-like structures by MDA-MB-231 cells on Matrigel. The number of tubes, branches, and loops are expressed as a % relative to WT control. Data shown are the mean of three independent experiments performed in quadruplicate. Error bars represent SEM.**p<.005 ***p<.001 ****p < .0001.

Conclusions and future directions: This experiment provides the first evidence that DDAH2 plays a key role in regulating VM. Future work is warranted to elucidate the molecular mechanisms by which DDAH2 contributes to tumor neovascularization and identify DDAH2 specific interactions with other VM modulators for the development of innovative anticancer treatments.

GSNOR safeguards S-phase entry and genomic stability in rhabdomyosarcoma cells: unraveling new molecular players

Paola Giglio[1], Gianmarco Matrullo[1], Giovanni Larotondo[1], Chiara Pecorari[2], Emiliano Maiani[3], Salvatore Rizza[2], Giuseppe Filomeni[1,2,4], and *Costanza Montagna*[1,3].

[1] *Department of Biology, University of Rome Tor Vergata, Rome, Italy* [2] *Redox Biology Group, Danish Cancer Society Research Center, Copenhagen, Denmark* [3] *UniCamillus-Saint Camillus, University of Health Sciences, Rome, Italy* [4] *Center for Healthy Aging, University of Copenhagen, Denmark*

Contact: costanzamontagna@gmail.com

Rhabdomyosarcoma (RMS) is a pediatric malignancy of muscles with a poor prognosis, underscoring an urgent need for therapy. RMS tumors arise from primitive muscle cells (myoblasts) that fail to exit the cell cycle, continue to proliferate indefinitely, and are unable to differentiate into myotubes. Recent advances in RMS biology point toward elevated levels of oxidative stress. Among free radicals, nitric oxide (NO) plays a role in cancer ontogenesis and progression. S-nitrosylation (SNO) is one of the main NO-induced posttranslational modifications, underling NO bioactivity. S-nitrosoglutathione reductase (GSNOR) is a denitrosylase that removes NO moiety from S-nitrosylated (SNO)-proteins, regulating the activity of many proteins that can be modified by SNO. GSNOR loss is associated with aging and malignancy, suggesting a role of SNO in neoplastic transformation and progression. Here, we show that GSNOR suppression inhibits RMS cell myogenic differentiation and increases the proliferation rate. In particular, we observed a higher percentage of cells in the S phase of the cell cycle in the absence of GSNOR, as well as a faster entry in the S phase. As a consequence of a deregulated cell cycle progression, GSNOR low cells display sufficient levels of replication stress and genomic instability that leads to accelerated cancer cell adaptation, resulting in faster cancer progression and aggressiveness. The oncogenic effect of GSNOR silencing in RMS points toward SNO signaling as a central regulator of RMS malignancy and will represent the starting point for future in vivo experiments in mouse models and (pre)clinical analyses.

Protein denitrosylation in tumor progression and therapy

Salvatore Rizza[1]
[1]*Redox Biology Group, Danish Cancer Society Research Center, 2100 - Copenhagen (DK)*

Contact: rizza@cancer.dk

S-nitrosylation, which is the posttranslational modification of proteins involving the attachment of nitric oxide (NO) to reactive cysteine residues, is a signaling mechanism frequently disrupted in cancer. This disruption can be attributed to either excessive production of NO or defective denitrosylation. Our research has identified a specific group of tumors characterized by the loss of S-nitrosoglutathione reductase (GSNOR), the most extensively studied denitrosylase[1]. We have found that GSNOR-deficient cancer cells exhibit a propensity for more aggressive behavior. A significant factor contributing to this phenotype is the activation of focal adhesion kinase-1 (FAK1) through S-nitrosylation[2]. FAK1 is a nonreceptor tyrosine kinase that senses the interaction between cells and the extracellular matrix (ECM) and transmits pro-survival signals into the cells. When cell adhesion to the ECM is incorrect, it triggers a form of apoptosis called "anoikis," that restricts misplaced readhesion of cells in inappropriate locations[3]. The hyperactivation of FAK1 confers anoikis-resistance to cancer cells and represents a pivotal step in metastasis. Consequently, FAK1 has emerged as an intriguing target for anticancer therapies.

We have demonstrated that mouse xenografts and organoids derived from GSNOR-deficient cancer cells are highly susceptible to FAK1 inhibitors[2]. This indicates that the expression level of GSNOR could be utilized to stratify cancer patients for treatment with FAK1 inhibitors, thereby enabling the development of personalized chemotherapy.

Furthermore, our studies indicate that GSNOR denitrosylation capabilities play a crucial role in various epithelia-derived cancers, with breast cancer (BC) being a representative example. In addition to its role in FAK1 activation, GSNOR silencing in BC leads to extensive rearrangement of cell signaling, interfering with cell adhesion, mitochondria physiology, and metabolism. These findings collectively highlight the multifaceted role of GSNOR and denitrosylation in tumor progression. Moreover, they offer new avenues for clinical interventions in cancer therapy.

References

1. Rizza S. and Filomeni G., Free Rad Biol Med, 2017, 110, 19–30.
2. Rizza S. et al, Cell Rep, 2023, 42(1), 111,997.
3. Paoli P. et al, BBA-Mol Cell Res, 2013, 1833(12), 3481–3498.

Blocking the soluble guanylyl cyclase/Thioredoxin 1 complex increases caspase-3 activity in Jurkat T cells

Chuanlong Cui[1,2], Ping Shu[2], Tanaz Sadeghian[1], Waqas Younis[2], Hong Li[3], *Annie Beuve*[2]

[1] *Rutgers School of Graduate Studies, Newark Health Science, Newark, NJ 07,103, USA*
[2] *Department of Physiology, Pharmacology and Neurosciences Rutgers New Jersey Medical School, Newark, NJ 07,103, USA*
[3] *Center for Advanced Proteomics Research, Department of Microbiology, Biochemistry and Molecular Genetics, Rutgers New Jersey Medical School, Newark, NJ 07,103, USA*
Contact: beuveav@njms.rutgers.edu

The NO-sensitive guanylyl cyclase (GC1, previously called sGC) is the major receptor for nitric oxide (NO). Upon binding of NO, GC1 catalytic activity is stimulated to produce the second messenger cGMP. Thioredoxin (Trx1) is an oxido-reductase and a major regulator of the cellular redox state. GC1 and Trx1 form a complex that mediates two NO signaling pathways as a function of the redox state of cells. Under physiological conditions, reduced Trx1 (rTrx1) supports the canonical NO-GC1-cGMP pathway by protecting GC1 activity from thiol oxidation. Under oxidative stress, the NO-cGMP pathway is disrupted by the S-nitrosation of GC1 (addition of a NO group to a cysteine). In turn, SNO-GC1 initiates transnitrosation cascades, using oxidized Trx1 (oTrx1) as a nitrosothiol relay. We designed an inhibitory peptide that blocked the interaction between GC1 and Trx1. This inhibitory peptide blocked the transfer of S-nitrosothiols from SNO-GC1 to oTrx1. In Jurkat T cells, oTrx1 transnitrosates procaspase-3, thereby inhibiting caspase-3 activity. Using the inhibitory peptide, we demonstrated that S-nitrosation of caspase-3 is the result of a transnitrosation cascade initiated by SNO-GC1 and mediated by oTrx1. Consequently, the peptide significantly increased caspase-3 activity in Jurkat cells, providing a promising therapy for some cancers.

Nitric oxide-modulated signaling: canonical and noncanonical signaling pathways

Sharon Glynn
University of Galway, Ireland

Contact: sharon.glynn@universityofgalway.ie]

During this educational session, we will discuss mechanisms both canonical versus non-canonical of nitric oxide signaling relevant to cancer research. This will include mechanisms such as cGMP signaling and s-nitrosation. We will examine the impact of NO concentrations and duration of exposure on cell signaling both inhibitory and activating. This will be an interactive session using Mentimeter to engage the audience, and compare and contrast people's knowledge and identify gaps in research. In particular, as an audience, we will identify the existing limitations in our knowledge and areas of need for further investigation. We will also discuss the pros and cons of different technologies for examining NO signaling, people's lived experiences in dealing with them moving from researching NO signaling *in vitro* to *in vivo* and finally into patients.

Nitric oxide in tumor biology: from stemness to metabolic reprogramming

Antonio Rodríguez-Ariza
Instituto Maimónides de Investigación Biomédica de Córdoba (IMIBIC), Córdoba, Spain

Contact: antonio.rodriguez.exts@juntadeandalucia.es

Nitric oxide (NO) plays a complex role in cancer biology, with both pro-tumorigenic and antitumorigenic effects. Over the past years, our laboratory has investigated the contributory role NO in diverse aspects of tumor development and progression. Experimental evidence supports a connection between NO metabolism and stem characteristics in certain tumors and we have shown that an inadequate balance between nitrosylation and denitrosylation promotes the expansion of cancer stem cells (CSCs) in breast cancer. Accordingly, NO-targeted therapy was found to effectively inhibit stemness and enhance the efficacy of antihormonal therapy in estrogen receptor-positive breast cancer cells. Besides, we have described that NO exerts autocrine and/or paracrine effects crucial for the generation and maintenance of cellular subpopulations of cancer stem cells (CSCs) in colorectal cancer (CRC). We have also studied the role of S-nitrosoglutathione reductase (GSNOR), an evolutionarily conserved denitrosylase enzyme with a potential tumor suppressor role. Our research revealed a strong correlation between reduced GSNOR expression in tumors and unfavorable survival outcomes in HER2+ breast cancer patients. Additionally, inhibiting GSNOR activity hampers the efficacy of the anti-HER2 drug trastuzumab. Our research has also elucidated that GSNOR deficiency correlates with an aggressive stem-like subtype of CRC, an immune-suppressed tumor microenvironment, unfavorable prognosis, and diminished survival rates in CRC patients. CRISPR-Cas9-mediated GSNOR gene knockout (KO) provided confirmation that GSNOR deficiency leads to increased tumorigenicity and immune evasion in CRC cells. Significantly, the increased tumorigenic potential observed in GSNOR-KO cells was linked to a metabolic shift from oxidative phosphorylation to aerobic glycolysis. Furthermore, we have offered evidence supporting the notion of targeting the metabolic vulnerabilities associated with this bioenergetic adaptation as a promising strategy for treating aggressive GSNOR-deficient CRC tumors.

Further research is required to understand the role of NO in cancer stemness and metabolic reprogramming. By uncovering the underlying mechanisms, targeted NO therapies can be developed to selectively inhibit cancer stemness and modify metabolic pathways, offering potential avenues for cancer treatment.

Funded by: Consejería de Universidad, Investigación e Innovación (P20-00,967) and Ministerio de Ciencia e Innovación (PID2019-105,256RB-I00).

Harnessing cardiac metabolism through receptor-mediated NO signaling

J-L Balligand, MD, PhD
Institut de Recherche Expérimentale et Clinique (IREC), University of Louvain Medical School, Brussels, Belgium

Contact: jl.balligand@uclouvain.be

Nitric oxide exerts pleiotropic effects on cardiac structure and function through receptor-mediated activation of the constitutive eNOS and nNOS expressed in cardiac myocytes. We and others showed that the third isotype of beta-adrenergic receptors (beta3AR) is coupled to the cardiac NOS and exerts protective effects against myocardial remodeling (e.g., hypertrophy, fibrosis) under hemodynamic or neurohormonal stress. This talk will cover recent observations illustrating the mechanism for this protective effect, particularly through the rewiring of cardiac metabolism. We will show that cardiac-specific moderate expression of the human *ADRB3* in a transgenic mouse model increases glucose uptake and preferentially directs its metabolism through the pentose phosphate pathway. The ensuing production of reducing equivalents (e.g., NADPH) underlies the antioxidant protection in the face of stress-induced ROS production. Lower metabolites are rerouted toward the lower glycolytic pathway for subsequent oxidation. Together with preserved mitochondrial respiration, this results in improved cardiac efficiency and myocardial function. Mechanistically, beta3AR-mediated NO production activates the transcription factor, NRF2 that coordinately drives the upregulation of key enzymes of the PPP as well as regulators of mitochondrial biogenesis to sustain these protective effects under chronic hemodynamic overload.

NRF2 interacts with distal enhancer and inhibits nitric oxide synthase 2 expression in *KRAS*-driven pancreatic cancer cells

Eros Di Giorgio[1], Ylenia Cortolezzis[1], Nicolò Gualandi[1], Valentina Rapozzi[1], Luigi E. Xodo[1]

[1] *Department of Medicine, Laboratory of Biochemistry, P.le Kolbe 4, 33,100 Udine, Italy*
Contact: eros.digiorgio@uniud.it

Nitric oxide is a pleiotropic free radical produced by three nitric oxide synthases (NOS1-3), of which inducible NOS2 is involved in tumor progression. Our RNA-seq, ChIP-seq, and qRT-PCR experiments combined with bioinformatic analyses demonstrated that NRF2 is a repressor of the *NOS2* gene by maintaining a distal enhancer located 22 kb downstream of the transcription start site (TSS) in an inactive state. Deletion of NRF2 leads to the activation of the enhancer, which exerts a pioneering function before its full activation. Specifically, NRF2 controls the expression of *NOS2* in response to intracellular oxidative stress and extracellular oxygen pressure. We found that abrogation of *NOS2* expression by siRNAs partially reduced the ability of WT Panc-1 cells to form 3D spheroids, but strongly reduced the formation of 3D spheroids by NRF2-depleted Panc-1 cells. Mechanistically, this effect correlates with the finding that NOS2 and nitric oxide stimulate epithelial-mesenchymal transition (EMT) in NRF2-depleted cells. We also observed that the knockdown of *NOS2* leads to the blockade of 3D Matrigel invasion of NRF2-depleted Panc-1 cells, demonstrating that a short circuit in the reciprocal regulation of *NOS2* and *NRF2* attenuates the malignancy of PDAC cells. Overall, we show here for the first time that: (i) NRF2 is a suppressor of *NOS2* in pancreatic cancer cells; (ii) a poised enhancer located 22 kb downstream TSS of *NOS2*, whose activation is triggered under conditions that simulate or stimulate NRF2 depletion, controls NOS2 expression; (iii) active NOS2 is required for NRF2-depleted Panc-1 cells to maintain malignancy and invasiveness.

hnRNP modifiers of TDP-43 pathology in Amyotrophic Lateral Sclerosis can act through the control of NOS1AP expression in neurons

S. Cappelli[1], A. Spalloni[2], L. Zangrando[1], C. Stuani[1], F. Feiguin[3], P. Longone[2] and *E. Buratti*[1]

[1] *International Centre for Genetic Engineering and Biotechnology, ICGEB Trieste, Italy;* [2] *Istituto Santa Lucia, Rome, Italy;* [3] *Cagliari University, Cagliari, Italy*
Contact: buratti@icgeb.org

Pathological aggregation of TDP-43 is principally associated with amyotrophic lateral sclerosis (ALS) but is also present in approximately 50% of all frontotemporal dementia patients and is a co-pathology in ~40% of Alzheimer's disease cases and in chronic traumatic encephalopathy (CTE). These observations suggest that TDP-43 pathological aggregation can occur in different brain areas/tissues and in many different pathological conditions, which can be differentially affected depending on the relative abundance and expression of other RNA-binding proteins that are present in the local context.

To address this issue, several years ago, we performed transcriptome analysis of SH-SY5Y cells silenced for DAZAP1, hnRNP Q, hnRNP D, hnRNP K and hnRNP U that were known to affect TDP-43 pathology in fly and human cell models of disease. After cross-comparing transcriptomic profiles of cells depleted by each of these factors, we identified seven commonly regulated transcripts: *CHPF2, IGF2, IRAK2, RNF112, UBE2E3, C1orf226*, and *NOS1AP*. Out of this list, NOS1AP (also known as CAPON) has recently emerged as an important player in brain physiology and pathophysiology. Several studies suggest that its interaction with nNOS contributes to NOS1AP-mediated excitotoxicity, the formation of neuronal processes, and probably schizophrenia. Most importantly, we observed a clear correlation between the reduction of NOS1AP and the inclusion of two previously characterized cryptic exons in different brain regions of patients with TDP-43 pathology. Then, using primary mouse neuronal cultures, we demonstrated that decrease of TDP-43 induced a drop in *NOS1AP* expression and this correlated with a significant decrease in several essential component of the synaptic network: PSD93, PSD95, SynGAP, and *Synapsin-3*. Finally, we observed that upregulation of NOS1AP in TDP-43 depleted SH-SY5Y cells can successfully rescue of many of these genes in the absence of TDP-43. In addition, downregulation of NOS1AP is also capable to rescue on its own the degenerative phenotype induced by TDP-43 overexpression in fly eyes.

Taken together, our data identify NOS1AP as capable of modulating the neurological dysfunctions associated with ALS, allowing a better understanding of the variability of patients affected by this disease, and making this protein a suitable candidate for the development of novel therapeutic strategies in the context of this pathology.

Dietary nitrate supplementation: effects on skeletal muscle and exercise performance in healthy humans

Andrew M. Jones

University of Exeter Medical School, St. Luke's Campus, Exeter, U.K.
Contact: a.m.jones@exeter.ac.uk

Nitric oxide (NO) is a gaseous signaling molecule that plays an important role in myriad physiological processes, including the regulation of vascular tone, neurotransmission, mitochondrial respiration, and skeletal muscle contractile function. NO may be produced via the canonical NO synthase-catalyzed oxidation of L-arginine and also by the sequential reduction of nitrate to nitrite and then NO. The body's nitrate stores can be augmented by the ingestion of nitrate-rich foods (primarily green leafy vegetables). NO bioavailability is greatly enhanced by the activity of bacteria residing in the mouth, which reduce nitrate to nitrite, thereby increasing the concentration of circulating nitrite, which can be reduced further to NO in regions of low oxygen availability. Recent investigations have focused on promoting this nitrate-nitrite-NO pathway to positively affect indices of cardiovascular health and exercise tolerance. It has been reported that dietary nitrate supplementation, for example, with beetroot juice, lowers blood pressure in hypertensive patients, and sodium nitrite supplementation improves vascular endothelial function and reduces the stiffening of large elastic arteries in older humans. Nitrate supplementation has also been shown to enhance skeletal muscle function and to improve exercise performance in some circumstances. Recently, it has been established that nitrate concentration in skeletal muscle is much higher than that in blood and that muscle nitrate stores are exquisitely sensitive to dietary nitrate supplementation and deprivation. This presentation will consider the possibility that nitrate represents an essential storage form of NO and discuss the integrated function of the oral microbiome, circulation, and skeletal muscle in nitrate-nitrite-NO metabolism, as well as the practical relevance for health and performance.

Effects of increased nitric oxide bioavailability on skeletal muscle function in aging

Simone Porcelli

University of Pavia, Italy
Contact: simone.porcelli@unipv.it

Elderly people are characterized by muscle weakness due to an overall loss of skeletal muscle tissue, an impairment of excitation-contraction coupling, and an intramuscular reduced sensitivity to calcium signaling. Moreover, aging seems to negatively affect plasma nitrate concentration. In young healthy adults, potential benefits of dietary nitrate supplementation on skeletal muscle contractile function and muscle fatigue have been recently suggested. This lecture will be focused on the effects of dietary nitrate supplementation on muscle contractile properties in elderly subjects. Results from a study conducted in elderly and young adults will be presented to show beneficial effects of dietary nitrate supplementation, in the form of beetroot juice, on muscle submaximal force production at different frequencies of stimulation, maximal voluntary isometric force, and fatigue resistance. Data obtained from a recent study on mice supplemented by nitrate-enriched water will be used to demonstrate the mechanistic bases of dietary nitrate supplementation on neuromuscular function and muscle force production.

Hydrogen sulfide donor AP123 restores endothelial nitric oxide via CREB-dependent pathway

Brancaleone V[1], Montanaro R[1], Torregrossa R[2,3], Vellecco V[4], Mitidieri E[4], Wood ME[3], d'Emmanuele di Villa Bianca R[4], Bucci M[4], Whiteman M[2], Cirino G[4],

[1]*Department of Science, University of Basilicata, Potenza, Italy;* [2]*University of Exeter Medical School, St. Luke's Campus, Exeter, England;* [3]*Biosciences, College of Life and Environmental Science, University of Exeter, England;* [4]*Department of Pharmacy, University of Naples Federico II, Naples, Italy*
Contact: vincenzo.brancaleone@unibas.it

Vascular impairment is associated to atherosclerosis and other inflammatory-based diseases. The mechanisms underlying these events have been widely disclosed and involve, among other factors, endothelial nitric oxide synthase (eNOS). In particular, eNOS activity is suppressed in hyperglycemic conditions, resulting in a reduced NO bioavailability [1, 2]. In addition, circulating levels of hydrogen sulfide (H_2S) are also suppressed by diabetic conditions, thus further undermining vascular homeostasis in terms of endogenous vasorelaxant response [3]. Here, we tested whether H_2S donor molecules, AP123 and AP39, could interfere with the impairment in NO signaling associated to hyperglycemia.

Thus, we used an *in vitro* model of hyperglycemia, where bovine aortic endothelial cells (BAECs) were grown in high glucose (HG, 50 mM) environment for 3 h. AP123 or AP39 (1 nM-1 μM) was added in HG condition (time 0 or +1 h) and NO/eNOS/cAMP response element-binding protein (CREB) pathway activity was evaluated. In addition, we used PKA and PI3K inhibitors, KT5720 and Wortmannin, respectively, to define the underlying mechanism. Finally, we used an *ex vivo* model mimicking hyperglycemic condition to evaluate the relevance of NO/eNOS/CREB in aorta exposed to HG environment.

AP39 or AP123 similarly restored NOx levels independently from administration time. Nonetheless, at lower concentration, AP39 and AP123 response were significantly different. In particular, AP39 induced an "on/off-like" effect, while AP123 modulated NOx levels in a concentration-dependent manner. Western blot analysis showed that AP39 was unable to modulate activation and expression of eNOS, suppressed by HG conditions. Conversely, AP123 protected BAEC from eNOS downregulation triggered by HG environment and this effect involved the activation of CREB. Indeed, this mechanism is affected by the administration of Wortmannin, thus highlighting a key role for PI3K. *Ex vivo* data showed that AP123 could also restore vascular function as for Ach-induced vasodilatory activity.

Overall, these data highlight a different modulatory mechanism between AP39 and AP123. Indeed, AP123 is more efficient in restoring eNOS/NO signaling through CREB activation via PI3K. These results suggest that H_2S is a crucial vasculoprotective mediator in diabetes and that specific H_2S-releasing molecules may be useful in counteracting the detrimental effects of "H_2S deficiency" associated to hyperglycemia within the vasculature.

References

1. Bucci M et al, Arterioscler Thromb Vasc Biol. 2004 Apr; 24(4):721−726
2. Bucci M et al, Vascul Pharmacol. 2008 Aug-Sep; 49(2−3):84−90
3. Brancaleone V et al, Br J Pharmacol. 2008 Nov; 155(5):673−680

Dissecting the Synergistic Effects of Nitric Oxide Synthase and PI3K Inhibitors in Metaplastic Breast Cancer

Jenny Chang
Houston Methodist Hospital, Texas, USA

Contact: jcchang@houstonmethodist.org

Metaplastic breast cancer (MpBC) is a rare, highly lethal breast cancer that affects <5% of all breast cancers. MpBC is often diagnosed at a later stage with more advanced disease, exhibits a worse prognosis in comparison to other subtypes of triple-negative breast cancer (TNBC), and has a median survival rate of mere 8 months after metastasis. MpBCs are characterized by hyper-activation of the phosphoinositide 3-kinase and protein kinase B (PI3K/AKT) pathway, expression of epithelial-mesenchymal transition (EMT)/cancer stem cells (CSCs) markers, and high levels of nitric oxide (NO). Therefore, we hypothesized that the synergistic inhibition of the nitric oxide synthase (NOS) pathway together with PI3K can eliminate tumor growth and metastases.

Our hypothesis is supported by robust preclinical and clinical data supporting the effects of NOS and/or PI3K inhibitors in MpBC. We have recently published in Science Translational Medicine a phase 1/2 clinical trial with NOS inhibitor L-NMMA and chemotherapy in metastatic TNBC patients (https://doi.org/10.1126/scitranslmed.abj5070). Overall, the regimen was well tolerated, with an overall response rate of 45.8% for the whole cohort. Notably MpBC patients enrolled in this trial also demonstrated significant benefit. Of the 13 MpBC patients in this study, one patient had complete response (CR), two patients had partial response, seven patients had stable disease, and three patients showed disease progression. In addition, our preclinical studies with MpBC cell lines and patient-derived xenografts (PDX) showed that inhibition of NOS pathways in MpBC may eliminate or significantly delay the growth of MpBC, and that these effects are augmented by simultaneous inhibition of PI3K. When treated with NOS and PI3K inhibitors, the MpBC cell lines showed significantly inhibited proliferation at 72 h. The in vivo studies similarly showed reduced tumor growth in taxane-treated mice co-treated with NOS + PI3K pathway inhibitors. Immunostaining analysis revealed that MpBC PDX tumors had elevated co-expression of iNOS and pAkt (60% vs 23%, $p=0.04$) relative to TNBC PDX tumors. Transcriptomic analysis found gene sets associated with EMT reversal, such as the formation of cornified envelope (NES = 2.04 Nom $p<0.00$) and keratinization pathway (NES = 2.06, Nom $p<0.00$) were enriched pathways in MpBC PDX tumors that responded to combination therapy. Together, our results support that combining NOS and PI3K inhibitors may be a novel therapeutic strategy to treat MpBC. This combination therapy will next be tested in a recently funded multicenter Phase 2 clinical trial (NIH U01 grant) to determine whether dual inhibition of PI3K/AKT using Alpelisib and NOS inhibition using L-NMMA combined with nab-paclitaxel will increase the objective response rate and survival in metastatic MpBC patients.

Decoding S-nitrosylation: Theoretical aspects and analytical approaches

Giuseppe Filomeni
Redox Biology, Danish Cancer Society Research Center, Copenhagen, Denmark; Department of Biology, Tor Vergata University, Rome, Italy.

Contact: giufil@cancer.dk

S-nitrosylation is a nitric oxide (NO)-induced, reversible modification that underlies the majority of NO bioactivity. It relies on the chemical addition of an NO moiety to a cysteine residue, thus generating an S-nitrosothiol (SNO). Protein S-nitrosylation induces structural modifications, altering protein function (e.g., protein—protein interactions, localization, and activity) and the capacity to undergo additional posttranslational modifications. It is highly specific and only occurs on specific cysteine residues that meet a number of criteria, such as proximity to the source of NO and reactivity, which is primarily dependent on the nature of neighboring residues that influence cysteine *pKa*.

NO-mediated signaling appears to be compartmentalized and spatially regulated, both because of the localization of NOS and the nature of chemical reactions involving NO, including S-nitrosylation. Recent research reveals, however, that S-nitrosylation-mediated signaling can also propagate far from sites where NO is generated *via* a thiol/nitrosotiol exchange reaction called *trans*-nitrosylation. This strengthens the biological relevance of S-nitrosylation as a redox mechanism of signal transduction in the cell. In fact, many studies of the last few years argue for S-nitrosylation being a crucial regulator of cell and tissue physiology, and the imbalance of protein SNOs has recently been proposed to be implicated in aging and several human diseases, such as neurodegeneration and cancer.

Here, we will provide a detailed overview of the molecular mechanisms through which S-nitrosylation takes place and propagates in cells, and discuss the effects of S-nitrosylation on the regulation of cell homeostasis and response to stress. We will also illustrate the strengths and limitations of the methodology currently in use to analyze S-nitrosylation, attempting to resolve any apparent discrepancies in the literature that have even called into question the stability and physiological role of S-nitrosylation over the last few years.

Widespread cysteine persulfidation in activated macrophages as a protective mechanism against oxidative-inflammatory stress

Moran Benhar[1]

[1]*Faculty of Medicine, Technion-Israel Institute of Technology, Haifa, Israel*
Contact: benhar@technion.ac.il

A hallmark of immune-inflammatory responses is the generation of reactive oxygen and nitrogen species by innate immune cells, in particular macrophages. How macrophages cope with excess oxidant production and associated redox stress is not fully understood. Recent evidence implicates reactive sulfur species (RSS) in inflammatory responses; however, how RSS affect macrophage function and its ability to cope with oxidative-inflammatory stress remains elusive. Herein, we investigated endogenous pathways of RSS biogenesis and clearance in macrophages, and in particular, explored how hydrogen sulfide (H_2S) and thiol persulfidation influences macrophage oxidative-inflammatory response. We report that classical activation of mouse or human macrophages with lipopolysaccharide and interferon-γ (LPS/IFN-γ) triggers a marked production of H_2S/RSS, leading to a widespread increase in protein persulfidation. We find that endogenous H_2S/persulfidation levels are governed by the activity of the cystine importer xCT, the H_2S-generating enzyme cystathionine γ-lyase, and the sulfide-metabolizing enzyme sulfide quinone oxidoreductase (SQOR). High levels of H_2S/persulfidation, due to LPS/IFN-γ stimulation or SQOR inhibition, are associated with a state of increased resistance to oxidative stress. Furthermore, enhanced persulfidation correlates with attenuated activation of the macrophage inflammasome and diminished inflammatory cell death. These findings shed light on the metabolism and effects of RSS in macrophages and point to an important role for persulfides in enabling macrophages to cope with oxidative-inflammatory stress.

Modulation of Oxidative Stress and Inflammation by the Nitrate-Nitrite-NO Pathway in Cardiometabolic and Renal Disorders

Mattias Carlstrom

Dept. of Physiology and Pharmacology, Karolinska Institutet, Stockholm, Sweden
Contact: mattias.carlstrom@ki.se

Nitric oxide (NO) plays a pivotal role in cardiovascular system, but also in the regulation of metabolic and renal functions. Reduced NO bioavailability, linked to compromised NO synthase (NOS) function and oxidative stress, has been associated with cardiometabolic and renal disorders, which often coexist. Treatment strategies that dampen oxidative stress and restore NO homeostasis have been suggested to be associated with therapeutic effects. Dietary nitrate supplementation is one such approach, which may boost the nitrate-nitrite-NO pathway.

Using translational research approaches, we and other groups have showed that inorganic nitrate treatment can ameliorate the development of aging- and diet-induced cardiometabolic dysfunctions as well as kidney disease. Underlying mechanisms include increased NO bioactivity, reduction of NADPH oxidase activity, modulation of mitochondria function, activation of AMPK, which together has been linked to reduced oxidative stress but also dampening of inflammation and fibrosis.

In summary, stimulation of the nitrate-nitrite-NO pathway has been coupled with significant therapeutic value in several experimental disease models. Future clinical studies are aimed at investigating the potential therapeutic value in patients with cardiovascular, metabolic, and renal disease.

A Novel Immunoregulatory Role for iNOS in Itaconate Production Mediated by Protein-Protein Interaction

Marina Diotallevi[a,b], Thomas Nicol[a,b], Faseeha Ayaz[a,b], Ganna O Krasnoselska[c,d], Svenja Hester[e], Roman Fischer[e], Benedikt Kessler[e], Raymond J Owens[d,e,f], Keith M. Channon[a,b] and *Mark J. Crabtree*[a,b,g]

[a]*BHF Centre of Research Excellence, Division of Cardiovascular Medicine, Radcliffe Department of Medicine, John Radcliffe Hospital, University of Oxford, UK;* [b]*Wellcome Trust Centre for Human Genetics, University of Oxford, UK* [c]*Structural Biology, Rosalind Franklin Institute, Harwell Science Campus, UK;* [d]*Protein Production UK, The Research Complex at Harwell, Science Campus, Didcot, UK;* [e]*Target Discovery Institute, Nuffield Department of Medicine, University of Oxford, UK;* [f]*Division of Structural Biology, The Wellcome Centre for Human Genetics, University of Oxford, UK;* [g]*Department of Biochemical Sciences, School of Biosciences & Medicine, University of Surrey, Guildford, GU2 7XH, United Kingdom.*
Contact: mark.crabtree@surrey.ac.uk

Itaconate, a TCA cycle-derived metabolite produced by Immunoresponsive Gene 1 (IRG1, Acod1), is one of the most abundant metabolites in activated immune cells and has pivotal roles in the inflammatory response, metabolic regulation, and redox signaling. We recently discovered that inflamed macrophages lacking inducible Nitric Oxide Synthase (iNOS) (an inducible enzyme generating high levels of nitric oxide (NO) during inflammation) or its cofactor tetrahydrobiopterin (BH4), produced markedly increased amounts of itaconate in comparison with wild-type activated macrophages, by a mechanism independent of IRG1 expression.

To further understand the role of iNOS in mediating itaconate production, and unravel the subsequent immuno-regulatory functions of iNOS, we used bone marrow−derived macrophages cultured from WT, iNOS KO, and Gch1 KO mice (GTP Cyclohydrolase I). Following activation with LPS and interferon γ ($M^{LPS/IFN\gamma}$), we uncover a strong correlation between the presence of iNOS/NO and a striking decrease in the production of itaconate over time.

We reveal that this dynamic inhibition of itaconate production occurs by an iNOS-dependent mechanism that will be presented, with iNOS effectively acting as a brake to control itaconate production and ultimately macrophage polarization state.

Nitric oxide plays a relevant role during cell resistance to Sorafenib treatment in liver cancer cells

Thaissa Horne[1,*], Joana Pieretti[1,*], Patricia de la Cruz-Ojeda[2,3], Elena Navarro-Villarán[1,3], Joaquín Dopazo Blazquez[4,5,6,7], Jordi Muntané[1,3,8]

[1]Institute of Biomedicine of Seville (IBiS), Hospital University Virgen del Rocío/CSIC/University of Seville, Seville, Spain. [2]FUNctional GEnomics of Solid Tumors-FunGeST, Sorbonne Université-INSERM-Université de Paris, Centre de Recherche des Cordeliers, Paris, France [3]Biomedical Research Center for Hepatic and Digestive Diseases (CIBERehd), Madrid, Spain. [4]Clinical Bioinformatics Area, Fundación Progreso y Salud (FPS). CDCA, Hospital Virgen del Rocio, Seville, Spain. [5]Computational Systems Medicine, Institute of Biomedicine of Seville (IBiS), Hospital Virgen del Rocio, Seville, Spain. [6]Bioinformatics in Rare Diseases (BiER), Centro de Investigación Biomédica en Red de Enfermedades Raras (CIBERer), FPS, Hospital Virgen del Rocío, Seville, Spain. [7]FPS/ELIXIR-Es, Hospital Virgen del Rocío, Seville, Spain. [8]Department of Medical Physiology and Biophysics, University of Seville, Seville, Spain.

**All authors have contributed equally to the study*

Background: Sorafenib is a tyrosine kinase inhibitor for the treatment of patients at advanced stage of hepatocellular carcinoma (HCC) who cannot benefit from immunotherapy. The induction of cell death and cell cycle arrest is associated with mitochondrial dysfunction and oxidative/nitrosative stress in cancer cells. The administration of Sorafenib is widely associated with the induction of cell resistance. **Aim:** To characterize the role of oxidative and nitrosative stress in Sorafenib-resistant cells. Methods: We have established a Sorafenib-resistant HepG2 cell line, hereby called R-HepG2. An inhibitor of nitric oxide synthase, L-NAME (5 mM), was administered to Sorafenib (10 μM)-treated R-HepG2 and HepG2 cells. Different parameters related to cell death and proliferation, cell respiration, and oxidative/nitrosative stress were assessed using different experimental approaches. Furthermore, we identified altered pattern of nitrosylated proteins using biotin-switch assay coupled to MS spectrometry. **Results:** Our data showed that R-HepG2 cells presented a different morphology from HepG2 and a higher IC_{50} in Sorafenib-treated cells. R-HepG2 also had higher proliferation rate and lower caspase-3 activity compared to parental Sorafenib-treated HepG2 cells. Interestingly, untreated R-HepG2 demonstrated a basal higher oxygen consumption rate (OCR) than HepG2, but Sorafenib treatment drastically decreased the OCR for both cells. Western blot and proteomics data indicated different S-nitrosylation patterns for R-HepG2 and HepG2 cells. The administration of L-NAME abolished the impact of R-HepG2 vs HepG2 in cell proliferation and death. Superoxide anion and nitric oxide generation were observed in Sorafenib-treated cells. The expression of GSNOR and TRX was increased in R-HepG2 compared to HepG2. **Conclusions:** (1) Nitric oxide plays a relevant role during the induction of cell proliferation and prevention of cell death in Sorafenib-resistant cells. (2) The induction of Sorafenib resistance is associated with a profound alteration of the pattern of S-nitrosylated proteins.

Using AI to discover new cancer therapies

Apostolos Zaravinos[1,2]

[1] *Department of Life Sciences, School of Sciences, European University Cyprus, 2404 Nicosia, Cyprus*; [2] *Cancer Genetics, Genomics and Systems Biology Group, Basic and Translational Cancer Research Center (BTCRC), 1516 Nicosia, Cyprus*

Contact: a.zaravinos@euc.ac.cy

The target audience for this educational workshop is expected to have some background knowledge in drug discovery and development. The workshop aims to introduce attendees to the emerging field of artificial intelligence (AI)-enabled drug discovery and demonstrate how AI techniques can be used to identify new drug targets. Furthermore, the workshop will delve into the role of AI in drug repurposing, as well as in preclinical and clinical drug development. Ethical considerations and challenges associated with AI-enabled drug discovery will also be discussed. The workshop will emphasize emerging trends and future directions in AI research in drug discovery, and highlight opportunities for collaboration between AI experts and drug discovery scientists. The session will be interactive and include hands-on exercises, which will be instrumental in helping attendees gain practical experience in applying AI techniques to drug discovery problems. Attendees will be provided with real-world examples of AI-enabled drug discovery, which will be used to illustrate key concepts and reinforce learning outcomes. Overall, the workshop is intended to equip attendees with the necessary knowledge and skills to effectively leverage AI in drug discovery, and encourage further research in this rapidly evolving field.

Applications of Artificial Intelligence techniques for therapeutic designs targeting modulation of gene expression

Alexandre Ferreira Ramos

Laboratory of Enzymology, Department of Cellular Biology, Biological Sciences Institute, University of Brasilia, Brasilia 70,910-900, DF, Brazil
Contact: alex.ramos@usp.br

Gene editing technologies reached a turning point toward epigenetic modulation for cancer treatment. Gene networks are complex systems composed of multiple nontrivially coupled elements capable of reliably processing dynamical information from the environment despite unavoidable randomness. However, this functionality is lost when the cells are in a diseased state. Hence, gene-editing-based therapeutic design can be viewed as a gene network dynamics modulation toward a healthy state. Enhancement of this control relies on mathematical models capable of effectively describing the regulation of stochastic gene expression. We use a two-state stochastic model for gene expression to investigate treatment response with a switching target gene. We show the necessity of modulating multiple gene-expression-related processes to reach a heterogeneity-reduced specific response using epigenetic-targeting cancer treatment designs. Our approach can be used as an additional tool for developing epigenetic-targeting treatments.

CYB5R3 and cGMP Regulation

Adam C. Straub, PhD
University of Pittsburgh

Contact: astraub@pitt.edu

Introduction: Cytochrome B5 reductase 3 (CYB5R3) is a heme reductase ubiquitously expressed in the cardiovascular system. We discovered a novel role for smooth muscle cell (SMC) CYB5R3 in reducing oxidized heme iron (Fe^{3+}) of soluble guanylate cyclase (sGC) back to its reduced form (Fe^{2+}) to enable nitric oxide signaling for cyclic guanosine monophosphate (cGMP) production and blood vessel relaxation. Our prior work has shown mice deficient in SMC CYB5R3 have higher systemic and pulmonary artery pressures. While there are over 40 human *CYB5R3* variants known to date, one coding variant, CYB5R3 T117S, results in decreased function (~50) and occurs with high frequency (0.23 allele frequency) in persons of African ancestry. We hypothesized that the T117S variant may increase the risk of stroke and poor cardiovascular outcomes. **Methods:** We studied transient middle cerebral artery occlusion (50 min) in mice with CRISPR-Cas9 generated T117S *CYB5R3* (T117S), Cre/Lox induced endothelial cell (EC)-specific CYB5R3 knockout (EC-KO), Cre/Lox induced SMC-specific CYB5R3 knockout (SMC-KO), and wild-type mice, all adult males on a C57Bl6 background. Cerebral infarct volume (mm^3), cerebral blood flow, and vascular function were determined 24 hr later using TTC-stain and ImageJ. **Results:** Infarct volume was greater in T117S versus wild-type controls, whereas cerebral blood flow was impaired, which was replicated in the SMC-KO but not the EC-KO. **Conclusion:** CYB5R3 T117S and SMC-KO of CYB5R3 causes worse stoke in mice, thus determining T117S carrier status in patients may help predict stroke risk and enable sGC targeted precision therapy approaches to stroke management.

Nitric oxide-mediated posttranslational protein modifications in neurodegenerative disease

Michael Heneka[1,2]

[1] *Luxemburg Centre for Systems Biomedicine, University of Luxemburg, L4367 Belvaux, Luxemburg*
[2]*Divison of Infectious Diseases and Immunology, University of Massachusetts, Medical School, Worcester, MA 01,605, USA*
Contact: michael.heneka@uni.lu

The accumulation of neurotoxic or misfolded amyloid beta peptides and/or neurofibrillary tangle formation represent key pathological hallmarks of neurodegenerative diseases including Alzheimer's disease. Evidence from preclinical, translational, genetic, and pathological studies suggests that the activation of distinct innate immune pathways represents an important factor that contributes to disease progression and chronicity. Microglia play a pivotal role in this immune response and their activation can result in NOS_2 induction and subsequent longstanding nitric oxide production. Nitric oxide reacts with molecular oxygen thereby forming peroxynitrite, which in turn can cause posttranslational modifications at tyrosine and cysteine residues. Peroxynitrite-driven changes of the beta-amyloid peptide changes its propensity to aggregate and aggravates the peptides suppression of hippocampal LTP. The latter may also be negatively regulated by NO-mediated posttranslational modifications of synaptic proteins that we have investigated at the cell culture level, in preclinical murine models of Alzheimer's disease and human postmortem tissue.

Interactions between persulfides, nitrite and nitric oxide in cardiovascular health and disease

Melanie Madhani
Institute of Cardiovascular Sciences, University of Birmingham, Birmingham, United Kingdom

Contact: m.madhani@bham.ac.uk

It is well established that nitric oxide, nitrite, and hydrogen sulfide have important physiological functions, and much of this signaling occurs in integrated fashion. However, the exact nature of the chemistry associated with the biological activity of hydrogen sulfide is not yet firmly established. A potential mechanism of hydrogen sulfide is the formation of persulfides from oxidized thiol species. Recent advances in the development of analytical methods have revealed that persulfides, specifically cysteine persulfide and related polysulfide species are endogenously produced in mammalian tissues. Endogenous formation of persulfides likely involves multiple pathways including interaction with other oxidants, enzymatic and nonenzymatic routes depending on cell types and tissue. Our studies demonstrate a fundamental role for reactive persulfides/polysulfides in regulation of redox-dependent signal transduction and cytoprotection from oxidative stress via their antioxidant activities in cardiovascular health and disease. Furthermore, these results highlight the interaction between persulfides, nitrite and nitric oxide, which may play a vital role in the development of therapeutic strategies to combat against cardiovascular disease.

References

1. Griffiths et al, Redox Biology, 2023, 60, 102,605
2. Feelisch et al, Cardiovascular Research, 2020, 116 (1), 51–62

Regions of hypoventilation as determined by the 4Dx scanner suggests that regional and not global inflammation impacts lung function in asthma

Mamta Sumi[1], Rosemary Westcott[1], Eric Stuehr[1], Chaitali Ghosh[1], Dennis J. Stuehr[1], Kewal Asosingh[1], Serpil C. Erzurum[1] and Arnab Ghosh[1*]

[1]*Department of Inflammation and Immunity, Lerner Research Institute, The Cleveland Clinic, Cleveland, Ohio-44,195.*

**Address of Correspondence to: Arnab Ghosh, Department of Inflammation and Immunity/NC22, Lerner Research Institute, The Cleveland Clinic, 9500 Euclid Ave., Cleveland, OH- 44,195, USA.*

Phone: 216-445-6950, Fax: 216-444-9329,

Contact: ghosha3@ccf.org

Asthma is characterized by airway remodeling and hyperreactivity. Our earlier studies determined that the NO-sGC-cGMP pathway plays a significant role in human lung bronchodilation implying that sGC maturation in its heme-containing form is essential for normal bronchomotor tone. However, this bronchodilation is dysfunctional in asthma due to high NO levels which cause sGC to become heme-free and desensitized to its natural activator, NO. Since asthma impacts the various lung segments/lobes differentially, we aimed to map regions of lungs which were hypoventilated using the 4Dx scanner and determine whether such regions also coincided with molecular signatures of sGC dysfunction. Acquiring of the 4Dx scanner by the Cleveland Clinic allowed us to use this technology for quantitative measures of lung function in translational models of mouse asthma with the ultimate aim to improve patient treatments and evaluate potential therapies. Using 4Dx scans on naïve and asthmatic mouse (OVA), we find that naturally hypoventilated lung regions in naïve mice are further aggravated upon developing asthma in the mice (OVA model). We further demonstrate that regions of lung hypoventilation in the inflamed OVA lungs can be tracked by upregulated expression of HO1 and these regions in-turn overlap with regions of heme-free sGC as evidenced by a decreased sGC-$\alpha 1\beta 1$ heterodimer and an increased response to heme-independent sGC activator, BAY 60-2770 relative to less hypoventilated regions. We also find that NO generated from iNOS upregulation in the hypoventilated regions has a higher impact in developing heme-free sGC as increasing iNOS activity correlates linearly with elevated heme-independent sGC activation. Since HO1 interacts with iNOS to modulate its activity, these events may depend on the regulation of iNOS activation. Among the regions of the lung segments, areas S1/S2, IF2, L1, L3, and A1 which are distributed in the four lung lobes were most prone to hypoventilation relative to regions L2, L4, IF1, IF3, and IF4, which seemed less hypoventilated. Recognition of these specific hypoventilated regions enhances our understanding of the inflamed lung in asthma and suggest that regional and not global inflammation impacts lung function in an inflammatory disease such as asthma.

Alpha hemoglobin in endothelium

Brant E. Isakson
Robert M. Berne Cardiovascular Research Center
University of Virginia School of Medicine

Contact: bei6n@virginia.edu

Nitric oxide (NO) is well known to be highly reactive in physiological and pathological processes. Thus, regulation of nitric oxide (NO) is a crucial component in normal physiological function. Here we discuss how expression of endothelial alpha hemoglobin in the systemic resistance arteries acts a potent regulator of NO delivery in mice and humans. We describe the unique localization of alpha hemoglobin in the endothelial cells and how it's spatial localization informs it's function. We also will describe the development of novel mouse reagents to target alpha globin including Hba1 floxed and Hba1$^{D36\text{-}39}$ to understand the physiological and pathological role of the globin, as well as therapeutic applications of the HbaX peptide used to disassociate alpha hemoglobin and eNOS. We also provide new data to demonstrate how endothelial cell regulation of iron is a key component to the alpha hemoglobin expression in in endothelial cells. Last, we put this work in the context of human pathology, including sickle cell disease.

NO-mediated neuroinflammatory pathways as treatment targets in neurodegeneration

Joern Steinert
School of Life Science, University of Nottingham, Nottingham, UK

Contact: joern.steinert@nottingham.ac.uk

Many neurodegenerative diseases associated with protein misfolding (e.g., Alzheimer's and Parkinson's disease) exhibit enhanced oxidative and nitrergic stress conditions following the activation of diverse neuroinflammatory pathways. The underlying activation of microglia within the central nervous system is responsible for the release of pro-inflammatory molecules. Additional neuroinflammatory signaling is mediated by the release of nitric oxide (NO) following expression of the inducible NO synthase in activated microglia. This prolonged activation constitutes a potent contribution to redox stress. NO and its related nitrergic species in particular are responsible for inducing posttranslational protein modifications, such as S-nitrosylation and 3-Nitrotyrosination, both of which may render cellular proteins dysfunctional. A further contributory factor to redox stress in disease is associated with dysregulation of a metal homeostasis include that of manganese (Mn), zinc (Zn), in addition to redox-sensitive metals, such copper (Cu) and iron (Fe). Their regulation is mediated by divalent metal transporter 1 (DMT1), a multimetal transporter with a primary role in iron trafficking while the ZRT/IRT-like protein-14 can mediate cellular uptake of iron, zinc, and manganese. In many diseases, nonenzymatic and irreversible glycation signaling has been implicated as a further underlying pathway that promotes protein misfolding via the generation of advanced glycation end-products (AGEs) followed by the activation of specific receptors recognizing AGEs (RAGE). However, the direct interactions between both, NO-mediated redox and AGE signaling pathways, are poorly understood. We studied various aspect of redox stress in neurodegeneration and assessed the therapeutic potential of suppressing NO signaling during early disease progression. Utilizing the prion disease mouse model, we characterized hippocampal markers of oxidative and nitrergic stress and redox metals and analyzed neuronal function and protein misfolding, detailing the progression of neurodegeneration. Increased neuroinflammatory signaling was observed in diseased mice characterized by enhanced nitrergic stress and redox metal dys-homeostasis, all associated with a decline in hippocampal neuronal function. Daily in vivo administration of the NOS inhibitor L-NAME prevented the functional degeneration of hippocampal neurons in prion-diseased mice and reduced 3-nitrotyrosination of triose-phosphate isomerase (TPI), an enzyme involved in the formation of disease-associated glycation and AGE formation. This work concludes that NO-mediated posttranslational modifications of TPI may enhance glycation signaling which contributes to cytotoxicity and thus illustrates an interaction between glycation and NO signaling.

The design of photoactivated therapeutics for controlled nitric oxide release

Gregory, Giles[1], Houman, Alimoradi[1], Sweta, Kumari[1], and Carol, Bussey[2].

[1]*Department of Pharmacology & Toxicology, School of Biomedical Sciences, University of Otago, New Zealand.*
[2]*Manaaki Mānawa Centre for Heart Research, Faculty of Medical and Health Sciences, University of Auckland, New Zealand.*
Contact: gregory.giles@otago.ac.nz

Hypothesis: We hypothesized that metabolically stable nitric oxide (NO) donors could be activated *in situ* via applied light (photoactivation), allowing the spatiotemporal control of NO fluxes for therapeutic applications.

Introduction: NO-based therapeutics can potentially be used to treat a range of pathologies including cancers, inflammatory disorders, and infectious diseases.[1] However clinically proven NO donors spontaneously degrade via metabolic pathways, limiting their therapeutic utility due to off-target vasodilatory action.

Aims: We set out to design NO donors based upon sterically hindered S-nitrosothiol groups. We aimed to demonstrate (**1**) that these compounds were stable to metabolism but could be photodegraded via applied light. (**2**) That they could be photoactivated *in vivo* to release NO, and (**3**) that controlled NO release could have therapeutic efficacy.

Experimental design: NO release measurements were conducted via oxy-myoglobin oxidation or an NO sensing electrode. The anticancer effects of NO release were assessed against A549 and HeLa cell lines. Tissue vasorelaxation measurements were conducted in a rat aortic ring organ bath apparatus coupled to an isometric force transducer. *In vivo* systemic and localized hemodynamic responses were assessed in Sprague Dawley rats. Drug action was compared to the existing NO donors sodium nitroprusside and S-nitrosoglutathione.

Results: We developed first, second, and third generation photoresponsive NO-releasing systems that displayed controllable half-lives ranging from 10 min to ~135 days depending upon photostimulus.[2–4] A range of anticancer effects were observed between 0.1 and 100 μM NO. In *ex vivo* aortic tissue lead compound tDodSNO display photoactive NO release, with an eightfold change in vasorelaxation EC_{50} under irradiation (from 8 to 1 μM).[5] *In vivo* tDodSNO (0.1−3.0 mg/kg) displayed low systemic effects, while targeted photoactivation resulted in localized vasodilatory responses, with increases to mean (26.0% ± 7.3%) and maximum (53.2% ± 10.4%) blood flow and decreases to vascular resistance (27.1% ± 3.9%).[6] Over this dosing range, tDodSNO did not induce met-Hb formation in blood, or display any signs of toxicity.[6]

Conclusions: These data demonstrate that drugs based upon a metabolically stable S-nitrosothiol motif can be photoactivated *in vivo* to release NO, and that such agents cause less systemic side effects than existing NO donors. Our findings support the use of S-nitrosothiols to enable the targeted release of NO for therapeutic applications.

References

1. Serafim R.A.M. *et al.*, Current Medicinal Chemistry, 2012, 19, 386−405.
2. Giles N.M. *et al.*, Chemical Biology and Drug Design, 2012, 80 (3), 471−478.
3. Alimoradi H. *et al.*, Free Radical Biology and Medicine, 2019, 130, 297−305.
4. Alimoradi H. *et al.*, Nanotechnology, 2022, 33, 195,101.
5. Kumari S. *et al.*, European Journal of Pharmacology, 2014, 737, 168−176.
6. Giles G.I. *et al.*, Nitric Oxide, 2022, 129, 53−62.

Recent developments of selective iNOS inhibitors as anticancer agents

Cristina Maccallini[1], Rosa Amoroso[1], M. Encarnación Camacho[2], M. Dora Carrión[2], Marialucia Gallorini[1], Belén Rubio-Ruiz[2,3]

[1]*Department of Pharmacy, University "G. d'Annunzio" of Chieti-Pescara, Via dei Vestini 31, 66,100, Chieti, Italy;* [2]*Department of Medicinal and Organic Chemistry, Faculty of Pharmacy, Campus Cartuja s/n, University of Granada, 18,071, Granada, Spain;* [3] *GENYO, Centre for Genomics and Oncological Research, Pfizer/University of Granada/Andalusian Regional Government, PTS Granada, Avda. Ilustraci on 114, 18,016, Granada, Spain*
Contact: cristina.maccallini@unich.it

Introduction: Nitric oxide and its production by the inducible Nitric Oxide Synthase (iNOS) is an established mechanism critical to tumor promotion or suppression [1]. In different cancers, such as the colon, breast, prostate, and cervical cancers, as well as in melanoma and leukemia, high iNOS expression has correlated relatively well with poor patient survival; therefore, it can be considered as a druggable target to counteract those cancers progression [2]. Encouraging results have been obtained from some iNOS inhibitors such as 1400 W or L-NAME in reducing triple-negative breast cancer, glioma, and melanoma growth, but these compounds are usually used at very high doses, typically in the mM range, due to their unfavorable pharmacokinetics [3–5].

Aims: To develop new potent and selective iNOS inhibitors endowed with an improved action profile as anticancer agents.

Experimental design: The leading molecular structure of 1400 W or BYK 191,023 was properly modified introducing ionizable and/or lipophilic moieties. The NOS L-citrulline assay was used to screen compounds activity. Different cell lines (C6 rat glioma cells, MDA-MB-231 cells, CRL-9855 monocytes, and *in vitro* differentiated macrophages) were adopted to evaluate the biological effects of the most promising iNOS inhibitors.

Results: A phenylamidine-based iNOS inhibitor was able to impair cancer cells proliferation and migration with improved potency with respect to 1400 W. Currently, biological evaluations of selected iNOS inhibitors showing an imidazole core are ongoing.

Conclusions: The adopted medicinal chemistry approach led to the disclosure of new effective iNOS inhibitors. The preliminary observed anticancer and immunomodulatory effects pave the way for the preclinical development of the selected compounds, and their further evaluation in disease models.

References

1. Khan, F.H.; et al, Int. J. Mol. Sci. 2020, 21, 9393
2. Liao, W. et al, BioMed Research International 2019, 6,304,851
3. Granados-Principal, S. et al, Breast Cancer Res. 2015, 17, 1–16
4. Maccallini, C. et al, Eur. J. Med. Chem., 2018, 152, 53–64
5. Sikora, A.G. et al, Clin. Cancer Res. 2010, 16, 1834–1844

Nitric oxide donating molecules and nanosystems: advances and perspectives on the current chemical tools

Greta Varchi

Italian National Research Council - Institute of Organic Synthesis and Photoreactivity - Via Piero Gobetti, 101 40,129 Bologna – Italy
Contact: greta.varchi@isof.cnr.it

Nitric oxide (NO) has emerged as a promising molecule for cancer treatment due to its ability to induce cell apoptosis, inhibit angiogenesis, and enhance the antitumor immune response. However, the direct use of NO is limited due to its short half-life and rapid degradation. Therefore, researchers have developed nitric oxide donating molecules (NODs) and nanosystems to overcome these limitations and enhance the therapeutic potential of NO. (1) Recent advances in NO-donating systems have shown great promise in cancer treatment, especially when combined with other therapies such as chemotherapy, radiotherapy, and immunotherapy. For example, NO-releasing nanoparticles have been developed for the targeted delivery of NO to tumor cells, enhancing the cytotoxic effects of chemotherapy drugs, reducing their side effects, and avoiding chemoresistance. (2) Furthermore, NO donors have been used to enhance the immune response against cancer cells, improving the efficacy of immunotherapy. For instance, NO-releasing nanoparticles have been shown to stimulate the activity of immune cells and enhance their antitumor activity. Additionally, the combination of NO-releasing compounds with immune checkpoint inhibitors has been shown to enhance the immune response against cancer cells and improve treatment efficacy. (3) One of the major advantages of NO-donating systems in cancer treatment is their ability to provide targeted and controlled delivery of NO to specific tumor cells, reducing systemic toxicity and improving treatment efficacy. Indeed, these systems can be designed to release NO in response to specific stimuli, such as changes in pH, temperature, light, and ultrasounds, allowing for precise control over NO release. (4) The seminar will cover the recent advances and challenges in the clinical translation of NO donating systems for cancer treatment, including the development of molecules with high NO release capacity and improved biocompatibility, as well as their potential in combination with other therapies such as chemotherapy, radiotherapy, and immunotherapy for targeted and controlled delivery of NO to specific tumor cells.

References

1. Vong LB, Nagasaki Y. Nitric Oxide Nano-Delivery Systems for Cancer Therapeutics: Advances and Challenges. *Antioxidants* (Basel). **2020**; 9:791.
2. Kim J, Yung BC, Kim WJ, Chen X. Combination of nitric oxide and drug delivery systems: tools for overcoming drug resistance in chemotherapy. *J Control Release.* **2017**; 263:223–230.
3. Kim J, Francis DM, Sestito LF, Archer PA, Manspeaker MP, O'Melia MJ, et al. Thermosensitive hydrogel releasing nitric oxide donor and anti-CTLA-4 micelles for anti-tumor immunotherapy. *Nat Commun.* **2022**; 13:1479.
4. Ji X, Zhong Z. External stimuli-responsive gasotransmitter prodrugs: Chemistry and spatiotemporal release. *Journal of Controlled Release.* **2022**; 351:81–101.

Preclinical models to assess the pharmacological properties of NO derivatives

Lucia Morbidelli[1] and Valerio Ciccone[1]

[1]*Dept. Life Sciences, University of Siena, Via A. Moro 2 53,100 Siena (Italy)*
Contact: lucia.morbidelli@unisi.it

The field of pharmacology of nitric oxide (NO) has various clinical perspectives as cardiovascular and metabolic disorders, neuro-vascular and neurodegenerative diseases, muscular-skeletal disorders, ocular, respiratory, and a large series of inflammation related pathologies, and, on top of these, infective and neoplastic diseases. Depending on the pathology and pathophysiological mechanisms, different pharmacological approaches can be developed and used, either to stimulate NO synthases (NOSs), improve NO availability, and activate downstream NO-related pathways or, on the opposite, to downregulate NOS, scavenge NO, and inhibit NO-related signaling.

In either conditions, preclinical tools and models are needed to allow molecule or therapeutic strategy screening, evaluating the relative potency and efficacy in a simple, relatively cheap, and not time-consuming manner.

Different models have been developed during time, starting from the classic isolated organs maintained in balanced buffers and measuring their typical functional responses as relaxation/contraction. The introduction of cell cultures, genetic manipulation, and molecular studies has allowed to go in deep detail on the cellular and molecular mechanisms controlling upstream and downstream NO signaling. The lesson will present (i) the molecular tools to measure the levels and intracellular localization of NO and redox species, (ii) the cell types and cellular models (2D and 3D) available to study functional aspects of NO-related strategies as well as the control of cell−cell interaction (1), (iii) the protocols to evaluate the activation and cell localization of receptor(s), NOS isoforms, and postreceptor signaling candidates. By the use of selective biochemical inhibitors of specific signaling pathways, the involvement of intracellular messengers can be assessed and verified in functional responses both on cell cultures and in isolated organs (2).

Finally, animal models reproducing human diseases or their symptoms can be used to assess the efficacy of synthetic strategies, NO delivery systems, and potential drugs or drug combinations in controlling pathology progression or in inducing disorder regression (3). Efficacy studies on experimental models, together with safety assessment, indeed allow to obtain predictive information for the proper design of safe human clinical trials.

References

3. Ciccone V. et al, Cancers 2022, 14(17), 4240.
4. Koika et al, Vascul. Pharmacol. 2010, 53(5–6), 215–222.
5. Monti M, et al, Pharmacol. Res. 2016,107, 352–335

Hybrid and composite porous materials as vehicles to deliver therapeutic nitric oxide

Moisés L. Pinto[1]

[1] *CERENA. Departamento de Engenharia Química, Instituto Superior Técnico, Universidade de Lisboa, Lisboa, Portugal*
Contact: moises.pinto@tecnico.ulisboa.pt.

Nanoporous materials with the capacity to store and release nitric oxide (NO) have been explored for therapeutic purposes in recent years [1,2]. Localized administration of NO can be a strategy to treat several human diseases, avoiding several side effects of systemic administration of NO donors. Nanoporous solid materials can lead to interesting therapeutic approaches for local delivery [1]. However, they need to be properly designed to achieve a balance between several key characteristics, namely: (1) high stability in biological media, (2) slow and controlled release kinetics in liquid phase, and (3) low toxicity at the relevant concentrations for biological effects.

The structure and the chemical environment in the pores are the most important aspects to give materials a slow delivery. The NO storage within the pores is normally assured by a coordination of the NO to unsaturated metal centers that offer covalent bonding of the NO. The release is usually achieved by a displacement of the NO by water molecules when in contact with biological media. Thus, the pore size and hydrophilic/hydrophobic balance of the pores needs to be tuned to achieve an efficient loading but also a slow release. The stability of the structure in biological media needs to be assured to guarantee that the release occurs within the expected timeframe. On the other hand, the toxicity of the material can be influenced by the chemical nature, particle size, and stability. Different types of nanoporous materials for NO delivery will be highlighted in this presentation with a special focus on some breakthroughs related with the control of biological functions [3–5]. The materials are normally obtained as powders, and their therapeutical application in this form is often not possible or very difficult. Incorporation in polymer matrixes is a possible strategy to overcome these problems and achieve a composite with NO storage/release properties. Additionally, this approach may enhance biocompatibility if biocompatible polymers are used.

These composites have also the important benefits of slowing, in some cases, the release rate of NO in liquid phase [6–8] and allow to produce shaped composites more suitable for therapeutic applications than powdered forms. Before applications of these nanoporous NO carries in clinical environment, several important aspects still need to be validated. Yet, the results obtained with some materials are very encouraging and motivate further research for several application areas.

Acknowledgments

Fundação para a Ciência e a Tecnologia (FCT) is acknowledge for funding to projects UIDB/04,028/2020 & UIDP/04,028/2020 (CERENA)

References

1. R.V. Pinto, S. Carvalho, F. Antunes, J. Pires, M.L. Pinto, ChemMedChem, 2022, 17 (1), e202100429.
2. M.L. Pinto, R.V. Pinto, Nanoporous Materials: New Generation of Nitric Oxide Donors. In L. Morbidelli, B. Bonavida (Eds.), Therapeutic Application of Nitric Oxide in Cancer and Inflammatory Disorders, 2019 (pp. 277–305). Academic Press.
3. R.V. Pinto, S. Wang, S.R. Tavares, J. Pires, F. Antunes, A. Vimont, G. Clet, M. Daturi, G. Maurin, C. Serre, M.L. Pinto, Angew. Chem. Int. Ed., 2020, 59 (13), 5135–5143.
4. M.L. Pinto, J. Rocha, J.R.B. Gomes, J. Pires, J. Am. Chem. Soc., 2011, 133(16), 6396–6402.
5. R.V. Pinto, F. Antunes, J. Pires, V. Graça, P. Brandão, M.L. Pinto Acta Biomaterialia, 2017, 51, 66–74.
6. R.V Pinto, S. Carvalho, F. Antunes, J. Pires, M.L. Pinto, ACS App. Nano Mater., 2022 5(7), 9230–9240.
7. S. Carvalho, R.V. Pinto, J. Pires, J. Rocha, F. Antunes, M.L. Pinto, Microporous Mesoporous Mater. 2023, 350, 112,432.
8. M. Batista, M., M.L. Pinto, F. Antunes, J. Pires, S. Carvalho, Materials, 2021, 14(21), 6701.

Nitric oxide consumption by intravascular hemolysis predicts pulmonary artery capacitance and right ventricle systolic function after cardiopulmonary bypass: results from a prospective cohort study

Emanuele Rezoagli[1,2] Simone Redaelli[1,3] Lorenzo Berra[4,5]

[1]*School of Medicine and Surgery, University of Milano-Bicocca, Monza, Italy;* [2]*Department of Emergency and Intensive Care, Fondazione IRCCS San Gerardo dei Tintori, Monza, Italy* [3]*Department of Anesthesia, Critical Care and Pain Medicine, Beth Israel Deaconess Medical Center, Harvard Medical School, Boston, MA, USA;* [4]*Department of Anesthesia, Critical Care and Pain Medicine, Massachusetts General Hospital, Harvard Medical School, Boston, MA, USA;* [5]*Respiratory Care Department, Massachusetts General Hospital, Boston, MA, USA*
Contact: emanuele.rezoagli@unimib.it

Introduction: Cardiopulmonary bypass (CPB) during cardiac surgery has been associated with hemolysis that results in increased plasma hemoglobin and consequent nitric oxide depletion [1]. This may translate into worsened pulmonary hemodynamics after CPB.

Objectives/Hypotheses: We hypothesized that CPB-related hemolysis is associated with lower pulmonary artery compliance and worse right ventricle systolic function.

Aims: Assessment of nitric oxide consumption, pulmonary artery capacitance, and right ventricle function at 12 h after CPB. Degree of association between nitric oxide consumption with pulmonary artery capacitance and right ventricle function after CPB.

Experimental design: Patients undergoing cardiac surgery with CPB at Massachusetts General Hospital, USA, between June 2014 and March 2015 were prospectively included. Plasma nitric oxide consumption was assessed before and after CPB, at 15 min, 4 and 12 h. Pulmonary artery capacitance (i.e., stroke volume-to-pulmonary artery pulse pressure ratio), and right ventricle function index (right ventricle end systolic pressure-to-cardiac output ratio)—a surrogate of right ventricle systolic function [2]—were measured at 15 minutes, 4 and 12 hours after CPB by a pulmonary artery catheter. Mixed-effects models for repeated measures and post-hoc multiple comparisons with Dunnett's correction were applied. Correlation between changes in nitric oxide consumption and hemodynamics between 15 min and 4 h after CPB was assessed by the Pearson correlation coefficient.

Results: Forty-six patients (26.1% female, median [IQR] age 66 [57 to 75] years, median [IQR] body mass index: 29.1 [26.6 to 33.0] kg/m^2) were included in this subanalysis. Pulmonary artery capacitance was the lowest immediately after the end of CPB and gradually increased over time. Similarly, the right ventricle function index was the highest 15 min after the end of CPB and decreased at the subsequent time points. The improvement of these hemodynamic parameters paralleled the decrease in nitric oxide consumption over the same time points. The nitric oxide consumption reduction between 15 min and 4 hours after CPB significantly correlated with the improvement in pulmonary artery capacitance ($r = 0.46$, $p = 0.002$) and right ventricle function index ($r = -0.79$, $p < 0.001$).

Conclusions: In this study, patients undergoing CPB showed a gradual increase of PAC and a gradual decrease of RVFI since 15 min until 12 h follow-up. The change in nitric oxide consumption—a robust biomarker of intravascular hemolysis—was significantly associated with the change in PAC and RVFI between 4 h and 15 min after CPB.

References

1. Rezoagli E, Ichinose F, Strelow S, Roy N, Shelton K, Matsumine R, Chen L, Bittner EA, Bloch DB, Zapol WM, Berra L. Pulmonary and Systemic Vascular Resistance after Cardiopulmonary Bypass: Role of Hemolysis, Journal of Cardiothoracic and Vascular Anesthesia, 201,731(2):505–515.
2. Couture EJ, Moses K, Monge García MI, Potes C, Haddad F, Grønlykke L, Garcia F, Paster E, Pibarot P, Denault AY. Longitudinal Validation of Right Ventricular Pressure Monitoring for the Assessment of Right Ventricular Systolic Dysfunction in a Large Animal Ischemic Model. Crit Care Explor. 2023 18; 5(1):e0847.

Effects of NO-supplementation Flexovital® on a novel reno-cardio-metabolic disease model to study diabetic nephropathy

L. Carvalho[1], M. Shimari[1], A. Boeder[1], M. M. O. Monteiro[2], K. Piraisoody[3], D. D. Guimarães[1], T. Schiffer[1], Z. Zhuge[1], M. Cai[1], C. Hagberg[1], P. Nordlander[3], B. Fellström[3], D. C. Andersson[1], J. O. Lundberg[1], E. Weitzberg[1], M. Carlström[1]

[1] *Department of Physiology and Pharmacology, Karolinska Institutet, Stockholm, Sweden*
[2] *Department of Biotechnology, Federal University of Paraiba, João Pessoa, Brazil* [3] *Department of Medical Sciences, Uppsala University, Uppsala, Sweden*
Contact: lucas.carvalho@ki.se

Abstract not available

Oxidative stress induces Cysteines-mediated tetramerization of the mitochondrial chaperone TRAP1

Claudio Laquatra[1,*], Fiorella Faienza[2,*], Matteo Castelli[3], Giorgio Colombo[3], Andrea Rasola[1], Giuseppe Filomeni[2,4,5]

[1], Department of Biomedical Sciences, University of Padova, Viale G. Colombo 3, I-35,131, Padova, Italy. [2], Department of Biology, University of Rome Tor Vergata, Via della Ricerca Scientifica I-00,133, Rome, Italy[3], Department of Chemistry, University of Pavia, via Taramelli 12, I-27,100, Pavia, Italy[4], Redox Biology, Danish Cancer Society Research Center, Strandboulevarden 49, DK-2100, Copenhagen, Denmark[5], Center for Healthy Aging, University of Copenhagen, Blegdamsvej 3B, DK-2200 Copenhagen, Denmark.

*These authors equally contributed to this work
Contact: claudio.laquatra@unipd.it

Expression of the Hsp90-family chaperone TRAP1 is restricted to mitochondria and increased in most tumor types. TRAP1 rapidly tunes metabolism with the changing environmental conditions and shields tumor cells from oxidative stress injury, ensuring survival under a wide array of pathophysiological condition. Here, we aim at connecting structural alterations elicited on TRAP1 by redox dyshomeostases to changes on its function and on downstream biological outputs.

By applying a stepwise approach that starts from molecular dynamics analyses and ends up with tumorigenesis assays, we have found that posttranslation modification of Cysteines (Cys) residues are crucial in regulating the activity of TRAP1 under oxidative stress. We report that TRAP1 is S-nitrosylated at Cys501, a redox modification associated with decreased ATPase activity and enhanced proteasomal degradation of the chaperone, whereas a nitrosylation-resistant C501S variant is more active and confers protection to cell death.

Moreover, we have observed that TRAP1 forms redox-sensitive tetramers via disulfide bonds involving Cys261 and Cys573. In cancer cells, TRAP1 tetramerization is induced by many oxidative agents, whereas the expression of a double Cys261-Cys573R mutant abrogates this effect and enhances TRAP1 ATPase activity. The expression of Cys261-Cys573R TRAP1 markedly increases mitochondrial ROS levels while reduces growth and invasion of an aggressive sarcoma cell model.

Our data indicate that oxidative stress causes posttranslation modifications in Cys residues of TRAP1 that controls its activity and structural rearrangements, thus modulating its pro-neoplastic functions.

Conference concluding remarks and remarks

Benjamin Bonavida[1], Luigi Xodo[2] & Valentina Rapozzi[2]
[1] *UCLA, Los Angeles, CA;* [2]*Department of Medicine, University of Udine, P.le Kolbe 4, 33100 Udine, Italy*

Contact: bbonavida@mednet.ucla.edu

As we draw to a close this remarkable scientific conference, we would like to take a moment to reflect in the insightful discussions, groundbreaking research, and valuable collaborations that have unfolded over the past 2 days. This conference has truly been a testament to the remarkable progress and dedication of the scientific community. Throughout the conference, we have witnessed an exceptional range of topics being presented and explored and witnessed advancements in biochemical, molecular mechanisms and therapeutic applications of NO in various diseases. The breadth and depth of knowledge presented at this conference have been truly inspiring. We will briefly reflect on the main themes that emerged as significant in this conference and highlight groundbreaking research and translational applications. In addition, we will also highlight emerging areas of research that were presented and discussed. We will also briefly discuss the main challenges that we experienced and address various means to overcome these challenges. Importantly, we will suggest interdisciplinarity collaborations to address complex scientific challenges. We will also highlight the importance of ongoing collaborations and knowledge sharing. The above will be followed by an open discussion by all participants

Posters' abstracts

An arginine to citrulline conversion controls the epigenome and the proteome of cancer cells leading to the acquisition of resistance

Francesca Agostini[1], Annalisa Ferino[1], Ylenia Cortolezzis[1], Valentina Rapozzi[1], Eros Di Giorgio[1], Luigi E. Xodo[1]

[1]*Laboratory of Biochemistry, Department of Medicine, University of Udine, Italy*
Contact: francesca.agostini@spes.uniud.it

KRAS mutations are frequently observed in pancreatic ductal adenocarcinoma (PDAC), in colorectal adenocarcinoma (CRC) and in nonsmall cell lung cancer (NSCLC). These tumors were reported to be KRAS-addicted, but differently from what was thought at the beginning, a relevant part of KRAS dependency is metabolic. We have recently demonstrated that the blockage of KRAS signaling achieved by knocking-out a down-stream target, the antioxidant transcription factor NRF2, or by forcing the degradation of KRAS with the KRAS G12C degrader LC-2, causes a metabolic reprogramming of PDAC cells, which switch off glycolysis and become auxotrophic to arginine. Arginine boosts the activity of creatine kinase and in turn the synthesis of phosphocreatine, thus feeding the energy demand of cancer cells. In PDAC cells, the treatment with neoadjuvant FOLFIRINOX therapy caused a similar burst in arginine auxotrophy. Interestingly, the dependence on arginine is coupled to the transcriptional induction of Nitric Oxide Synthase (NOS) 2, Peptidyl Arginine Deiminase (PADI) 1, and PADI3. These are different enzymes which all catalyze the conversion of arginine into citrulline. NOS2 is regulated by a distal enhancer through the engagement of NRF2 and acts on free arginine, influencing its availability for the synthesis of phosphocreatine. Differently, PADIs work on peptidyl-arginine residues of proteins. In this way, PADIs can influence the activity of targeted enzymes as protein citrullination modifies the secondary and tertiary structure, and can also modify the state of the chromatin by histone citrullination, which promotes gene expression by creating a relaxed chromatin active for the engagement of histone acetyltransferases. In this project, we aim to evaluate the contribution of NOS2, PADI1, and PADI3 enzymes in controlling the metabolic resetting of cancer cells that have acquired resistance to KRAS inhibition, NRF2 blockage, and FOLFIRINOX therapy and that are all characterized by a reexpression of these enzymes. It remains also to be clarified how PADIs are regulated in this context and why only PADI1 and PADI3 are upregulated even though all the PADI family's members (PADI1,2,3,4,6) are encoded by a genic cluster (1p36.13) comprised in the same topological-associated domain. We found that hypoxia triggers the expression of PADI1, PADI3, and NOS2 in PDAC and CRC cancer cells. Inhibition of PADI with BB-Cl amidine perturbed the viability of cancer cells. CRC and PDAC patients with higher mRNA expression levels of PADI1 and PADI3 were characterized by a worse prognosis. As future experiments, we aim to: (a) identify the signature of genes under the control of NOS2, PADI1, and PADI3 by evaluating the transcriptome in cells treated with 1400W or BB-Cl amidine inhibitors and in knocked-down cells; (b) define the citrullinome by identifying the citrullinated proteins in hypoxia and normoxia; (c) identify the direct transcriptional targets of PADI1 and PADI3 by obtaining the H3R8cit profile in PADI1 and PADI3 knocked-down cells in respect to control cells; (d) identify the epigenetic and post-transcriptional mechanisms that control the expression of PADI1 and PADI3; (e) quantify the impact of NOS2 up-regulation in perturbing arginine metabolism. Even though we are only at the beginning of this project, a better comprehension of the role of NOS2 and PADIs is of particular interest because it will allow to understand more deeply the molecular mechanisms that sustain the metabolism of cancer cells. This could also allow the identification of new and more specific targets for therapy.

Nitric Oxide and Tumor Development: Effects in Cell-Induced Platelet Aggregation in Colorectal Cancer

Fernanda LT Aquino[1], Fernando A Fonseca[2], Isabelle T Almeida[1], Maytê S Toledo[1], Ana Caroline S Teodoro[1], Ricardo C Cintra[1], Hugo P Monteiro[1]

[1] *Department of Biochemistry-Center for Cellular and Molecular Therapy-Universidade Federal de São Paulo – São Paulo – Brazil.* [2] *ABC Medical School, Santo André, São Paulo, Brazil*
Contact: ufal.fernanda@gmail.com

Background: Nitric oxide (NO) is a signaling free radical produced by normal and tumor cells, with growing evidence on its role in tumor initiation and development. NO potentially regulates epithelial-mesenchymal transition (EMT)[1] and tumor cell-induced platelet aggregation (TCIPA)[2], indicating the importance of fine intracellular regulation of NO levels in tumor cells[3].

Methods and Results: We investigated the role of NO/NOS2 in tumor development using two human colorectal cancer (CRC) cell lines: SW480 derived from the primary tumor and SW620 from lymph node metastasis. NOS2 expression was manipulated by permanently knocking it down in SW620 cells (SW620-SH2) or overexpressing it in SW480 cells (SW480–NOS2+). SW620-SH2 cells exhibited lower NO/NOS2 levels and displayed morphological and EMT marker expression resembling SW480 cells. Abnormalities in the coagulation system, particularly TCIPA, contribute to CRC patient mortality. Expression levels of TCIPA-associated proteins (fibronectin and ICAM), tissue factor, and PDI were determined in SW480, SW620, and SW480–NOS2+ cell lines. SW480 cells exhibited high expression levels of tissue factor, ICAM, and fibronectin, effectively promoting TCIPA. SW620 cells do not express these proteins and therefore do not promote TCIPA. SW480–NOS2+ cells demonstrated lower tissue factor expression as compared to SW480 cells and reduced thrombus formation capabilities. Non-significant changes on PDI expression levels were observed in all cell lines.

Conclusion: Collectively, NO/NOS2 levels in CRC cell lines regulate EMT, determine tissue factor expression, and are directly associated with TCIPA. We propose that in addition to ICAM and fibronectin, TCIPA may occur by the combined actions of NOS2, tissue factor, and PDI. Our findings emphasize the importance of NO/NOS2 in the regulation of EMT and TCIPA in CRC cell lines. Understanding the intricate molecular mechanisms underlying these processes may lead to the characterization of new potential therapeutic targets for CRC treatment.

References

1. Baritaki, S. et al. Cell Cycle, 2010, Volume 9, Pages 4931–4940. 2 43
2. Radomkski M.W. et al. Cancer Res, Volume 51, Pages 6073–6078. 44
3. Monteiro, H. P. et al. Nitric Oxide Biol. Chem., 2019, Volume 89, Pages 1–13.

Functional analysis of Thioredoxin interaction protein (Txnip) in the redox biology of human colon, breast and melanoma tumors

Ricardo C Cintra[1], Paulo V Sousa[1], Letícia T Barbosa[1], Igor R Nascimento[1], Fernanda LT Aquino[1], Hugo P Monteiro[1]

[1]*Department of Biochemistry - Center for Cellular and Molecular Therapy- Universidade Federal de São Paulo — São Paulo — Brazil.*
Contact: ricardo.cintra@unifesp.br

Background: In recent years, it has been evidenced that increased levels of reactive oxygen species (ROS) and nitric oxide (NO) are fundamental in tumor establishment and progression. Generation of these reactive species has been attributed to the inflammatory and hypoxic nature of the tumor microenvironment, in addition to the intrinsic increase in cellular metabolism[1]. Consequently, for tumor cells, antioxidant protection is essential for their survival. The Thioredoxin (Trx)/Trx Reductase antioxidant system has been associated with a protective role in tumor growth, survival, treatment resistance, angiogenesis, and in the promotion of the epithelial-mesenchymal transition[2]. On the other hand, the expression of the protein interacting with Trx, known as Txnip, has been reported to decrease in human colon, breast, and melanoma tumors. Txnip has been suggested as tumor suppressor protein[3]. By inhibiting the nuclear translocation of Trx, Txnip promotes an imbalance on redox homeostasis and favors apoptosis[4].

Aim: Based on the aforementioned information, the aim of this study is to explore the potential clinical value and biological role of Txnip expression in human colon, breast, and melanoma tumors, correlating it with clinical data and pathological aspects. **Methods and Results**: The first step of this project utilized transcriptome data from TCGA and revealed that the expression of Txnip is reduced in tumor tissue compared to normal tissue, accompanied by hypermethylation in its promoter region, suggesting a potential epigenetic process controlling its expression. Additionally, in a melanoma model, decreased Txnip expression emerged as an independent negative prognostic factor for this tumor model. Another noteworthy observation was the negative correlation between Txnip expression and Trx expression.

Prospects: The second part of this study will employ primary and metastatic tumor cell lines from colon, breast, and melanoma to perform molecular characterization and establish Txnip overexpression models based on the CRISPR/Cas9 system. Phenotypic assays investigating migration, clonogenicity, and resistance to genotoxic approaches in these cellular models will provide insights into the functions of Txnip, which holds potential as a therapeutic target for these tumor types.

References

1. González, R. et al, Antioxidants & Redox Signaling, 2018, Volume 29, Pages 1312—1332
2. Monteiro, H.P. et al, Nitric Oxide Biol. Chem., 2019, Volume 89, Pages 1—13
3. Guo, X. et al, Cancer Cell International, 2022, Volume 22, Pages 1—16.
4. Ogata F.T. et al Antioxidants & Redox Signaling, 2022, Volume 36, Pages 1036—1050

The role of NO/RNS in the acquired resistance of pancreatic cancer cells to gemcitabine

Mariachiara Gani[1], Eros Di Giorgio[1], Marina Comelli[1], Raffaella Picco[1], Jordi Muntanè[2], Luigi E. Xodo[1], Valentina Rapozzi[1]

[1] *Department of Medicine, University of Udine, P.le Kolbe 4, 33,100 Udine, Italy;* [2] *Institute of Biomedicine of Seville (IBiS), Seville, Spain*
Contact: mariachiara.gani@spes.uniud.it

Pancreatic ductal adenocarcinoma (PDAC) is among the most malignant tumors of the gastrointestinal tract, and genomic alterations play a key role in its progression, in addition to lifestyle, smoking, and other factors. Metastasis of PDAC cells is considered one of the main causes of patient death and poor prognosis. In addition, an increase in the aggressiveness of cancer cells has been associated with the development of drug resistance.

Gemcitabine (GEM) is the first-line therapy for PDAC. It is a deoxycytidine-analog prodrug that requires cellular uptake and intracellular phosphorylation for activation. Once activated, GEM acts as a potent inhibitor of DNA synthesis. However, resistance of GEM is a major obstacle to achieving satisfactory clinical results.

To propose new treatment options for the vast majority of advanced PDAC patients facing treatment resistance, it is essential to better understand the mechanisms of acquired chemoresistance.

Like other chemotherapeutic agents (e.g., cisplatin, paclitaxel, doxorubicin), GEM increases the levels of reactive oxygen and nitrogen species (ROS/RNS) in the tumor. As low levels of ROS and RNS promote cancer progression, the aim of this work is to evaluate their impact in the development of resistance as well as identify the role of these radical species.

Therefore, we developed a PDAC cell line BxPC-3 resistant to GEM (BxPC-3/GEM Res) by repeatedly treating BxPC-3 cells with low doses of GEM for months.

We obtained a more aggressive and resistant tumor population compared to the parental BxPC-3 cell line. RNA-Seq analyses confirmed the occurrence of epithelial-mesenchymal transition (EMT), hypoxia state, and accentuated glycolytic metabolism in BxPC-3 RES/GEM cells compared to the parental cell line.

During the development of GEM resistance, there are no changes in NRF2 levels, but there was an increase in iNOS (NOS2) levels accompanied by an increase in levels of the denitrosylated enzyme GSNOR, suggesting a protective role in tumor progression. To evaluate the role of protein nitrosylation, we performed a biotin switch assay coupled with mass spectrometry. Preliminary data suggest that in pancreatic cancer cells, upregulation of both NOS2 and GSNOR is associated with the development of resistance and higher cancer cell aggressiveness.

Further analyses are underway to understand the role of S-nitrosylation/denitrosylation in resistant pancreatic cancer cells

GSNOR preserves the G1/S checkpoint in rhabdomyosarcoma: implication in cell-cycle control

Giovanni Larotondo[1#*], Paola Giglio[1#], Gianmarco Matrullo[1], Emiliano Maiani[2], Giuseppe Filomeni[1,3,4], and Costanza Montagna[1,2].

[1]*Department of Biology, University of Rome Tor Vergata, Rome, Italy;*[2] *UniCamillus-Saint Camillus, University of Health Sciences, Rome, Italy;*[3] *Redox Biology Group, Danish Cancer Society Research Center, Copenhagen, Denmark;*[4] *Center for Healthy Aging, University of Copenhagen, Denmark*
[#] Shared first-authors
Contact: glarotondo98@gmail.com

S-nitrosylation is the posttranslational modification (PTM) of the free radical nitric oxide (NO), and it is based on the chemical addition of NO moiety to a cysteine residue (becoming in such a way an *S*-nitrosothiol—SNO). The maintenance of the SNO-proteins level is mediated by denitrosylases, and *S*-nitrosoglutathione reductase (GSNOR) is the most characterized. By regulating SNO-proteins' quality and quantity, GSNOR has an active role in different cellular processes. In skeletal muscle, GSNOR promotes the myogenic differentiation of myoblast into myotubes. Rhabdomyosarcoma (RMS) is a childhood malignancy, deriving from undifferentiated and proliferative muscle cells. Coherently with its role in skeletal muscle, we decided to investigate the role of GSNOR in RMS cell lines. Here we show that GSNOR deficiency reduces myogenic differentiation in RMS cells. In addition, GSNOR knock-down cells show an increase in the S phase by EdU staining and a faster progression of the cell cycle after releasing synchronized cells in the G1/S as well as the S phase. As a result, the same cells showed increased phosphorylation of H2A.X on serine 139, a marker of replication stress (RS) and genomic instability. To further investigate the safeguard mechanism underlying the G1/S checkpoint, we are generating RMS GSNOR knock-out cells by CRISPR-CAS9 technology. Uncontrolled proliferation and RS are among the leading causes of cancer. These results point out the role of aberrant S-nitrosylation in driving RMS malignancy and aggressiveness and highlight GSNOR as a tumor suppressor.

Potential role of micro-RNAs in Breast Cancer biology through S-nitrosoglutathione reductase (GSNOR) downregulation

Gianmarco Matrullo[1], Valeria Fiorentini[1], Costanza Montagna[1], Giuseppe Filomeni[123], Salvatore Rizza[2]

[1]*Department of Biology, University of Rome "Tor Vergata", Rome 00,133, Italy;* [2]*. Redox Biology, Danish Cancer Society Research Center, Copenhagen 2100, Denmark;* [3]*. Center for Healthy Aging, Copenhagen University - 2200 Copenhagen, Denmark.*
Contact: gianmarco.matrullo@gmail.com

S-nitrosylation is the posttranslational addition of nitric oxide (NO) to a protein's free thiols, generating an S-nitrosylated protein (PSNOs) (Benhar M. et al, 2005). Various PSNOs have been found to impact cancer related pathways, such as proliferation, migration, epithelial-mesenchymal transition (EMT), and programmed cell death (Rizza and Filomeni, 2020). NO and PSNOs levels are regulated by several enzymes, such as NO-Synthases (NOSs), and different denitrosylases, the most characterized one being the S-nitrosoglutathione reductase (GSNOR) (Rizza and Filomeni, 2017). We found GSNOR to be strongly deregulated in breast cancer (BC). GSNOR ablation can significantly alter cellular homeostasis, playing a crucial role in chemoresistance, tumor onset, and progression. This suggests an important role of GSNOR in this class of tumors. However, the mechanisms underlying GSNOR downregulation and its relevance in BC biology are still unclear. However, in silico analyses indicate a potential contribution of micro-RNAs (miRNAs) in decreasing GSNOR expression. Our analysis points toward miR-23a, a well-known cancer biomarker, and several others miRNAs to be found strongly deregulated in BC, suggesting that GSNOR decrease might represent one of the mechanisms through which these miRNAs exert their pro-tumoral activity.

Here, we provide evidence on the driving role of mi-RNAs and GSNOR downregulation in BC, with the final goal of proposing new predictive and therapeutical targets of BC.

References

1. Benhar M, Forrester M, Stamler J. Protein denitrosylation: enzymatic mechanisms and cellular functions. Nat Rev Mol Cell Biol. 2009. 10, 721—732.
2. Rizza S, Filomeni G. Exploiting S-nitrosylation for cancer therapy: facts and perspectives. Biochem J. 2020. 477, 3649—3672.
3. Rizza S, Filomeni G, Chronicles of a reductase: Biochemistry, genetics and physio-pathological role of GSNOR. Free Radical Biol Med. 2017. 110, 19—30

AKR1A1 drives metabolic reprogramming and redox adaptive response

Chiara Pecorari[1], Salvatore Rizza[1], Simone Cardaci[2], Yongluon Luo[3], Thomas Sauter[4], Giuseppe Filomeni[1,5,6]

[1] *Redox Biology Group, Danish Cancer Society Research Center, Copenhagen, Denmark;*[2] *Cancer Metabolism Unit, Division of Genetics and Cell Biology, IRCCS San Raffaele Scientific Institute, Milan, Italy;* [3] *Department of Biomedicine, Aarhus University, Denmark;*[4] *Life Sciences Research Unit, University of Luxembourg, Belvaux, Luxembourg;*[5] *Department of Biology, University of Rome Tor Vergata, Rome, Italy;*[6] *Center for Healthy Aging, Copenhagen University, Denmark*
Contact: chp@cancer.dk

Renal and hepatocellular carcinomas (RCC and HCC, respectively) are two of the leading causes of death worldwide. Standard treatment for these cancer types is surgical resection followed by nonspecific chemotherapy, a procedure poorly effective due to late diagnosis, chemoresistance, and recurrence. Recent researches indicated that S-nitrosylation, a nitric oxide-induced protein posttranslational modification, may contribute to the metabolic reprogramming of RCC and HCC, a crucial step during tumor development and progression. In this context, we discovered that the enzyme denitrosylase aldo-keto reductase 1A1 (AKR1A1), which neutralizes S-nitrosylation, is under-expressed in RCC and HCC and that AKR1A1 expression correlates with patient's prognosis. We gathered compelling evidence that AKR1A1 downregulation causes significant metabolic changes in RCC and HCC cellular models, including: (i) a decreased glycolytic rate; (ii) an increased pentose phosphate pathway flux; and (iii) elevated levels of methylglyoxal (MGO).

As a result of this metabolic reprogramming, AKR1A1-deficient RCC and HCC cell lines exhibit a higher antioxidant response and develop a more invasive and aggressive phenotype. We believe that definition of the many-faced roles of AKR1A1 in RCC and HCC will bring to light new exploitable paths in the treatment of these diseases.

Zinc-based nanoparticles, but not silicon-based nanoparticles, accumulates in mitochondria promoting cell death in liver cancer cells

Joana Pieretti[1,2,*], Thaissa Horne[1,*], Amedea Seabra[2], Jordi Muntané[1,3,4]

[1]*Institute of Biomedicine of Seville (IBiS), Hospital University Virgen del Rocío/CSIC/University of Seville, Seville, Spain.* [2]*Center for Natural and Human Sciences (CCNH), Federal University of ABC (UFABC), Santo André, SP, Brazil* [3]*Biomedical Research Center for Hepatic and Digestive Diseases (CIBERehd), Madrid, Spain.* [4]*Department of Medical Physiology and Biophysics, University of Seville, Seville, Spain.*

All authors have contributed equally to the study

Background: Conventional cancer treatments still face difficulties related to poor targetability, nonspecificity, and increasing tumor resistance[1]. Thus, nanotechnology has gained space in the field of medicine providing an alternative strategy for improving therapeutic efficiency[2]. Among several nanoparticles (NPs), metal oxide NPs, such as zinc and silicon oxides (ZnO and SiO_2), consist in a potential strategy for an efficient drug delivery. **Aim:** To identify molecular mechanisms of ZnO and SiO_2 NPs proapoptotic or antiproliferative effects in differentiated hepatoblastoma cells (HepG2) and mesenchymal liver cancer cell (SNU449). **Methods:** We have determined the dose-dependent induction of cell cytotoxicity by ZnO and SiO_2 NPs (5–50 μg/mL) by Trypan blue in HepG2 and SNU449 cells (24 h). We have determined NPs intracellular localization using transmission electron microscopy (TEM), caspase-3/8 activities, and cell respiration (Mito Stress, Seahorse, Agilent). **Results:** SiO_2 NPs did not reduce cell viability in none of cell lines. On the other hand, ZnO NPs reduced cell viability of both cell lines. Interestingly, SNU449 appeared to be more susceptible to cytotoxicity than HepG2 (IC_{50} of 27.4 ± 1.4 μg/mL and 41.8 ± 0.4 μg/mL, respectively). SiO_2 NPs tended to be localized in lysosomes in both cell lines, while ZnO NPs demonstrated a higher accumulation in mitochondrial fraction in HepG2, and a nonspecific localization in SNU449 cells. As expected, SiO_2 NPs did not impact cells respiration, while ZnO NPs decreased oxygen consumption rate in both cell lines. Moreover, ZnO NPs induced caspase-3 and -8 in liver cancer cells. In contrast, SiO_2 NPs did not affect caspase activities. **Conclusion:** SiO_2 demonstrated no cytotoxic profile against HCC cells, promoting no significant alterations in the parameters evaluated. By contrast, ZnO NPs demonstrated a high cytotoxic potential against both HCC cells, directly impacting in cells respiration, which might be related directly to ZnO NPs accumulation in mitochondria.

References

1. Timin, AS. et al., Journal of Controlled Release, 2022, 344, Page 1–11
2. Tian, H. et al., Journal of Hematology & Oncology, 2022, 15, Page 132

Oxidative stress induces structural and functional changes of the mitochondrial chaperone TRAP1

Claudio Laquatra[1], *, Fiorella Faienza[2], *, Matteo Castelli[3], Giorgio Colombo[3], Giuseppe Filomeni[2,4,5], *Andrea Rasola*[1]

[1] *Department of Biomedical Sciences, University of Padova, viale G. Colombo 3, I—35,131, Padova, Italy.* [2] *Department of Biology, University of Rome Tor Vergata, Via della Ricerca Scientifica I—00,133, Rome, Italy.*[3] *Department of Chemistry, University of Pavia, via Taramelli 12, I-27,100, Pavia, Italy.* [4] *Redox Biology, Danish Cancer Society Research Center, Strandboulevarden 49, DK-2100, Copenhagen, Denmark.*[5] *Center for Healthy Aging, University of Copenhagen, Blegdamsvej 3B, DK-2200 Copenhagen, Denmark.*
*These authors equally contributed to this work
Contact: andrea.rasola@unipd.it

Expression of the Hsp90-family chaperone TRAP1 is restricted to mitochondria and increased in most tumor types. TRAP1 rapidly tunes metabolism with the changing environmental conditions and shields tumor cells from oxidative stress injury, ensuring survival under a wide array of pathophysiological condition. Here, we aim at connecting structural alterations elicited on TRAP1 by redox dyshomeostases to changes on its function and on downstream biological outputs.

By applying a stepwise approach that starts from molecular dynamics analyses and ends up with tumorigenesis assays, we have found that posttranslation modification of Cysteines (Cys) residues are crucial in regulating the activity of TRAP1 under oxidative stress. We report that TRAP1 is S-nitrosylated at Cys501, a redox modification associated with decreased ATPase activity and enhanced proteasomal degradation of the chaperone, whereas a nitrosylation-resistant C501S variant is more active and confers protection to cell death.

Moreover, we have observed that TRAP1 forms redox-sensitive tetramers via disulfide bonds involving Cys261 and Cys573. In cancer cells, TRAP1 tetramerization is induced by many oxidative agents, whereas the expression of a double Cys261-Cys573R mutant abrogates this effect and enhances TRAP1 ATPase activity. The expression of Cys261-Cys573R TRAP1 markedly increases mitochondrial ROS levels while reduces growth and invasion of an aggressive sarcoma cell model.

Our data indicate that oxidative stress causes posttranslation modifications in Cys residues of TRAP1 that controls its activity and structural rearrangements, thus modulating its pro-neoplastic functions.

Alphabetical list of speakers

Oral presentations

(*Continued*)

Poster presentations

Objective of this volume

The objective of this volume is to publish selected topics that were presented at the Third International Conference *"Therapeutic Applications of Nitric Oxide in Cancer and Inflammatory-Related Disorders"* held in Udine, Italy, September 3–5, 2023. This volume was to publish selected topics presented at the conference and to make them accessible to the scientists at large. Several contributors were invited speakers at the conference. As the field on the role of NO in various diseases is rapidly evolving, the contributions are updated with the most recent developments with areas covering fundamental mechanisms of NO-mediated biological effects, novel synthetic NO donors with therapeutic potential, and future perspectives and directives for the application of NO and derivatives in the treatment of human diseases.

Highlights

The Third International Conference *"Therapeutic Applications of Nitric Oxide in Cancer and Inflammatory-Related Disorders"* was held on September 3–5th, 2023, in Udine, Italy. This conference was attended by scientists from over the world and the speakers presented up-to-date research and clinical applications of the rapidly emerging field of nitric oxide and its derivatives in both cancer and inflammatory diseases. Briefly, there were presentations on the following: (1) The general properties of nitric oxide–mediated signaling pathways and their potential roles in the development of targeted therapeutics; (2) The new developments of the contrasting effects of nitric oxide in various cancers and a presentation on the complex role of protein nitrosylation and denitrosylation in tumor progression and response to therapeutics; (3) The important role mediated by NO in the metabolism of cancer cells and the heart; (4) The importance of NO in the regulation of neurons in the brain, in skeletal muscle function as a function of age and hyperglycemia; (5) Several presentations on NO and oxidative inflammatory stress, cardio and renal disorders, in asthma and the role of iNOS in macrophages; (6) NO-mediated effects in neurodegenerative diseases; and (7) Several presentations on the emerging field of the therapeutic applications of NO in various diseases including cancer. An important addition to the conference was the introduction of various educational sessions by leaders in the NO field. These included (1) Methods to accurately measure and determine the levels of NO in biological systems; (2) Analyses of both the canonical and noncanonical signaling pathways modulated by NO; (3) The S-nitrosylation system in cells, its detection, and its biological importance; (4) The recent advances of the application of artificial intelligence (AI) in clinical settings and in the discovery of new therapeutics; and (5) The current approaches in the synthesis and the delivery of novel NO-containing complexes for therapeutic purposes. Overall, this conference was highly productive and allowed open discussions and critiques, as well as encouraging scientific collaborations.

The proceedings of this conference were followed by the development of a book through the sponsorship of Elsevier/Academic Press. Below, a brief introduction of the contents in this volume by each contributor.

PART I

General introduction

CHAPTER

1

Nitric oxide pathophysiology and applications in patients undergoing major cardiac surgery

Davide Signori[1,2], Mattias Carlström[3], Lorenzo Berra[4,5] and Emanuele Rezoagli[1,6]

[1]School of Medicine and Surgery, University of Milano-Bicocca, Monza, Italy; [2]Department of Anesthesia and Intensive Care Medicine, Seriate, Italy; [3]Department of Physiology and Pharmacology, Karolinska Institute, Stockholm, Sweden; [4]Department of Anesthesia, Critical Care and Pain Medicine, Massachusetts General Hospital and Harvard Medical School, Boston, MA, United States; [5]Respiratory Care Department, Massachusetts General Hospital and Harvard Medical School, Boston, MA, United States; [6]Department of Anesthesia and Intensive Care Medicine, Fondazione IRCCS San Gerardo dei Tintori, Monza, Italy

Abbreviations

AKI Acute kidney injury
CPB Cardiopulmonary bypass
CSA-AKI Cardiac surgery-associated acute kidney injury
eNOS Endothelial Nitric Oxide Synthase
Hb Hemoglobin
Hp Haptoglobin
iNO Inhaled nitric oxide
IRI Ischemia−reperfusion injury
KDIGO Kidney Disease: Improving Global Outcomes
NO Nitric oxide
ppm Parts per million

Therapeutic Applications of Nitric Oxide in Cancer and Inflammatory Disorders
https://doi.org/10.1016/B978-0-443-22354-9.00001-2

Introduction

Cardiac surgery—associated acute kidney injury (CSA-AKI) is a common complications of cardiac surgery and is associated with increased mortality and morbidity (Zakeri et al., 2005). CSA-AKI is characterized by an abrupt deterioration in kidney function following surgery, which is evident from a reduction in glomerular filtration rate, reduced urine output, and/or an increase in serum creatinine from baseline (Mao et al., 2013). Diagnosis is made based on the Kidney Disease: Improving Global Outcomes (KDIGO) criteria for AKI staging that have become the new consensus definition of AKI (Kdigo Clinical Practice Guidelines for Acute Kidney Injury, 2023; Luo et al., 2014). Any patient who had cardiac surgery in the past week and fulfilled the KDIGO criteria for AKI has CSA-AKI.

Due to differences in patients baseline characteristics and in surgery type, reported CSA-AKI incidences widely vary between studies with ranges between 3.1% and 42% (Hobson et al., 2009; Jiachang Hu et al., 2016; Lagny et al., 2015). Moreover, AKI is associated with a prolonged hospital stay, more postoperative complications, increased risk of developing chronic kidney disease, increased early and late mortality, and significant healthcare costs for the society (Hobson et al., 2015).

Preoperative risks factors for the development of CSA-AKI include female sex, advanced age, the presence of comorbidities (particularly preexisting chronic kidney disease, diabetes mellitus, hypertension, chronic heart failure), and obesity (Coleman et al., 2011). Intra- and postoperative factors largely impact on the development of CSA-AKI factors such as the type and complexity of cardiac surgical procedure (Shaw et al., 2008, pp. p55—p60) and the use of nephrotoxic agents such as contrast media or aminoglycoside antibiotics (Casanova et al., 2022). Also, cardiopulmonary bypass (CPB) itself and its duration may be associated with an increased risk of CSA-AKI (Fuhrman & Kellum, 2017; Yi et al., 2016).

The therapy for CSA-AKI in clinical practice is still based on supportive and preventive measure (avoidance of nephrotoxins, optimization of systemic hemodynamics, and the limitation of both fluid depletion and fluid overload) since the identification of the biological and pathophysiological variability is complex and actually requires greater characterization (Wang & Bellomo, 2017; Yu et al., 2023). However, these strategies have not proven to be very effective. Given the magnitude of this global clinical problem and all its complications, further large trials assessing underlying pathophysiological mechanism and investigating new specific therapies for the prevention and treatment of CSA-AKI remain a priority in the field.

In this chapter, we focus on the pathogenesis of CSA-AKI paying particular attention on the role of NO dysregulation as a biological mechanism of injury and as a possible therapeutic target.

Pathophysiology of CSA-AKI and NO dysregulation

NO is not just a simple endogenous selective vasodilator (Ignarro et al., 1987; Palmer et al., 1987; Premont et al., 2020), but plays an important role in the whole cardiovascular system. This small molecule, classically derived enzymatically from endothelial NO synthase (eNOS), exerts a protective role on endothelial homeostasis by reducing smooth muscle proliferation

(Garg & Hassid, 1989), inhibits platelet aggregation (Radomski et al., 1993), modulates leukocyte binding to endothelium (Gao et al., 2018), and dampens oxidative stress by modulating NADPH oxidase activity and oxidative phosphorylation in mitochondria. These functions, and in particular its antiinflammatory and antioxidative properties, can contribute to its favorable regulation of cell bioenergetics and apoptosis (Moncada & Erusalimsky, 2002). However, the NOS pathway is significantly impaired in conditions characterized by low oxygen tension, particularly during ischemia, as this enzymatic system for NO generation relies not only on cofactors and substrates but also on molecular oxygen for its proper functioning. In conditions like this, and alternative system termed the nitrate–nitrite–NO system may serve as a back-up. Nitrate and nitrite were long considered as inert end-products of NO metabolism, but accumulating research over the past 2 decades have showed that these anions can undergo reductive bioactivation and again for bioactive NO and other bioactive nitrogen oxide species. Intriguingly, this pathway may be boosted via our daily diet since nitrate is found in high levels in certain food sources, especially in green leafy vegetables and beets. Presently, there are many experimental studies that have showed favorable renal and cardiovascular, as well as metabolic, effects of dietary supplementation with inorganic nitrate in IR-induced kidney disease (Carlström, 2021).

Since NO has homeostatic and protective roles on endothelium, several studies tried to assess the putative beneficial effect of NO on different organs that may be potentially prone to develop injury consequent to ischemia. Although research about the role of NO in human and mammals is profound and dates decades back, to date its complete understanding is still far from being achieved, especially for its exploitation in the therapeutic field. In fact, NO exerts its action both by direct activation on its cellular receptor (soluble guanylate cyclase), and above all with S-nitrosylation phenomena of countless proteins, a phenomenon which is now recognized as a key posttranslational modification and as a major key player in the NO bioactivity (Premont et al., 2020). Moreover, NO-like bioactivity can be transduced by NO-ferroheme species, which can be transferred between proteins and directly exert NO pathways without intermediacy of free NO, thus avoiding rapid NO scavenging (DeMartino et al., 2023; Kleschyov et al., 2023). This complexity accounts for the efficacy of both NO-donors and inhaled NO (iNO) in organs, which develop ischemia. NO is rapidly inactivated by hemoglobin (Hb)-mediated oxidation in the circulation and may not reach the target organ. However, reactive NO molecule can undergo oxidative modification(s), resulting in the formation of more long-lived nitrogen oxides that via nitrosation or nitration reactions can alter protein functions, which in turn may account for the preconditioning effects of NO itself (Nagasaka et al., 2018).

The relation of NO dysregulation and CSA-AKI is close even if the pathophysiology underlying CSA-AKI remains not completely understood. Its mechanisms are multiple, involves several organ systems, and often present alternately and in variable combinations (Wang & Bellomo, 2017). Factors related to the development of AKI include the following: the role of renal hypoperfusion that depends on the use of CPB (which guarantees only continuous flow, sometimes with acute hemodilution and hypothermia), the CPB-induced hemolysis, the low cardiac output syndrome, the hemorrhagic complications, and the duration of the aortic cross-clamp (Wang & Bellomo, 2017; Yu et al., 2023). Renal vasoconstriction and subsequent ischemia are associated with the level of hemolysis in cardiac surgery since free-hemoglobin (free-Hb) is a potent scavenger of NO (Hu et al., 2021; Vermeulen Windsant et al., 2011). Also,

NO depletion is directly related to the level of systemic vasoconstriction (Rezoagli et al., 2017). Furthermore, free-Hb also causes direct damage to the renal parenchyma (Schaer, Buehler et al., 2013) via inflammation and oxidative stress. Surgical injury to tissues and the exposure of blood to the CPB pump and circuit is associated with increase of inflammatory cytokines and complement activation (Baliga et al., 1999; Stoner et al., 2007; Zhang et al., 2015). After the cessation of the ischemic insult, ischemia–reperfusion injury (IRI) may occur and may contribute to AKI. IRI damage is mediated by multiple factors such as excessive production of reactive oxygen species (ROS) leading to oxidative stress (Stoner et al., 2007), which further induces inflammation by upregulating proinflammatory cytokines. Cytokines activate neutrophils, macrophages, and lymphocytes and recruit them into the renal parenchyma, causing parenchymal infiltration and thus contributing to AKI (Ascon et al., 2009; Yang et al., 2011). Activation of immune cells like neutrophils and macrophages also leads to further ROS generation (from NADPH oxidase), leading to a vicious cycle with progressive oxidative stress and reduced NO bioavailability and mitochondrial dysfunction (PMID: 21629295 & 38061207) (Carlström et al., 2024; Drummond et al., 2011). Moreover, IRI-associated mitochondrial ROS generation causes the opening of mitochondrial permeability transition pores, which is associated with activation of NADPH oxidase and cell injury or death (Crompton, 1999; Schiffer et al., 2020).

Beyond the direct effect of renal hypoperfusion as a potential cause of CSA-AKI because of the CPB, we identified three specific pathways of renal damage in the field of AKI:

- renal vasoconstriction induced by free-Hb;
- direct renal damage from free-Hb and catalytic iron; and
- IRI.

We then highlighted the studies that have proven promising in the stratification and treatment of this disease through NO biological pathways. A brief summary of what is highlighted in the following paragraphs is reported in Table 1.1.

Renal vasoconstriction

Renal ischemia during cardiac surgery plays a central role in the pathogenesis of CSA-AKI (Wang & Bellomo, 2017). The intraoperative ischemic damage is currently only partially avoidable, and its prevention is largely related to the improvement of surgical techniques and organ support therapies. Kidneys often suffer secondary ischemic insults linked to various stimuli such as low cardiac output in postcardiac surgery patients (Hoffman et al., 2003). In particular, one of the fundamental mechanisms is covered by CPB-induced hemolysis (Signori et al., 2022). Hemolysis determines a massive release of free-Hb that causes systemic and renal vasoconstriction and possibly ischemia (Hu et al., 2021; Kam et al., 1996; Vermeulen Windsant et al., 2011). During intravascular hemolysis, free-Hb in plasma is able to rapidly bind NO, thus affecting vasomotor tone and consequently organ perfusion (Helms & Kim-Shapiro, 2013). This occurs by the dioxygenation reaction of plasma oxy-Hb (Fe^{2+}) with NO to form bioinactive nitrate and met-Hb (Fe^{3+}) (Doyle & Hoekstra, 1981) (see Fig. 1.1). The depletion of endogenous NO induced by hemolysis produces dose-dependent systemic and pulmonary vasoconstriction (Minneci et al., 2005). Similarly, hemolysis during CPB determines a massive

TABLE 1.1 Pathophysiological mechanism proposed therapeutic interventions and their actual evidences about the treatment of CSA-AKI.

Pathophysiological mechanism	Proposed therapeutic intervention	Actual highest level of evidence
1. Renal vasoconstriction and ischemia consequent to NO scavenging by plasma free-Hb	Exogenous NO expedites the transition of the highly unstable plasma Oxy-Hb to its reduced and inert form Met-Hb, which is unable to deplete NO from the vasculature	iNO reduces AKI and transition to stage 3 CKD in human undergoing cardiac surgery under with CPB. *Lei C* et al. *Am J Respir Crit Care Med. 2018* (Lei et al., 2018).
2. Direct free-Hb and catalytic iron tubular injury	Free-Hb elimination via binding to plasma protein Hp. Within the Hb/Hp complex, the free-Hb vasoactivity is controlled, its redox reactivity is stabilized, and Hb itself is protected against oxidative damage.	Intraoperative administration of Hp during cardiac surgery in humans was associated with lower risk of AKI. *Kubota k* et al. *Anesth Analg. 2017* (Kubota et al., 2017).
3. eNOS uncoupling and NO depletion in IRI	Exogenous NO supplementation may restore endothelial physiology and reduce IRI damage	Supplementation of nitrite and nitrate in an animal renal IRI model reduces AKI. *Tripatara* et al. *J Am Soc Nephrol. 2007* (Tripatara et al., 2007) and *Yang* et al. *Redox Biol 2017* (T. Yang et al., 2017). More consistence evidence are available for iNO benefits in animal studies of cardiac arrest (Minamishima et al., 2011; Miyazaki & Ichinose, 2020), myocardial infarction (Hataishi et al., 2006; Liu et al., 2007) and stroke (Li et al., 2013; Terpolilli et al., 2012). In human trials dietary nitrate treatment reduced the risk for contrast-induced nephropathy, helped stabilizing postprocedure kidney function, and reduced the risk for major adverse cardiac events in at-risk patients undergoing coronary angiography for non-ST-elevation acute coronary syndrome (according to trial findings presented at the European Society of Cardiology Congress 2023) (Beirne et al., 2021).

release of free-Hb that causes systemic and renal vasoconstriction and possibly ischemia (Hu et al., 2021; Kam et al., 1996; Vermeulen Windsant et al., 2011). In 2016, Rezoagli et al. (2017) found that a prolonged duration of CPB ($\geq$140 min) was associated with higher levels of plasma free-Hb, NO consumption, and thus both systemic and pulmonary vasoconstriction. The interplay between hemolysis during CPB and hemodynamics was also observed in a recent reanalysis of the same study by Rezoagli and colleagues who observed a significant adjusted association between the levels of NO consumption and both pulmonary vascular compliance and right ventricle function (Rezoagli et al., 2024). Also, CPB induces endothelial dysfunction, a condition associated with reduced NO bioavailability (Wessel et al., 1993). The

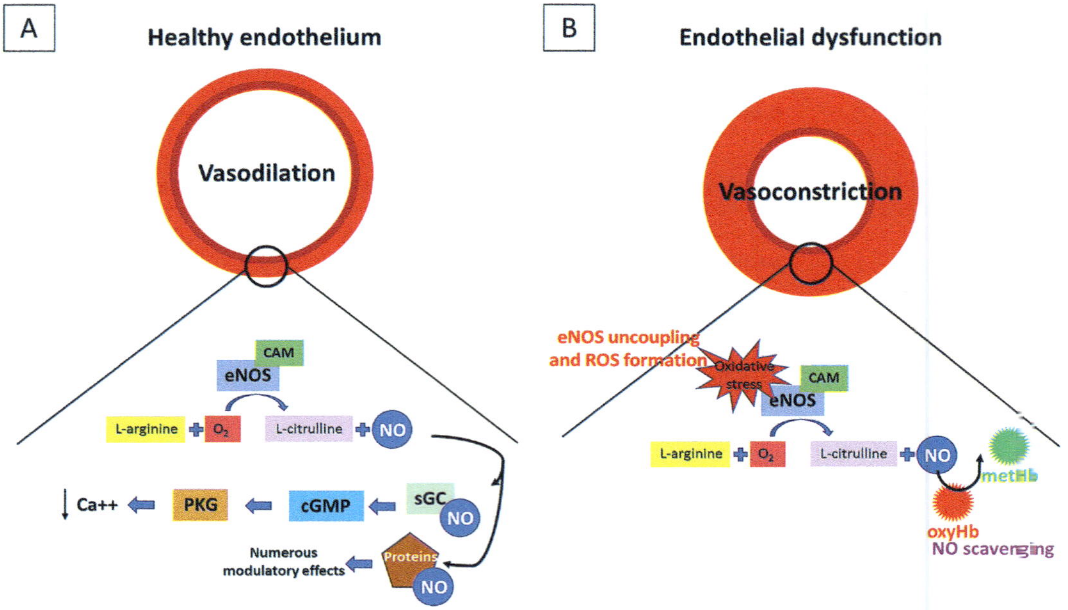

FIGURE 1.1 NO biosynthesis and eNOS uncoupling. Endogenous NO is produced by NOS by the oxidation of L-arginine to L-citrulline + NO (NADPH- and BH4-dependent reaction). NO is one of the end-products of the reaction. Most of the effects of NO in the cardiovascular system are mediated by the activation of sGC, which catalyzes the formation of the second messenger cGMP from GTP. The activation of GMP-dependent PKG leads to vascular relaxation and protein function modulation via nitration and nitrosylation (A). Several circumstances may alter eNOS activity causing the reduction of NO levels and triggering the production of superoxide instead of NO, a process defined as "eNOS uncoupling". For example, oxidative stress consequent to IRI causes eNOS uncoupling and NO scavenging by free oxy-Hb reduces NO bioavailability (B). *BH4*, tetrahydrobiopterin; *cGMP*, cyclic guanosine monophosphate; *GTP*, guanosine-5′-triphosphate; *IRI*, ischemia-reperfusion injury; *NO*, nitric oxide; *NOS*, nitric oxide synthase; *oxy-Hb*, oxyhemoglobin; *PKG*, Protein Kinase G; *sGC*, soluble guanylate cyclase. *Adapted from Signori, D., Magliocca, A., Hayashida, K., Graw, J. A., Malhotra, R., Bellani, G., Berra, L., Rezoagli, E. (1975). Inhaled nitric oxide: role in the pathophysiology of cardio-cerebrovascular and respiratory diseases.* Intensive Care Medicine Experimental, *10(1):28. https://doi.org/10.1186/s40635-022-00455-6.*

enhanced NO consumption and reduced NO synthesis after CPB were reported to play a role in organ damage, particularly in CSA-AKI (Vermeulen Windsant et al., 2014). This process contributes to the reduction of renal organ blood flow because of vasoconstriction. All these evidences have led researchers to investigate the possibility of correcting the described mechanisms (Redaelli et al., 2022). A recent systematic review and meta-analysis found that NO was associated with a reduced risk of AKI when administered from the beginning of CPB (Hu et al., 2019). Lei et al. (2018) published a prospective, randomized controlled interventional trial comparing treatment with 80 ppm iNO versus placebo (nitrogen) during CPB (trough the membrane lung) and during the first 24 h in the postoperative period (trough inhalation) in adult patients undergoing valvular cardiac surgery. The authors reported that iNO delivered during hemolysis expedites the transition from the highly unstable plasma Oxy-Hb to its reduced and inert form Met-Hb, which is unable to deplete NO from the vasculature.

Moreover, they reported a 14% reduction of AKI, and lowered transition to stage 3 chronic kidney disease at 90 days and 1 year. Since this study has some limitation particularly related to the study population (young age and homogeneous disease and type of cardiac surgery), a further clinical trial is currently ongoing to evaluate whether administration of 80 ppm iNO during CPB and for 24 h after cardiac surgery reduces the risk of AKI in patients with endothelial dysfunction (NCT02836899) (Berra, 2022). Similar trials are ongoing in neonates who undergo cardiac surgery due to congenital heart diseases (https://clinicaltrials.gov/study/NCT04216927; Savorgnan, 2021).

Direct free-hemoglobin and catalytic iron injury

Hemolysis induced by the CPB accounts for a widely accepted pathophysiological process of CSA-AKI, as widely described in the previous paragraph. However, vasoconstriction from NO scavenging is not the only mechanism of renal damage. Actually, the free-Hb is filtered by the kidney, where it may directly determine tubule barrier deregulation and oxidative damage (Deuel et al., 2016; Leaf et al., 2015). Also, dysregulated iron metabolism plays a particularly important pathogenic role in AKI in experimental animal models (Baliga et al., 1993, 1996; Paller, 1988) since, even if catalytic iron and free-Hb levels correlated with each other after CPB, an association with CSA-AKI was observed with the former but not the latter. This discordance suggests the importance of redox activity in the pathogenesis of AKI (Leaf et al., 2015). However, to date, limited data are available on iron in the context of human AKI.

The adverse changes and injury produced by free-Hb and catalytic iron may be prevented by haptoglobin (Hp) treatment as reported in preclinical in vitro and animal studies (Baek et al., 2012; Boretti et al., 2009; Deuel et al., 2016; Schaer et al., 2016). Thereby, free-Hb is eliminated by binding to plasma protein Hp and then transported into cells of monocytic lineage (Ascenzi et al., 2005). Moreover, within the Hb/Hp complex, the free-Hb vasoactivity is controlled, its redox reactivity is stabilized, and Hb itself is protected against oxidative damage. Moreover, even if the current theories regarding Hp protection on direct oxidative toxicity were not supported by strong evidence, it is accepted that Hp exert a protective role also against free-Hb direct toxicity. The protective role of Hp against endothelium-directed Hb toxicity is based primarily on the Hb structure stabilization, thus preventing unfolding and aggregation of the protein, and on Hb catalytic iron combination with superoxide ions to form the very reactive and injurious hydroxyl radical (Halliwell & Gutteridge, 1990; Schaer, Deuel et al., 2013). The protective role of Hp in patients undergoing cardiac surgery has been investigated to date incompletely. Specifically, patients with the Hp 2-2 phenotype (with inferior ability to bind Hb compared to other wild-type Hp) have risk of CSA-AKI and is associated with decreased short- and long-term survival after cardiac surgery in patients with diabetes mellitus (Feng et al., 2017). Furthermore, in a single-center retrospective observational study, intraoperative administration of Hp during cardiac surgery was associated with a lower risk of AKI (Kubota et al., 2017). Interestingly, in the presence of endothelial dysfunction such as in models of diabetes mellitus or hyperlipidemia in mice, Hp was not able to prevent vasoconstriction (Graw et al., 2017). This mechanism may be attributed to the reduction of endogenous NO scavenging. Because endothelial dysfunction enhances vasoconstriction due to NO scavenging by cell-free Hb—or it prevents vasodilation, this may suggest that, in contrast to supplementation of NO, sequestration of cell-free Hb by Hb-binding molecules is not effective to prevent vasoconstriction (Ascon et al., 2009; Halliwell & Gutteridge, 1990).

Other scavenging molecules may have a role in reducing direct damage from hemolysis, however current evidence is still marginal. Of note, hemopexin is suggested thanks to its extraordinary affinity for free-heme group, which is bound and transported into hepatocytes and macrophages (Hvidberg et al., 2005). Hemopexin has a debated protective role against hemolysis and severe inflammation and infection (Lin et al., 2015) and it may work as a potential therapeutic intervention in future trials. Similar considerations are open for hepcidin, which is known to bind iron (Wunderer et al., 2020) and about the possibility of clearing free-Hb from plasma via renal replacement therapy techniques (Hulko et al., 2015), strategies that open future directions in CSA-AKI prevention/treatment.

Ischemia–reperfusion injury

Intra- and postoperative renal ischemia is frequent in cardiac surgery patients. It is a consequence of various factors such as acute ischemic conditions consequent to aortic cross-clamp and renal arteries embolization or prolonged settings with renal oxygen supply/demand mismatch such as low-flow low-pressure CPB, hemolysis, hemodilution, and postoperative low-cardiac output states (Lameire et al., 2008; Lannemyr et al., 2017; Ranucci, 2007; Ranucci et al., 2005; Wang & Bellomo, 2017; Yu et al., 2023), but also contrast-media-induced nephropathy in patients undergoing cardiac catheterization (Morcos et al., 2019). Even CPB alone seems to cause endothelial dysfunction (Wessel et al., 1993). In IRI, endothelial dysfunction appears to play a pivotal role in determining the severity of the injury and the organ's ability to recover (Perkins et al., 2012; Signori et al., 2022). This process is largely deleterious and has been linked to endothelial dysfunction, ROS increase, and other cardiovascular diseases (Egea et al., 2017). Endothelial dysfunction is a disorder characterized by an imbalance between vasodilating, antimitogenic, and antithrombogenic molecules, in particular NO (Flammer & Lüscher, 2010; Virdis et al., 2010). This condition may alter endothelial NOS activity, causing the decrease of NO levels and triggering the production of superoxide instead of NO, a process which is defined as "eNOS uncoupling" (Tejero et al., 2019) (see Fig. 1.1). This mechanism is particularly important in the development of AKI since kidneys are particularly sensitive to NO dysregulation in IRI (Goligorsky et al., 2002). Therefore, numerous studies attempted to evaluate the effectiveness of exogenous NO supplementation in IRI. Preclinical evidence of iNO beneficial effect on IRI was reported on neurological outcomes and survival after cardiac arrest (Minamishima et al., 2011; Miyazaki & Ichinose, 2020), on infarction size and cardiac function after myocardial infarction (Hataishi et al., 2006; Liu et al., 2007), and on infarction size in stroke (Li et al., 2013; Terpolilli et al., 2012). Interestingly, a key study on iNO in humans has shown benefits in survival after cardiac arrest, even if neurological outcome did not improve (Patel et al., 2021). However, the precise role of NO supplementation—inhaled or in NO-donors form—in renal IRI has been poorly documented. A study conducted in children undergoing cardiac surgery for congenital heart disease evaluated the effect of gaseous NO delivery to the membrane oxygenator of the CPB circuit. The authors observed an optimized fluid balance, brain natriuretic peptide, and diuretic usage in the NO group. However, no difference in plasma creatinine was detected between the two groups. The authors concluded that these findings support the hypothesis that exogenous NO improved either hemodynamic profile or renal function (Checchia et al., 2013). Given the positive impacts associated with inorganic nitrate supplementation in models of cardiorenal disease models, it's important to mention the promising findings which have emerged from a recently completed large phase 2 study (NCT03627130) (Beirne et al., 2021). According to trial findings presented at the European Society of Cardiology Congress 2023, dietary nitrate treatment reduced the risk for contrast-induced nephropathy (CIN), helped stabilizing postprocedure kidney function, and reduced the risk for major adverse cardiac events (MACE) in at-risk patients undergoing coronary angiography for nonST-elevation acute coronary syndrome.

Collaterally, it is of interest to search for alternative endogenous mechanisms of NO production or enhance its downstream signaling. Indeed, in physiology, the reduction of direct NO synthesis may be partly replaced by alternative synthetic processes of NO production that are established in conditions of hypoxia and ischemia (by NO donors or inorganic nitrate/nitrite). Of particular importance is the retroconversion of nitrates and nitrites into bioactive NO, i.e., the nitrate—nitrite—NO pathway, which can serve as a back-up system for NO generation (Carlström, 2021; Zweier et al., 1995). As already described above, this biological process has proven (especially in disease conditions with low oxygen) to be an effective therapeutic target in reducing development and/or progression of AKI in animal models of kidney IRI via different administration routes (nitrate orally or nitrite intravenously or topically to the kidney capsule) (Tripatara et al., 2007; Yang et al., 2017; Zhang et al., 2021). Furthermore, the dietary supplementation of nitrate seems like a reasonable option for patients with endothelial dysfunction even considering their lower availability as compared with healthy volunteers (Checchia et al., 2013).

Conclusions

The role of NO dysregulation in CSA-AKI is supported by multiple scientific evidences, although—currently—these are limited to preclinical studies in animal models and in small human population studies. Renal vasoconstriction induced by free-Hb NO scavenging, direct damage of free-Hb and catalytic iron on the renal tubule, and endothelial dysfunction induced by IRI appear to be mechanisms most closely linked to the development of CSA-AKI. Therapies including inhaled NO or dietary supplementation with nitrate together with small molecules aimed at scavenging free-Hb and its derivative products including iron are open to further research in the prevention and treatment of CSA-AKI.

Conflict of interest statement

No potential conflicts of interest were disclosed.

Acknowledgments

None.

References

Ascenzi, P., Bocedi, A., Visca, P., Altruda, F., Tolosano, E., Beringhelli, T., & Fasano, M. (2005). Hemoglobin and heme scavenging. *IUBMB Life, 57*(11), 749–759. https://doi.org/10.1080/15216540500380871

Ascon, M., Ascon, D. B., Liu, M., Cheadle, C., Sarkar, C., Racusen, L., Hassoun, H. T., & Rabb, H. (2009). Renal ischemia-reperfusion leads to long term infiltration of activated and effector-memory T lymphocytes. *Kidney International, 75*(5), 526–535. https://doi.org/10.1038/ki.2008.602

Baek, J. H., D'Agnillo, F., Vallelian, F., Pereira, C. P., Williams, M. C., Jia, Y., Schaer, D. J., & Buehler, P. W. (2012). Hemoglobin-driven pathophysiology is an in vivo consequence of the red blood cell storage lesion that can be attenuated in Guinea pigs by haptoglobin therapy. *Journal of Clinical Investigation, 122*(4), 1444–1458. https://doi.org/10.1172/JCI59770

Baliga, R., Ueda, N., & Shah, S. V. (1993). Increase in bleomycin-detectable iron in ischaemia/reperfusion injury to rat kidneys. *Biochemical Journal, 291*(3), 901–905. https://doi.org/10.1042/bj2910901

Baliga, R., Zhang, Z., Baliga, M., & Shah, S. V. (1996). Evidence for cytochrome P-450 as a source of catalytic iron in myoglobinuric acute renal failure. *Kidney International, 49*(2), 362–369. https://doi.org/10.1038/ki.1996.53

Baliga, R., Ueda, N., Walker, P. D., & Shah, S. V. (1999). Oxidant mechanisms in toxic acute renal failure. *Drug Metabolism Reviews, 31*(4), 971–997. https://doi.org/10.1081/DMR-100101947

Beirne, A. M., Mitchelmore, O., Palma, S., Andiapen, M., Rathod, K. S., Hammond, V., Bellin, A., Cooper, J., Wright, P., Antoniou, S., Yaqoob, M. M., Naci, H., Mathur, A., Ahluwalia, A., & Jones, D. A. (2021). NITRATE-CIN study: Protocol of a randomized (1:1) single-center, UK, double-blind placebo-controlled trial testing the effect of inorganic nitrate on contrast-induced nephropathy in patients undergoing coronary angiography for acute coronary syndromes. *Journal of Cardiovascular Pharmacology and Therapeutics, 26*(4), 303–309. https://doi.org/10.1177/1074248421000520

Berra. (2022). *Prevention of acute kidney injury by nitric oxide in prolonged cardiopulmonary bypass. A double blind controlled randomized trial in cardiac surgical patients with endothelial dysfunction.*

Boretti, F. S., Buehler, P. W., D'Agnillo, F., Kluge, K., Glaus, T., Butt, O. I., Jia, Y., Goede, J., Pereira, C. P., Maggiorini, M., Schoedon, G., Alayash, A. I., & Schaer, D. J. (2009). Sequestration of extracellular hemoglobin within a haptoglobin complex decreases its hypertensive and oxidative effects in dogs and Guinea pigs. *Journal of Clinical Investigation, 119*(8), 2271–2280. https://doi.org/10.1172/JCI39115

Carlström, M., Rannier Ribeiro Antonino Carvalho, L., Guimaraes, D., Boeder, A., & Schiffer, T. A. (2024). Dimethyl malonate preserves renal and mitochondrial functions following ischemia-reperfusion via inhibition of succinate dehydrogenase. *Redox Biology, 69*. https://doi.org/10.1016/j.redox.2023.102984

Carlström, M. (2021). Nitric oxide signalling in kidney regulation and cardiometabolic health. *Nature Reviews Nephrology, 17*(9), 575–590. https://doi.org/10.1038/s41581-021-00429-z

Casanova, A. G., Sancho-Martínez, S. M., Vicente-Vicente, L., Ruiz Bueno, P., Jorge-Monjas, P., Tamayo, E., Morales, A. I., & López-Hernández, F. J. (2022). Diagnosis of cardiac surgery-associated acute kidney injury State of the art and perspectives. *Journal of Clinical Medicine, 11*(15). https://doi.org/10.3390/jcm11154576

Checchia, P. A., Bronicki, R. A., Muenzer, J. T., Dixon, D., Raithel, S., Gandhi, S. K., & Huddleston, C. B. (2013). Nitric oxide delivery during cardiopulmonary bypass reduces postoperative morbidity in children - a randomized trial. *The Journal of Thoracic and Cardiovascular Surgery, 146*(3), 530–536. https://doi.org/10.1016/j.jtcvs.2012.09.100

Coleman, M. D., Shaefi, S., & Sladen, R. N. (2011). Preventing acute kidney injury after cardiac surgery. *Current Opinion in Anaesthesiology, 24*(1), 70–76. https://doi.org/10.1097/ACO.0b013e3283422ebc

Crompton, M. (1999). The mitochondrial permeability transition pore and its role in cell death. *Biochemical Journal, 341*(2), 233–249. https://doi.org/10.1042/bj3410233

DeMartino, A. W., Poudel, L., Dent, M. R., Chen, X., Xu, Q., Gladwin, B. S., Tejero, J., Basu, S., Alipour, E., Jiang, Y., Rose, J. J., Gladwin, M. T., & Kim-Shapiro, D. B. (2023). Thiol-catalyzed formation of NO-ferroheme regulates intravascular NO signaling. *Nature Chemical Biology, 19*(10), 1256–1266. https://doi.org/10.1038/s41589-023-01413-3

Deuel, J. W., Schaer, C. A., Boretti, F. S., Opitz, L., Garcia-Rubio, I., Baek, J. H., Spahn, D. R., Buehler, P. W., & Schaer, D. J. (2016). Hemoglobinuria-related acute kidney injury is driven by intrarenal oxidative reactions triggering a heme toxicity response. *Cell Death & Disease, 7*(1). https://doi.org/10.1038/cddis.2015.392

Doyle, M. P., & Hoekstra, J. W. (1981). Oxidation of nitrogen oxides by bound dioxygen in hemoproteins. *Journal of Inorganic Biochemistry, 14*(4), 351–358. https://doi.org/10.1016/S0162-0134(00)80291-3

Drummond, G. R., Selemidis, S., Griendling, K. K., & Sobey, C. G. (2011). Combating oxidative stress in vascular disease: NADPH oxidases as therapeutic targets. *Nature Reviews Drug Discovery, 10*(6), 453–471. https://doi.org/10.1038/nrd3403

Efficacy of nitric oxide administration during cardiopulmonary bypass in neonates at reducing the risk of acute kidney injury. https://clinicaltrials.gov/study/NCT04216927.

Egea, J., Fabregat, I., Frapart, Y. M., Ghezzi, P., Görlach, A., Kietzmann, T., Kubaichuk, K., Knaus, U. G., Lopez, M. G., Olaso-Gonzalez, G., Petry, A., Schulz, R., Vina, J., Winyard, P., Abbas, K., Ademowo, O. S., Afonso, C. B., Andreadou, I., Antelmann, H., … Daiber, A. (2017). European contribution to the study of ROS: A summary of the findings and prospects for the future from the COST action BM1203 (EU-ROS). *Redox Biology, 13*, 94–162. https://doi.org/10.1016/j.redox.2017.05.007

Feng, C., Naik, B. I., Xin, W., Ma, J. Z., Scalzo, D. C., Thammishetti, S., Thiele, R. H., Zuo, Z., & Raphael, J. (2017). Haptoglobin 2-2 phenotype is associated with increased acute kidney injury after elective cardiac surgery in patients with diabetes mellitus. *Journal of the American Heart Association, 6*(10). https://doi.org/10.1161/JAHA.117.006565

Flammer, A. J., & Lüscher, T. F. (2010). Human endothelial dysfunction: EDRFs. *Pflugers Archiv European Journal of Physiology, 459*(6), 1005–1013. https://doi.org/10.1007/s00424-010-0822-4

Fuhrman, D. Y., & Kellum, J. A. (2017). Epidemiology and pathophysiology of cardiac surgery-associated acute kidney injury. *Current Opinion in Anaesthesiology, 30*(1), 60–65. https://doi.org/10.1097/ACO.0000000000000412

Gao, F., Lucke-Wold, B. P., Li, X., Logsdon, A. F., Xu, L. C., Xu, S., LaPenna, K. B., Wang, H., Talukder, M. A. H., Siedlecki, C. A., Huber, J. D., Rosen, C. L., & He, P. (2018). Reduction of endothelial nitric oxide increases the adhesiveness of constitutive endothelial membrane ICAM-1 through Src-mediated phosphorylation. *Frontiers in Physiology, 8*. https://doi.org/10.3389/fphys.2017.01124

Garg, U. C., & Hassid, A. (1989). Nitric oxide-generating vasodilators and 8-bromo-cyclic guanosine monophosphate inhibit mitogenesis and proliferation of cultured rat vascular smooth muscle cells. *Journal of Clinical Investigation, 83*(5), 1774–1777. https://doi.org/10.1172/JCI114081

Goligorsky, M. S., Brodsky, S. V., & Noiri, E. (2002). Nitric oxide in acute renal failure: NOS versus NOS. *Kidney International, 61*(3), 855–861. https://doi.org/10.1046/j.1523-1755.2002.00233.x

Graw, J. A., Yu, B., Rezoagli, E., Warren, H. S., Buys, E. S., Bloch, D. B., & Zapol, W. M. (2017). Endothelial dysfunction inhibits the ability of haptoglobin to prevent hemoglobin-induced hypertension. *American Journal of Physiology - Heart and Circulatory Physiology, 312*(6), H1120–H1127. https://doi.org/10.1152/ajpheart.00851.2016

Halliwell, B., & Gutteridge, J. M. C. (1990). Role of free radicals and catalytic metal ions in human disease: An overview. *Methods in Enzymology, 186*(C), 1–85. https://doi.org/10.1016/0076-6879(90)86093-B

Hataishi, R., Rodrigues, A. C., Neilan, T. G., Morgan, J. G., Buys, E., Shiva, S., Tambouret, R., Jassal, D. S., Raher, M. J., Furutani, E., Ichinose, F., Gladwin, M. T., Rosenzweig, A., Zapol, W. M., Picard, M. H., Bloch, K. D., & Scherrer-Crosbie, M. (2006). Inhaled nitric oxide decreases infarction size and improves left ventricular function in a murine model of myocardial ischemia-reperfusion injury. *American Journal of Physiology - Heart and Circulatory Physiology, 291*(1), H379–H384. https://doi.org/10.1152/ajpheart.01172.2005

Helms, C., & Kim-Shapiro, D. B. (2013). Hemoglobin-mediated nitric oxide signaling. *Free Radical Biology and Medicine, 61*, 464–472. https://doi.org/10.1016/j.freeradbiomed.2013.04.028

Hobson, C. E., Yavas, S., Segal, M. S., Schold, J. D., Tribble, C. G., Layon, A. J., & Bihorac, A. (2009). Acute kidney injury is associated with increased long-term mortality after cardiothoracic surgery. *Circulation, 119*(18), 2444–2453. https://doi.org/10.1161/circulationaha.108.800011

Hobson, C., Ozrazgat-Baslanti, T., Kuxhausen, A., Thottakkara, P., Efron, P. A., Moore, F. A., Moldawer, L. L., Segal, M. S., & Bihorac, A. (2015). Cost and mortality associated with postoperative acute kidney injury. *Annals of Surgery, 261*(6), 1207–1214. https://doi.org/10.1097/SLA.0000000000000732

Hoffman, T. M., Wernovsky, G., Atz, A. M., Kulik, T. J., Nelson, D. P., Chang, A. C., Bailey, J. M., Akbary, A., Kocsis, J. F., Kaczmarek, R., Spray, T. L., & Wessel, D. L. (2003). Efficacy and safety of milrinone in preventing low cardiac output syndrome in infants and children after corrective surgery for congenital heart disease. *Circulation, 107*(7), 996–1002. https://doi.org/10.1161/01.CIR.0000051365.81920.28

Hu, J., Chen, R., Liu, S., Yu, X., Zou, J., & Ding, X. (2016). Global incidence and outcomes of adult patients with acute kidney injury after cardiac surgery: A systematic review and meta-analysis. *Journal of Cardiothoracic and Vascular Anesthesia, 30*(1), 82–89. https://doi.org/10.1053/j.jvca.2015.06.017

Hu, J., Spina, S., Zadek, F., Kamenshchikov, N. O., Bittner, E. A., Pedemonte, J., & Berra, L. (2019). Effect of nitric oxide on postoperative acute kidney injury in patients who underwent cardiopulmonary bypass: A systematic review and meta-analysis with trial sequential analysis. *Annals of Intensive Care, 9*(1). https://doi.org/10.1186/s13613-019-0605-9

Hu, J., Rezoagli, E., Zadek, F., Bittner, E. A., Lei, C., & Berra, L. (2021). Free hemoglobin ratio as a novel biomarker of acute kidney injury after on-pump cardiac surgery: Secondary analysis of a randomized controlled trial. *Anesthesia & Analgesia, 132*(6), 1548–1558. https://doi.org/10.1213/ANE.0000000000005381

Hulko, M., Kunz, M., Yildirim, M., Homeyer, S., Amon, O., & Krause, B. (2015). Cell-free plasma hemoglobin removal by dialyzers with various permeability profiles. *Scientific Reports, 5*(1). https://doi.org/10.1038/srep16367

Hvidberg, V., Maniecki, M. B., Jacobsen, C., Højrup, P., Møller, H. J., & Moestrup, S. K. (2005). Identification of the receptor scavenging hemopexin-heme complexes. *Blood, 106*(7), 2572–2579. https://doi.org/10.1182/blood-2005-03-1185

Ignarro, L. J., Byrns, R. E., Buga, G. M., & Wood, K. S. (1987). Endothelium-derived relaxing factor from pulmonary artery and vein possesses pharmacologic and chemical properties identical to those of nitric oxide radical. *Circulation Research, 61*(6), 866–879. https://doi.org/10.1161/01.res.61.6.866

Kam, P., Hines, L., & O'Connor, E. (1996). Effects of cardiopulmonary bypass on systemic vascular resistance. *Perfusion, 11*(4), 346–350. https://doi.org/10.1177/026765919601100408

KDIGO clinical practice guidelines for acute kidney injury. (2023).

Kleschyov, A. L., Zhuge, Z., Schiffer, T. A., Guimarães, D. D., Zhang, G., Montenegro, M. F., Tesse, A., Weitzberg, E., Carlström, M., & Lundberg, J. O. (2023). NO-ferroheme is a signaling entity in the vasculature. *Nature Chemical Biology, 19*(10), 1267–1275. https://doi.org/10.1038/s41589-023-01411-5

Kubota, K., Egi, M., & Mizobuchi, S. (2017). Haptoglobin administration in cardiovascular surgery patients: Its association with the risk of postoperative acute kidney injury. *Anesthesia & Analgesia, 124*(6), 1771–1776. https://doi.org/10.1213/ane.0000000000002093

Lagny, M.-G., Jouret, F., Koch, J.-N., Blaffart, F., Donneau, A.-F., Albert, A., Roediger, L., Krzesinski, J.-M., & Defraigne, J.-O. (2015). Incidence and outcomes of acute kidney injury after cardiac surgery using either criteria of the RIFLE classification. *BMC Nephrology, 16*(1). https://doi.org/10.1186/s12882-015-0066-9

Lameire, N., Biesen, W. V., & Vanholder, R. (2008). Acute kidney injury. *The Lancet, 372*(9653), 1863–1865. https://doi.org/10.1016/S0140-6736(08)61794-8

Lannemyr, L., Bragadottir, G., Krumbholz, V., Redfors, B., Sellgren, J., & Ricksten, S. E. (2017). Effects of cardiopulmonary bypass on renal perfusion, filtration, and oxygenation in patients undergoing cardiac surgery. *Anesthesiology, 126*(2), 205–213. https://doi.org/10.1097/ALN.0000000000001461

Leaf, D. E., Rajapurkar, M., Lele, S. S., Mukhopadhyay, B., Rawn, J. D., Frendl, G., & Waikar, S. S. (2015). Increased plasma catalytic iron in patients may mediate acute kidney injury and death following cardiac surgery. *Kidney International, 87*(5), 1046–1054. https://doi.org/10.1038/ki.2014.374

Lei, C., Berra, L., Rezoagli, E., Yu, B., Dong, H., Yu, S., Hou, L., Chen, M., Chen, W., Wang, H., Zheng, Q., Shen, J., Jin, Z., Chen, T., Zhao, R., Christie, E., Sabbisetti, V. S., Nordio, F., Bonventre, J. V., Xiong, L., & Zapol, W. M. (2018). Nitric oxide decreases acute kidney injury and stage 3 chronic kidney disease after cardiac surgery. *American Journal of Respiratory and Critical Care Medicine, 198*(10), 1279–1287. https://doi.org/10.1164/rccm.201710-2150OC

Li, Y. S., Shemmer, B., Stone, E., A Nardi, M., Jonas, S., & Quartermain, D. (2013). Neuroprotection by inhaled nitric oxide in a murine stroke model is concentration and duration dependent. *Brain Research, 1507*, 134–145. https://doi.org/10.1016/j.brainres.2013.02.031

Lin, T., Maita, D., Thundivalappil, S. R., Riley, F. E., Hambsch, J., Van Marter, L. J., Christou, H. A., Berra, L., Fagan, S., Christiani, D. C., & Warren, H. S. (2015). Hemopexin in severe inflammation and infection: Mouse models and human diseases. *Critical Care, 19*(1). https://doi.org/10.1186/s13054-015-0885-x

Liu, X., Huang, Y., Pokreisz, P., Vermeersch, P., Marsboom, G., Swinnen, M., Verbeken, E., Santos, J., Pellens, M., Gillijns, H., Van de Werf, F., Bloch, K. D., & Janssens, S. (2007). Nitric oxide inhalation improves microvascular flow and decreases infarction size after myocardial ischemia and reperfusion. *Journal of the American College of Cardiology, 50*(8), 808–817. https://doi.org/10.1016/j.jacc.2007.04.069

Luo, X., Jiang, L., Du, B., Wen, Y., Wang, M., & Xi, X. (2014). A comparison of different diagnostic criteria of acute kidney injury in critically ill patients. *Critical Care, 18*(4). https://doi.org/10.1186/cc13977

Mao, H., Katz, N., Ariyanon, W., Blanca-Martos, L., Adýbelli, Z., Giuliani, A., Danesi, T. H., Kim, J. C., Nayak, A., Neri, M., Virzi, G. M., Brocca, A., Scalzotto, E., Salvador, L., & Ronco, C. (2013). Cardiac surgery-associated acute kidney injury. *Cardiorenal Medicine, 3*(3), 178–199. https://doi.org/10.1159/000353134

Minamishima, S., Kida, K., Tokuda, K., Wang, H., Sips, P. Y., Kosugi, S., Mandeville, J. B., Buys, E. S., Brouckaert, P., Liu, P. K., Liu, C. H., Bloch, K. D., & Ichinose, F. (2011). Inhaled nitric oxide improves outcomes after successful cardiopulmonary resuscitation in mice. *Circulation, 124*(15), 1645–1653. https://doi.org/10.1161/CIRCULATIONAHA.111.025395

Minneci, P. C., Deans, K. J., Zhi, H., Yuen, P. S. T., Star, R. A., Banks, S. M., Schechter, A. N., Natanson, C., Gladwin, M. T., & Solomon, S. B. (2005). Hemolysis-associated endothelial dysfunction mediated by accelerated NO inactivation by decompartmentalized oxyhemoglobin. *Journal of Clinical Investigation, 115*(12), 3409–3417. https://doi.org/10.1172/JCI25040

Miyazaki, Y., & Ichinose, F. (2020). Nitric oxide in post-cardiac arrest syndrome. *Journal of Cardiovascular Pharmacology, 75*(6), 508–515. https://doi.org/10.1097/fjc.0000000000000765

Moncada, S., & Erusalimsky, J. D. (2002). Does nitric oxide modulate mitochondrial energy generation and apoptosis? *Nature Reviews Molecular Cell Biology, 3*(3), 214–220. https://doi.org/10.1038/nrm762

Moroos, R., Kucharik, M., Bansal, P., Al Taii, H., Manam, R., Casale, J., Khalili, H., & Maini, B. (2019). Contrast-induced acute kidney injury: Review and practical update. *Clinical Medicine Insights: Cardiology, 13*. https://doi.org/10.1177/1179546819878680

Nagasaka, Y., Fernandez, B. O., Steinbicker, A. U., Spagnolli, E., Malhotra, R., Bloch, D. B., Bloch, K. D., Zapol, W. M., & Feelisch, M. (2018). Pharmacological preconditioning with inhaled nitric oxide (NO): Organ-specific differences in the lifetime of blood and tissue NO metabolites. *Nitric Oxide, 80*, 52–60. https://doi.org/10.1016/j.niox.2018.08.006

Paller, M. S. (1988). Hemoglobin- and myoglobin-induced acute renal failure in rats: Role of iron in nephrotoxicity. *American Journal of Physiology - Renal Fluid and Electrolyte Physiology, 255*(3), F539–F544. https://doi.org/10.1152/ajprenal.1988.255.3.f539

Palmer, R. M. J., Ferrige, A. G., & Moncada, S. (1987). Nitric oxide release accounts for the biological activity of endothelium-derived relaxing factor. *Nature, 327*(6122), 524–526. https://doi.org/10.1038/327524a0

Patel, J. K., Schoenfeld, E., Hou, W., Singer, A., Rakowski, E., Ahmad, S., Patel, R., Parikh, P. B., & Smaldone, G. (2021). Inhaled nitric oxide in adults with in-hospital cardiac arrest: A feasibility study. *Nitric Oxide, 115*, 30–33. https://doi.org/10.1016/j.niox.2021.07.001

Perkins, K. A. A., Pershad, S., Chen, Q., McGraw, S., Adams, J. S., Zambrano, C., Krass, S., Emrich, J., Bell, B., Iyamu, M., Prince, C., Kay, H., Teng, J. C. W., & Young, L. H. (2012). The effects of modulating eNOS activity and coupling in ischemia/reperfusion (I/R). *Naunyn-Schmiedeberg's Archives of Pharmacology, 385*(1), 27–38. https://doi.org/10.1007/s00210-011-0693-z

Premont, R. T., Reynolds, J. D., Zhang, R., & Stamler, J. S. (2020). Role of nitric oxide carried by hemoglobin in cardiovascular physiology: Developments on a three-gas respiratory cycle. *Circulation Research, 126*(1), 129–158. https://doi.org/10.1161/CIRCRESAHA.119.315626

Radomski, M. W., Vallance, P., Whitley, G., Foxwell, N., & Moncada, S. (1993). Platelet adhesion to human vascular endothelium is modulated by constitutive and cytokine induced nitric oxide. *Cardiovascular Research, 27*(7), 1380–1382. https://doi.org/10.1093/cvr/27.7.1380

Ranucci, M. (2007). Perioperative renal failure: Hypoperfusion during cardiopulmonary bypass? *Seminars in Cardiothoracic and Vascular Anesthesia, 11*(4), 265–268. https://doi.org/10.1177/1089253207311141

Ranucci, M., Romitti, F., Isgrò, G., Cotza, M., Brozzi, S., Boncilli, A., & Ditta, A. (2005). Oxygen delivery during cardiopulmonary bypass and acute renal failure after coronary operations. *The Annals of Thoracic Surgery, 80*(6), 2213–2220. https://doi.org/10.1016/j.athoracsur.2005.05.069

Redaelli, S., Magliocca, A., Malhotra, R., Ristagno, G., Citerio, G., Bellani, G., Berra, L., & Rezoagli, E. (2022). Nitric oxide: Clinical applications in critically ill patients. *Nitric Oxide, 121*, 20–33. https://doi.org/10.1016/j.niox.2022.01.007

Rezoagli, E., Ichinose, F., Strelow, S., Roy, N., Shelton, K., Matsumine, R., Chen, L., Bittner, E. A., Bloch, D. B., Zapol, W. M., & Berra, L. (2017). Pulmonary and systemic vascular resistances after cardiopulmonary bypass: Role of hemolysis. *Journal of Cardiothoracic and Vascular Anesthesia, 31*(2), 505–515. https://doi.org/10.1053/j.jvca.2016.06.009

Rezoagli, E., Redaelli, S., Bittner, E.A., Fumagalli, R., Ichinose, F., & Berra, L. (2024). Role of hemolysis on pulmonary arterial compliance and right ventricular systolic function after cardiopulmonary bypass. *Nitric Oxide, 146,* 24–30. https://doi.org/10.1016/j.niox.2024.03.003. PMID: 38521488.

Savorgnan, F. (2021). *A prospective randomized controlled trial of supplemental nitric oxide during cardiopulmonary bypass to reduce acute kidney injury after cardiac surgery.*

Schaer, C. A., Deuel, J. W., Bittermann, A. G., Rubio, I. G., Schoedon, G., Spahn, D. R., Wepf, R. A., Vallelian, F., & Schaer, D. J. (2013). Mechanisms of haptoglobin protection against hemoglobin peroxidation triggered endothelial damage. *Cell Death & Differentiation, 20*(11), 1569–1579. https://doi.org/10.1038/cdd.2013.113

Schaer, C. A., Deuel, J. W., Schildknecht, D., Mahmoudi, L., Garcia-Rubio, I., Owczarek, C., Schauer, S., Kissner, R., Banerjee, U., Palmer, A. F., Spahn, D. R., Irwin, D. C., Vallelian, F., Buehler, P. W., & Schaer, D. J. (2016). Haptoglobin preserves vascular nitric oxide signaling during hemolysis. *American Journal of Respiratory and Critical Care Medicine, 193*(10), 1111–1122. https://doi.org/10.1164/rccm.201510-2058OC

Schaer, D. J., Buehler, P. W., Alayash, A. I., Belcher, J. D., & Vercellotti, G. M. (2013). Hemolysis and free hemoglobin revisited: Exploring hemoglobin and hemin scavengers as a novel class of therapeutic proteins. *Blood, 121*(8), 1276–1284. https://doi.org/10.1182/blood-2012-11-451229

Schiffer, T. A., Lundberg, J. O., Weitzberg, E., & Carlström, M. (2020). Modulation of mitochondria and NADPH oxidase function by the nitrate-nitrite-NO pathway in metabolic disease with focus on type 2 diabetes. *Biochimica et Biophysica Acta - Molecular Basis of Disease, 1866*(8). https://doi.org/10.1016/j.bbadis.2020.165811

Shaw, A., Swaminathan, M., & Stafford-Smith, M. (2008). Cardiac surgery-associated acute kidney injury: Putting together the pieces of the puzzle. *Nephron Physiology, 109*(4), p55–p60. https://doi.org/10.1159/000142937

Signori, D., Magliocca, A., Hayashida, K., Graw, J. A., Malhotra, R., Bellani, G., Berra, L., & Rezoagli, E. (2022). Inhaled nitric oxide: Role in the pathophysiology of cardio-cerebrovascular and respiratory diseases. *Intensive Care Medicine Experimental, 10*(1). https://doi.org/10.1186/s40635-022-00455-6

Stoner, J. D., Clanton, T. L., Aune, S. E., & Angelos, M. G. (2007). O2 delivery and redox state are determinants of compartment-specific reactive O2 species in myocardial reperfusion. *American Journal of Physiology - Heart and Circulatory Physiology, 292*(1), H109–H116. https://doi.org/10.1152/ajpheart.00925.2006

Tejero, J., Shiva, S., & Gladwin, M. T. (2019). Sources of vascular nitric oxide and reactive oxygen species and their regulation. *Physiological Reviews, 99*(1), 311–379. https://doi.org/10.1152/physrev.00036.2017

Terpolilli, N. A., Kim, S. W., Thal, S. C., Kataoka, H., Zeisig, V., Nitzsche, B., Klaesner, B., Zhu, C., Schwarzmaier, S., Meissner, L., Mamrak, U., Engel, D. C., Drzezga, A., Patel, R. P., Blomgren, K., Barthel, H., Boltze, J., Kuebler, W. M., & Plesnila, N. (2012). Inhalation of nitric oxide prevents ischemic brain damage in experimental stroke by selective dilatation of collateral arterioles. *Circulation Research, 110*(5), 727–738. https://doi.org/10.1161/CIRCRESAHA.111.253419

Tripatara, P., Patel, N. S. A., Webb, A., Rathod, K., Lecomte, F. M. J., Mazzon, E., Cuzzocrea, S., Yaqoob, M. M., Ahluwalia, A., & Thiemermann, C. (2007). Nitrite-derived nitric oxide protects the rat kidney against ischemia/reperfusion injury in vivo: Role for xanthine oxidoreductase. *Journal of the American Society of Nephrology, 18*(2), 570–580. https://doi.org/10.1681/ASN.2006050450

Vermeulen Windsant, I. C., de Wit, N. C. J., Sertorio, J. T. C., van Bijnen, A. A., Ganushchak, Y. M., Heijmans, J. H., Tanus-Santos, J. E., Jacobs, M. J., Maessen, J. G., & Buurman, W. A. (2014). Hemolysis during cardiac surgery is associated with increased intravascular nitric oxide consumption and perioperative kidney and intestinal tissue damage. *Frontiers in Physiology, 5.* https://doi.org/10.3389/fphys.2014.00340

Vermeulen Windsant, I. C., Hanssen, S. J., Buurman, W. A., & Jacobs, M. J. (2011). Cardiovascular surgery and organ damage: Time to reconsider the role of hemolysis. *The Journal of Thoracic and Cardiovascular Surgery, 142*(1), 1–11. https://doi.org/10.1016/j.jtcvs.2011.02.012

Virdis, A., Ghiadoni, L., & Taddei, S. (2010). Human endothelial dysfunction: EDCFs. *Pflügers Archiv - European Journal of Physiology, 459*(6), 1015–1023. https://doi.org/10.1007/s00424-009-0783-7

Wang, Y., & Bellomo, R. (2017). Cardiac surgery-associated acute kidney injury: Risk factors, pathophysiology and treatment. *Nature Reviews Nephrology, 13*(11), 697–711. https://doi.org/10.1038/nrneph.2017.119

Wessel, D. L., Adatia, I., Giglia, T. M., Thompson, J. E., & Kulik, T. J. (1993). Use of inhaled nitric oxide and acetylcholine in the evaluation of pulmonary hypertension and endothelial function after cardiopulmonary bypass. *Circulation, 88*(5), 2128–2138. https://doi.org/10.1161/01.cir.88.5.2128

Wunderer, F., Traeger, L., Sigurslid, H. H., Meybohm, P., Bloch, D. B., & Malhotra, R. (2020). The role of hepcidin and iron homeostasis in atherosclerosis. *Pharmacological Research, 153.* https://doi.org/10.1016/j.phrs.2020.104664

Yang, L., Humphreys, B. D., & Bonventre, J. V. (2011). Pathophysiology of acute kidney injury to chronic kidney disease: Maladaptive repair. *Contributions to Nephrology, 174*, 149–155. https://doi.org/10.1159/000329385

Yang, T., Zhang, X. M., Tarnawski, L., Peleli, M., Zhuge, Z., Terrando, N., Harris, R. A., Olofsson, P. S., Larsson, E., Persson, A. E. G., Lundberg, J. O., Weitzberg, E., & Carlstrom, M. (2017). Dietary nitrate attenuates renal ischemia-reperfusion injuries by modulation of immune responses and reduction of oxidative stress. *Redox Biology, 13*, 320–330. https://doi.org/10.1016/j.redox.2017.06.002

Yi, Q., Li, K., Jian, Z., Xiao, Y. B., Chen, L., Zhang, Y., & Ma, R. Y. (2016). Risk factors for acute kidney injury after cardiovascular surgery: Evidence from 2,157 cases and 49,777 controls - a meta-analysis. *Cardiorenal Medicine, 6*(3), 237–250. https://doi.org/10.1159/000444094

Yu, Y., Li, C., Zhu, S., Jin, L., Hu, Y., Ling, X., Miao, C., & Guo, K. (2023). Diagnosis, pathophysiology and preventive strategies for cardiac surgery-associated acute kidney injury: A narrative review. *European Journal of Medical Research, 28*(1). https://doi.org/10.1186/s40001-023-00990-2

Zakeri, R., Freemantle, N., Barnett, V., Lipkin, G. W., Bonser, R. S., Graham, T. R., Rooney, S. J., Wilson, I. C., Cramb, R., Keogh, B. E., & Pagano, D. (2005). Relation between mild renal dysfunction and outcomes after coronary artery bypass grafting. *Circulation, 112*(9), I270–I275. https://doi.org/10.1161/CIRCULATIONAHA.104.522623

Zhang, G., Han, H., Zhuge, Z., Dong, F., Jiang, S., Wang, W., Guimarães, D. D., Schiffer, T. A., Lai, E. Y., Ribeiro Antonino Carvalho, L. R., Lucena, R. B., Braga, V. A., Weitzberg, E., Lundberg, J. O., & Carlstrom, M. (2021). Renovascular effects of inorganic nitrate following ischemia-reperfusion of the kidney. *Redox Biology, 39*. https://doi.org/10.1016/j.redox.2020.101836

Zhang, W. R., Garg, A. X., Coca, S. G., Devereaux, P. J., Eikelboom, J., Kavsak, P., McArthur, E., Thiessen-Philbrook, H., Shortt, C., Shlipak, M., Whitlock, R., & Parikh, C. R. (2015). Plasma IL-6 and IL-10 concentrations predict AKI and long-term mortality in adults after cardiac surgery. *Journal of the American Society of Nephrology, 26*(12), 3123–3132. https://doi.org/10.1681/ASN.2014080764

Zweier, J. L., Wang, P., Samouilov, A., & Kuppusamy, P. (1995). Enzyme-independent formation of nitric oxide in biological tissues. *Nature Medicine, 1*(8), 804–809. https://doi.org/10.1038/nm0895-804

PART II

Nitric oxide and cancer

CHAPTER

2

The emerging role of dimethylarginine dimethylaminohydrolase 2 in regulating vasculogenic mimicry in cancer

Arduino A. Mangoni[1,2], *Lashika Weerakoon*[1], *Vinitha N. Ragavan*[1,3], *Natalia Jarzebska*[3], *Roman N. Rodionov*[3] and *Sara Tommasi*[1,2]

[1]Discipline of Clinical Pharmacology, College of Medicine and Public Health, Flinders University, Adelaide, SA, Australia; [2]Department of Clinical Pharmacology, Flinders Medical Centre, Southern Adelaide Local Health Network, Adelaide, SA, Australia; [3]Department of Internal Medicine III, Technische Universität Dresden, Dresden, Germany

Abbreviations

ADMA Asymmetric N^G,N^G-dimethyl-L-arginine
DDAH1 Dimethylarginine dimethylaminohydrolase 1
DDAH2 Dimethylarginine dimethylaminohydrolase 2
eNOS Endothelial nitric oxide synthase
HIF-1α Hypoxia-inducible factor 1-alpha
iNOS Inducible nitric oxide synthase
NMMA N^G-monomethyl-L-arginine
nNOS Neuronal nitric oxide synthase
NO Nitric oxide
NOS Nitric oxide synthase
TGFβ Transforming growth factor beta
TNBC Triple-negative breast cancer
VEGF-A Vascular endothelial growth factor A

Therapeutic Applications of Nitric Oxide in Cancer and Inflammatory Disorders
https://doi.org/10.1016/B978-0-443-22354-9.00002-4

Introduction

Despite significant advances in diagnosis and treatment, particularly in relation to the discovery of targeted therapies, cancer remains a leading public health burden globally in terms of disability, premature mortality, and healthcare costs ("Cancer Incidence, Mortality, Years of Life Lost, Years Lived With Disability, and Disability-Adjusted Life Years for 29 Cancer Groups From 2010 to 2019: A Systematic Analysis for the Global Burden of Disease Study," 2019; Sung et al., 2021). These vexing issues have stimulated a significant body of research to identify new "druggable" molecular targets for cancer. The ongoing quest for new anticancer treatments is also focused on the mechanisms that regulate the blood supply of oxygen and nutrients to cancer cells. Angiogenesis, a type of neovascularization that involves the formation, sprouting, extension, and remodeling of preexisting blood vessels by endothelial cells, critically favors the development of vascular networks within solid tumors. These networks are an important target for antiangiogenic therapies (Cao et al., 2023; De Palma et al., 2017; Dudley & Griffioen, 2023; Giordo et al., 2022; Lopes-Coelho et al., 2021; Lugano et al., 2020). However, despite their initial promise, antiangiogenic treatments have not translated into a significant reduction in tumor progression and improved survival in several types of cancer. A possible reason for this lack of effect is the development of "resistance" to these agents (Dashtaki & Ghasemi, 2023; Itatani et al., 2018; Ribatti et al., 2021). One recognized factor causing resistance to antiangiogenic drugs is vasculogenic mimicry, an alternative mechanism of vascularization, which is also being increasingly studied as a possible "druggable" target in cancer (Belotti et al., 1912; Haibe et al., 2020; Wei et al., 2021).

A recently discovered regulator of vasculogenic mimicry is the enzyme dimethylarginine dimethylaminohydrolase (DDAH) (Hulin et al., 2020). Specifically, isoform 1 of DDAH (DDAH1) plays a critical physiological role in the biotransformation of the methylated arginines, asymmetric N^G,N^G-dimethyl-L-arginine (ADMA), and N^G-monomethyl-L-arginine (LNMMA), endogenous inhibitors of nitric oxide (NO) synthase (NOS) into inactive metabolites (Jarzebska et al., 2019). However, the role of the other identified isoform of DDAH, isoform 2 (DDAH2), in modulating vasculogenic mimicry has not been investigated until recently (Fig. 2.1).

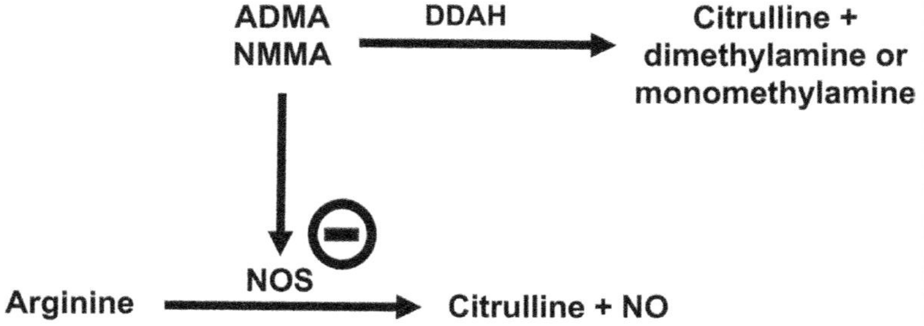

FIGURE 2.1 Methylated arginines and nitric oxide synthesis. *ADMA*, asymmetric N^G,N^G-dimethyl-L-arginine; *DDAH*, dimethylarginine dimethylaminohydrolase; *NMMA*, N^G-monomethyl-L-arginine; *NO*, nitric oxide; *NOS*, nitric oxide synthase.

This chapter describes the current knowledge regarding the pathophysiology and the clinical significance of vasculogenic mimicry in cancer, the role of DDAH1 and methylated arginines in modulating vasculogenic mimicry, and the results of recent studies investigating the biology of DDAH2 and its possible effects on vasculogenic mimicry.

Vasculogenic mimicry: Pathogenesis and clinical significance

Vasculogenic mimicry describes the formation of blood vessel-like structures and networks by cancer cells rather than endothelial cells (Folberg et al., 2006; Folberg & Maniotis, 2004; Maniotis et al., 1999). Given their nonendothelial origin, the newly formed structures are microscopically visible using the periodic acid-Schiff reagent over conventional endothelial cell markers, e.g., CD31 (Maniotis et al., 1999). Vasculogenic mimicry has been described in several solid cancers, including melanoma, breast, ovary, prostate, and lung cancer (Folberg et al., 2006; Maniotis et al., 1999; Shirakawa et al., 2001). Vasculogenic mimicry and canonical angiogenesis can coexist in cancer tissues (Shirakawa et al., 2002).

There has been an increasing interest in the identification of the molecular mechanisms underpinning vasculogenic mimicry. Current evidence suggests the important role played by transforming growth factor beta (TGFβ), stromal cell-derived factor 1, interleukin-6, and TGFβ, secreted by tumor-associated fibroblasts in the tumor microenvironment (Luo et al., 2020). Another critical step in the development of vasculogenic mimicry is represented by epithelial–mesenchymal transition. This transition is characterized by the reduced expression of cell adhesion molecules and the upregulation of vascular endothelial cadherin and fibronectin (Luo et al., 2020) and is further modulated by the hypoxia-inducible factor 1-alpha (HIF-1α), TGFβ (Shuixian Li et al., 2016), the Wnt and Notch signaling pathways (Krishna Priya et al., 2016; Wang, Huang, et al., 2018), and microRNA regulatory networks (Nieto et al., 2016). Several other molecular factors modulating vasculogenic mimicry have been proposed (Basu et al., 2006; Li et al., 2018; Liu et al., 2019; Park & Kim, 2019; Salinas-Vera et al., 2018; Shevde et al., 2010; Wagenblast et al., 2015). Hypoxia and HIF-1α signaling pathways and vascular endothelial growth factor A (VEGF-A) have also been shown to stimulate the transformation of cancer stem cells into endothelial cell-like structures, inducing vasculogenic mimicry (Wei et al., 2021), in triple-negative breast cancer (TNBC) (Liu, Sun, et al., 2013), glioblastoma (Mao et al., 2013), prostate (Wang, Sun, et al., 2018), and liver cancer (Zhao et al., 2018). The results of these studies have led to the identification of promising treatments targeting vasculogenic mimicry, particularly the inhibitors of the sarcoma family kinases (Li et al., 2016), the combination of acridine and metals (Pérez et al., 2017), and several natural compounds (Andonegui-Elguera et al., 2020; Belotti et al., 1912; Shuixian Li et al., 2016, 2017; Luo et al., 2020; Morales-Guadarrama et al., 2021; Tu et al., 2016; Wei et al., 2021; Xu et al., 2019). Importantly, conventional antiangiogenic drugs targeting VEGF-A do not significantly inhibit vasculogenic mimicry (Hendrix et al., 2003; van der Schaft et al., 2004). Instead, they have been shown to paradoxically stimulate vasculogenic mimicry in TNBC and melanoma, with a consequent increase in tumor aggressiveness (Pezzella & Ribatti, 2022; Schnegg et al., 2015; Sun et al., 2017).

The clinical significance of vasculogenic mimicry has been particularly well investigated in breast cancer, with studies reporting a significant association with metastasis, relapse, and mortality (Mitra et al., 2020; Yang et al., 2016). Vasculogenic mimicry is often observed in TNBC (Andonegui-Elguera et al., 2020), a type of cancer that affects 15%–20% of women with breast cancer (Borri & Granaglia, 2021; Marotti et al., 2017) and is associated with aggressive behavior, high relapse rate, and poor survival (Baser et al., 2012; Bianchini et al., 2016; Hsu et al., 2022; Li, Yang, et al., 2017).

Dimethylarginine dimethylaminohydrolase 1: Biology and role in vasculogenic mimicry

NO, an ubiquitous signaling molecule that regulates several physiological and pathophysiological processes, e.g., vascular tone, platelet function, inflammation, and immunity (Coleman, 2001; Khan et al., 2020; Morbidelli et al., 2003; Pacher et al., 2007), is synthesized from L-arginine by the family of NO synthases, including the endothelial (eNOS), neuronal (nNOS), and inducible isoform (iNOS) (Iyengar et al., 1987; Stuehr et al., 1991). There is increasing evidence that NO also regulates carcinogenesis and tumor progression (Khan et al., 2020; Korde Choudhari et al., 2013; Salimian Rizi et al., 2017). In particular, the excessive local tissue concentrations of NO, driven by iNOS, are associated with increased invasiveness and poor prognosis in several cancers, including breast cancer (Jadeski et al., 2000; Jin et al., 2015; Loibl et al., 2002).

The enzyme DDAH, which was traditionally thought to include two isoforms with similar enzymatic activity (DDAH1 and DDAH2), is responsible for the metabolic clearance of the methylated arginines, ADMA and NMMA (Jarzebska et al., 2019; Vallance & Leiper, 2004; Wadham & Mangoni, 2009), potent endogenous inhibitors of the three NOS isoforms (Asif et al., 2013; Bouteldja et al., 2013; Erre et al., 2019; Jarzebska et al., 2019; Lee et al., 2022; Mangoni et al., 2019, 2021; Rees et al., 1990; Shah et al., 2021; Vallance & Leiper, 2004; Wadham & Mangoni, 2009). However, the effects of modulating the enzymatic biotransformation of ADMA and NMMA have been primarily investigated in DDAH1 (Dayoub et al., 2008; Jarzebska et al., 2019; Kopaliani et al., 2021; Kozlova et al., 2023; Murphy et al., 2016).

Increasing evidence suggests an important pathophysiological role of DDAH1 in cancer. For example, studies have reported an increased expression of DDAH1 in several solid cancers, e.g., pancreas, breast, colon, ovary, prostate, and glioma (Hulin et al., 2020; Papaevangelou et al., 2018; Reddy et al., 2018). Our group has reported the overexpression of *DDAH1* in TNBC cell lines, i.e., MDA-MB-231, MDA-MB-453, and BT549 (Hulin et al., 2017). These cells also exhibited the capacity to undertake vasculogenic mimicry (Hulin et al., 2017). Notably, *DDAH1* knockdown was associated with a significant reduction in vasculogenic mimicry and cell migration without affecting cell proliferation and viability (Hulin et al., 2017).

Using small molecule arginine analogues developed by our group, compounds ZST316 and ZST152 (Murphy et al., 2016; Tommasi et al., 2015), the pharmacological inhibition of DDAH1 also significantly reduced vasculogenic mimicry by MDA-MB-231 and BT549 cells (Hulin et al., 2019). Similar to *DDAH1* knockdown, treatment with ZST316 and ZST152 did not influence TNBC cell viability or proliferation but reduced cell migration (Hulin et al., 2019). Notably, the DDAH1 substrate LNMMA was also able to significantly inhibit vasculogenic mimicry further supporting the role of DDAH1 and NO in modulating this alternative neovascularization pathway (Hulin et al., 2019).

Taken together, these experiments highlight the promising role of DDAH1 as a "druggable" target to suppress vasculogenic mimicry in TNBC and, potentially, exert anticancer effects in vivo. By contrast, the potential role of DDAH2 in modulating vasculogenic mimicry has not been investigated until recently.

Recent developments in understanding the biological role of DDAH2 and its possible effects in vasculogenic mimicry

DDAH2, initially described in 1999 (Leiper et al., 1999), is encoded by a gene that is located in a different human chromosome, 6p21.3, than DDAH1, 1p22 (Tran et al., 2000). With a 50% identity in their amino acid sequences, both isoforms are widely expressed in multiple tissues. However, while both genes are highly expressed in the endothelium, the kidney, and the liver, DDAH1 is relatively more expressed in the central nervous system, whereas DDAH2 is predominant in cardiac tissues, the lung, and the placenta (Kozlova et al., 2022; Leiper et al., 1999; Tran et al., 2000).

Our group has recently reported that the expression of *DDAH2* is significantly increased in TNBC cell lines (MDA-MB-231, MDA-MB-453, MFM223s, BT459) when compared to normal breast hTERT-MEC cells (personal data). Pending additional experiments to investigate the effects of DDAH2 downregulation, currently under way, these results suggest that this isoform may also regulate vasculogenic mimicry in TNBC (Fig. 2.2).

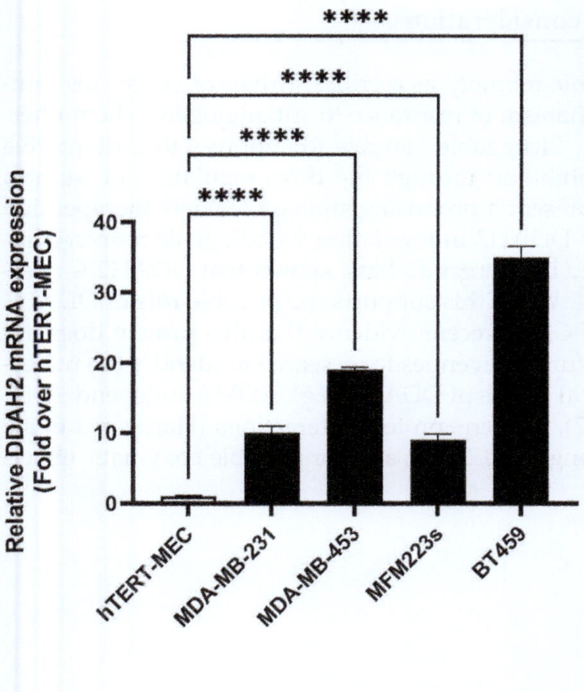

FIGURE 2.2 Expression of *DDAH2* in triple-negative breast cancer cell lines (MDA-MB-231, MDA-MB-453, MFM223s, BT459) and normal breast hTERT-MEC cells. Data are normalized to the housekeeping gene *18S* and presented relative to hTERT-MEC cells, which is set to a value of 1. Data are average of three independent experiments performed in duplicate. Error bars represent SEM. $^{*}P < .05$, $^{****}P < .0001$.

However, despite a significant body of evidence supporting the critical role of DDAH1 in metabolizing ADMA and NMMA (Tommasi et al., 2017; Liu, Fassett, et al., 2013; Zhang et al., 2013), there has been significant controversy regarding whether DDAH2 can effective catalyze the same reaction (Altmann et al., 2012; Hu et al., 2011). Notably, a landmark international consortium of research groups has recently demonstrated that DDAH2 is unable to metabolize ADMA using in silico, in vitro, cell culture, and murine models (Ragavan et al., 2023). Specifically, molecular docking and molecular dynamics studies demonstrated that ADMA is unable to bind efficiently to the predicted substrate binding site of DDAH2. Furthermore, simulations conducted using thermophoresis analysis showed that the binding of ADMA to recombinant human DDAH2 exhibits binding modes that are catalytically unproductive, transient, and unstable. In further experiments, recombinant human DDAH2, unlike recombinant human DDAH1, did not metabolize ADMA, and the downregulation of DDAH2 in human primary endothelial cells did not affect total DDAH activity, whereas the global homozygous deletion of *DDAH1* in mice caused the complete loss of total DDAH activity in tissue lysates. Finally, the homozygous loss of *DDAH2* expression in mice did not affect total DDAH activity in tissue lysates (Ragavan et al., 2023). While the results of this study resolve the controversy regarding the lack of enzymatic biotransformation of ADMA by DDAH2, they also raise significant questions regarding the possible mechanisms underpinning the reported physiological and pathophysiological effects of DDAH2 on angiogenesis, vascular permeability, immunity, and several cellular processes (Hasegawa et al., 2006; Lambden et al., 2015; Shiozawa et al., 2016; Vanella et al., 2011; Yuan et al., 2015), and its possible role in regulating vasculogenic mimicry in cancer.

General considerations

The increasing recognition of vasculogenic mimicry as a critical driver of cancer dissemination and poor prognosis as well as a mechanism of resistance to antiangiogenic chemotherapeutics has prompted the search for new "druggable" targets to suppress this alternative neovascularization pathway. While NO inhibition through the downregulation of isoform 1 of DDAH has recently been shown to represent a promising strategy for new therapeutics, the role of the second identified isoform of DDAH2 in regulating vasculogenic mimicry has been unclear. Recent experiments conducted by our group have shown that DDAH2 is overexpressed in TNBC cells, similar to DDAH1. While this supports the possible role of DDAH2 in regulating vasculogenic mimicry in TNBC, the recent evidence that this protein does not metabolize ADMA, unlike DDAH1, opens further avenues for research to identify the molecular mechanisms underpinning the biological effects of DDAH2, e.g., ADMA-independent effects on NO metabolism (Wang et al., 2007), protein–protein interactions (Hasegawa et al., 2006), effects on mitochondrial fission (Huang et al., 2021), and/or possible enzymatic effects on other substrates.

Conflict of interest statement

No potential conflicts of interest were disclosed.

References

Altmann, K. S., Havemeyer, A., Beitz, E., & Clement, B. (2012). Dimethylarginine-Dimethylaminohydrolase-2 (DDAH-2) does not metabolize methylarginines. *ChemBioChem, 13*(17), 2599–2604. https://doi.org/10.1002/cbic.201200499

Andonegui-Elguera, M. A., Alfaro-Mora, Y., Cáceres-Gutiérrez, R., Caro-Sánchez, C. H. S., Herrera, L. A., & Díaz-Chávez, J. (2020). An overview of vasculogenic mimicry in breast cancer. *Frontiers in Oncology, 10*. https://doi.org/10.3389/fonc.2020.00220

Asif, M., Soiza, R. L., McEvoy, M., & Mangoni, A. A. (2013). Asymmetric dimethylarginine: A possible link between vascular disease and dementia. *Current Alzheimer Research, 10*(4), 347–356. https://doi.org/10.2174/1567205011310040001

Baser, O., Wei, W., Henk, H. J., Teitelbaum, A., & Xie, L. (2012). Patient survival and healthcare utilization costs after diagnosis of triple-negative breast cancer in a United States managed care cancer registry. *Current Medical Research and Opinion, 28*(3), 419–428. https://doi.org/10.1185/03007995.2011.628649

Basu, G. D., Liang, W. S., Stephan, D. A., Wegener, L. T., Conley, C. R., Pockaj, B. A., & Mukherjee, P. (2006). A novel role for cyclooxygenase-2 in regulating vascular channel formation by human breast cancer cells. *Breast Cancer Research, 8*(6). https://doi.org/10.1186/bcr1626

Belotti, D., Pinessi, D., & Taraboletti, G. (1912). Alternative vascularization mechanisms in tumor resistance to therapy. *Cancers, 13*.

Bianchini, G., Balko, J. M., Mayer, I. A., Sanders, M. E., & Gianni, L. (2016). Triple-negative breast cancer: Challenges and opportunities of a heterogeneous disease. *Nature Reviews Clinical Oncology, 13*(11), 674–690. https://doi.org/10.1038/nrclinonc.2016.66

Borri, F., & Granaglia, A. (2021). Pathology of triple negative breast cancer. *Seminars in Cancer Biology, 72*, 136–145. https://doi.org/10.1016/j.semcancer.2020.06.005

Bouteldja, N., Woodman, R. J., Hewitson, C. L., Domingo, E., Barbara, J. A., & Mangoni, A. A. (2013). Methylated arginines and nitric oxide in end-stage renal disease: Impact of inflammation, oxidative stress and haemodialysis. *Biomarkers, 18*(4), 357–364. https://doi.org/10.3109/1354750X.2013.795608

Cancer incidence, mortality, years of life Lost, years lived with disability, and disability-adjusted life years for 29 cancer groups from 2010 to 2019: A systematic analysis for the global burden of disease study. *JAMA Oncology, 8*, (2019), 420–444.

Cao, Y., Langer, R., & Ferrara, N. (2023). Targeting angiogenesis in oncology, ophthalmology and beyond. *Nature Reviews Drug Discovery, 22*(6), 476–495. https://doi.org/10.1038/s41573-023-00671-z

Coleman, J. W. (2001). Nitric oxide in immunity and inflammation. *International Immunopharmacology, 1*(8), 1397–1406. https://doi.org/10.1016/S1567-5769(01)00086-8

Dashtaki, M. E., & Ghasemi, S. (2023). Anti-angiogenic drug resistance: Roles and targeting of non-coding RNAs (microRNAs and long non-coding RNAs). *Current Molecular Pharmacology, 16*(8), 855–869. https://doi.org/10.2174/1874467216666221206100135

Dayoub, H., Rodionov, R., Lynch, C., Cooke, J. P., Arning, E., Bottiglieri, T., Lentz, S. R., & Faraci, F. M. (2008). Overexpression of dimethylarginine dimethylaminohydrolase inhibits asymmetric dimethylarginine-induced endothelial dysfunction in the cerebral circulation. *Stroke, 39*(1), 180–184. https://doi.org/10.1161/STROKEAHA.107.490631

De Palma, M., Biziato, D., & Petrova, T. V. (2017). Microenvironmental regulation of tumour angiogenesis. *Nature Reviews Cancer, 17*(8), 457–474. https://doi.org/10.1038/nrc.2017.51

Dudley, A. C., & Griffioen, A. W. (2023). Pathological angiogenesis: Mechanisms and therapeutic strategies. *Angiogenesis, 26*(3), 313–347. https://doi.org/10.1007/s10456-023-09876-7

Erre, G. L., Mangoni, A. A., Castagna, F., Paliogiannis, P., Carru, C., Passiu, G., & Zinellu, A. (2019). Meta-analysis of asymmetric dimethylarginine concentrations in rheumatic diseases. *Scientific Reports, 9*(1). https://doi.org/10.1038/s41598-019-41994-5

Folberg, R., Arbieva, Z., Moses, J., Hayee, A., Sandal, T., Kadkol, S. H., Lin, A. Y., Valyi-Nagy, K., Setty, S., Leach, L., Chévez-Barrios, P., Larsen, P., Majumdar, D., Pe'er, J., & Maniotis, A. J. (2006). Tumor cell plasticity in uveal melanoma: Microenvironment directed dampening of the invasive and metastatic genotype and phenotype accompanies the generation of vasculogenic mimicry patterns. *American Journal of Pathology, 169*(4), 1376–1389. https://doi.org/10.2353/ajpath.2006.060223

Folberg, R., & Maniotis, A. J. (2004). Vasculogenic mimicry. *Acta Pathologica, Microbiologica et Immunologica Scandinavica, 112*(7–8), 508–525. https://doi.org/10.1111/j.1600-0463.2004.apm11207-0810.x

Giordo, R., Wehbe, Z., Paliogiannis, P., Eid, A. H., Mangoni, A. A., & Pintus, G. (2022). Nano-targeting vascular remodeling in cancer: Recent developments and future directions. *Seminars in Cancer Biology, 86*, 784–804. https://doi.org/10.1016/j.semcancer.2022.03.001

Haibe, Y., Kreidieh, M., El Hajj, H., Khalifeh, I., Mukherji, D., Temraz, S., & Shamseddine, A. (2020). Resistance mechanisms to anti-angiogenic therapies in cancer. *Frontiers in Oncology, 10*. https://doi.org/10.3389/fonc.2020.00221

Hasegawa, K., Wakino, S., Tanaka, T., Kimoto, M., Tatematsu, S., Kanda, T., Yoshioka, K., Homma, K., Sugano, N., Kurabayashi, M., Saruta, T., & Hayashi, K. (2006). Dimethylarginine dimethylaminohydrolase 2 increases vascular endothelial growth factor expression through Sp1 transcription factor in endothelial cells. *Arteriosclerosis, Thrombosis, and Vascular Biology, 26*(7), 1488–1494. https://doi.org/10.1161/01.atv.0000219615.88323.b4

Hendrix, M. J. C., Seftor, E. A., Hess, A. R., & Seftor, R. E. B. (2003). Vasculogenic mimicry and tumour-cell plasticity: Lessons from melanoma. *Nature Reviews Cancer, 3*(6), 411–421. https://doi.org/10.1038/nrc1092

Hsu, J. Y., Chang, C. J., & Cheng, J. S. (2022). Survival, treatment regimens and medical costs of women newly diagnosed with metastatic triple-negative breast cancer. *Scientific Reports, 12*(1). https://doi.org/10.1038/s41598-021-04316-2

Hu, X., Atzler, D., Xu, X., Zhang, P., Guo, H., Lu, Z., Fassett, J., Schwedhelm, E., Böger, R. H., Bache, R. J., & Chen, Y. (2011). Dimethylarginine dimethylaminohydrolase-1 is the critical enzyme for degrading the cardiovascular risk factor asymmetrical dimethylarginine. *Arteriosclerosis, Thrombosis, and Vascular Biology, 31*(7), 1540–1546. https://doi.org/10.1161/ATVBAHA.110.222638

Huang, S., Li, Z., Wu, Z., Liu, C., Yu, M., Wen, M., Zhang, L., & Wang, X. (2021). DDAH2 suppresses RLR-MAVS-mediated innate antiviral immunity by stimulating nitric oxide-activated, Drp1-induced mitochondrial fission. *Science Signaling, 14*.

Hulin, J. A., Gubareva, E. A., Jarzebska, N., Rodionov, R. N., Mangoni, A. A., & Tommasi, S. (2020). Inhibition of dimethylarginine dimethylaminohydrolase (DDAH) enzymes as an emerging therapeutic strategy to target angiogenesis and vasculogenic mimicry in cancer. *Frontiers in Oncology, 9*. https://doi.org/10.3389/fonc.2019.01455

Hulin, J. A., Tommasi, S., Elliot, D., Hu, D. G., Lewis, B. C., & Mangoni, A. A. (2017). MiR-193b regulates breast cancer cell migration and vasculogenic mimicry by targeting dimethylarginine dimethylaminohydrolase. *Scientific Reports, 7*(1). https://doi.org/10.1038/s41598-017-14454-1

Hulin, J. A., Tommasi, S., Elliot, D., & Mangoni, A. A. (2019). Small molecule inhibition of DDAH1 significantly attenuates triple negative breast cancer cell vasculogenic mimicry in vitro. *Biomedicine & Pharmacotherapy, 111*, 602–612. https://doi.org/10.1016/j.biopha.2018.12.117

Itatani, Y., Kawada, K., Yamamoto, T., & Sakai, Y. (2018). Resistance to anti-angiogenic therapy in cancer—alterations to anti-VEGF pathway. *International Journal of Molecular Sciences, 19*(4). https://doi.org/10.3390/ijms19041232

Iyengar, R., Stuehr, D. J., & Marletta, M. A. (1987). Macrophage synthesis of nitrite, nitrate, and N-nitrosamines: Precursors and role of the respiratory burst. *Proceedings of the National Academy of Sciences of the United States of America, 84*(18), 6369–6373. https://doi.org/10.1073/pnas.84.18.6369

Jadeski, L. C., Hum, K. O., Chakraborty, C., & Lala, P. K. (2000). Nitric oxide promotes murine mammary tumour growth and metastasis by stimulating tumour cell migration, invasiveness and angiogenesis. *International Journal of Cancer, 86*(1), 30–39. https://doi.org/10.1002/(SICI)1097-0215(20000401)86:1<30::AID-IJC5>3.0.CO;2-I

Jarzebska, N., Mangoni, A. A., Martens-Lobenhoffer, J., Bode-Böger, S. M., & Rodionov, R. N. (2019). The second life of methylarginines as cardiovascular targets. *International Journal of Molecular Sciences, 20*(18). https://doi.org/10.3390/ijms20184592

Jin, Z., Wang, W., Jiang, N., Zhang, L., Li, Y., Xu, X., Cai, S., Wei, L., Liu, X., Chen, G., Zhou, Y., Liu, C., Li, Z., Jin, F., Chen, B., & Wang, Y.-J. (2015). Clinical implications of iNOS levels in triple-negative breast cancer responding to neoadjuvant chemotherapy. *PLoS One, 10*(7). https://doi.org/10.1371/journal.pone.0130286

Khan, F. H., Dervan, E., Bhattacharyya, D. D., McAuliffe, J. D., Miranda, K. M., & Glynn, S. A. (2020). The role of nitric oxide in cancer: Master regulator or not? *International Journal of Molecular Sciences, 21*(24), 1–30. https://doi.org/10.3390/ijms21249393

Kopaliani, I., Jarzebska, N., Brilloff, S., Kolouschek, A., Martens-Lobenhoffer, J., Bornstein, S. R., Bode-Böger, S. M., Weiss, N., Deussen, A., & Rodionov, R. N. (2021). Transgenic overexpression of dimethylarginine dimethylaminohydrolase 1 protects from angiotensin II - induced cardiac hypertrophy and vascular remodeling. *Atherosclerosis, 331*. https://doi.org/10.1016/j.atherosclerosis.2021.06.281

Korde Choudhari, S., Chaudhary, M., Bagde, S., Gadbail, A. R., & Joshi, V. (2013). Nitric oxide and cancer: A review. *World Journal of Surgical Oncology, 11*(1). https://doi.org/10.1186/1477-7819-11-118

Kozlova, A. A., Ragavan, V. N., Jarzebska, N., Lukianova, I. V., Bikmurzina, A. E., Rubets, E., Suzuki-Yamamoto, T., Kimoto, M., Mangoni, A. A., Gainetdinov, R. R., Weiss, N., Bauer, M., Markov, A. G., Rodionov, R. N., & Bernhardt, N. (2022). Divergent dimethylarginine dimethylaminohydrolase isoenzyme expression in the central nervous system. *Cellular and Molecular Neurobiology, 42*(7), 2273–2288. https://doi.org/10.1007/s10571-021-01101-7

Kozlova, A. A., Rubets, E., Vareltzoglou, M. R., Jarzebska, N., Ragavan, V. N., Chen, Y., Martens-Lobenhoffer, J., Bode-Böger, S. M., Gainetdinov, R. R., Rodionov, R. N., & Bernhardt, N. (2023). Knock-out of the critical nitric oxide synthase regulator DDAH1 in mice impacts amphetamine sensitivity and dopamine metabolism. *Journal of Neural Transmission, 130*(9), 1097–1112. https://doi.org/10.1007/s00702-023-02597-7

Krishna Priya, S., Nagare, R. P., Sneha, V. S., Sidhanth, C., Bindhya, S., Manasa, P., & Ganesan, T. S. (2016). Tumour angiogenesis—origin of blood vessels. *International Journal of Cancer, 139*(4), 729–735. https://doi.org/10.1002/ijc.30067

Lambden, S., Kelly, P., Ahmetaj-Shala, B., Wang, Z., Lee, B., Nandi, M., Torondel, B., Delahaye, M., Dowsett, L., Piper, S., Tomlinson, J., Caplin, B., Colman, L., Boruc, O., Slaviero, A., Zhao, L., Oliver, E., Khadayate, S., Singer, M., Arrigoni, F., & Leiper, J. (2015). Dimethylarginine dimethylaminohydrolase 2 regulates nitric oxide synthesis and hemodynamics and determines outcome in polymicrobial sepsis. *Arteriosclerosis, Thrombosis, and Vascular Biology, 35*(6), 1382–1392. https://doi.org/10.1161/atvbaha.115.305278

Lee, T. F., Bersten, A. D., Heilbronn, L. K., Zinellu, A., Carru, C., Sotgia, S., Mangoni, A. A., & Burt, M. G. (2022). ADMA and homoarginine independently predict mortality in critically ill patients. *Nitric Oxide - Biology and Chemistry, 122–123*, 47–53. https://doi.org/10.1016/j.niox.2022.03.002

Leiper, J. M., Santa Maria, J., Chubb, A., MacAllister, R. J., Charles, I. G., Whitley, G. S. J., & Vallance, P. (1999). Identification of two human dimethylarginine dimethylaminohydrolases with distinct tissue distributions and homology with microbial arginine deiminases. *Biochemical Journal, 343*(1), 209–214. https://doi.org/10.1042/0264-6021:3430209

Li, G., Huang, M., Cai, Y., Ke, Y., Yang, Y., & Sun, X. (2018). miR-141 inhibits glioma vasculogenic mimicry by controlling EphA2 expression. *Molecular Medicine Reports, 18*(2), 1395–1404. https://doi.org/10.3892/mmr.2018.9108

Li, S., Meng, W., Guan, Z., Guo, Y., & Han, X. (2016). The hypoxia-related signaling pathways of vasculogenic mimicry in tumor treatment. *Biomedicine & Pharmacotherapy, 80*, 127–135. https://doi.org/10.1016/j.biopha.2016.03.010

Li, S., Zhang, Q., Zhou, L., Guan, Y., Chen, S., Zhang, Y., & Han, X. (2017). Inhibitory effects of compound DMBT on hypoxia-induced vasculogenic mimicry in human breast cancer. *Biomedicine & Pharmacotherapy, 96*, 982–992. https://doi.org/10.1016/j.biopha.2017.11.137

Li, X., Yang, J., Peng, L., Sahin, A. A., Huo, L., Ward, K. C., O'Regan, R., Torres, M. A., & Meisel, J. L. (2017). Triple-negative breast cancer has worse overall survival and cause-specific survival than non-triple-negative breast cancer. *Breast Cancer Research and Treatment, 161*(2), 279–287. https://doi.org/10.1007/s10549-016-4059-6

Liu, S., Ni, C., Zhang, D., Sun, H., Dong, X., Liang, C. N., Chen, X., Liu, C., Bai, F., Lin, J., Zhao, X., & Sun, X. (2019). S1PR1 regulates the switch of two angiogenic modes by VE-cadherin phosphorylation in breast cancer. *Cell Death & Disease, 10*.

Liu, T. J., Sun, B. C., Zhao, X. L., Zhao, X. M., Sun, T., Gu, Q., Yao, Z., Dong, X. Y., Zhao, N., & Liu, N. (2013). CD133+ cells with cancer stem cell characteristics associates with vasculogenic mimicry in triple-negative breast cancer. *Oncogene, 32*(5), 544–553. https://doi.org/10.1038/onc.2012.85

Liu, X., Fassett, J., Wei, Y., & Chen, Y. (2013). Regulation of DDAH1 as a potential therapeutic target for treating cardiovascular diseases. *Evidence-based Complementary and Alternative Medicine*, 1–6. https://doi.org/10.1155/2013/619207

Loibl, S., Von Minckwitz, G., Weber, S., Sinn, H. P., Schini-Kerth, V. B., Lobysheva, I., Nepveu, F., Wolf, G., Strebhardt, K., & Kaufmann, M. (2002). Expression of endothelial and inducible nitric oxide synthase in benign and malignant lesions of the breast and measurement of nitric oxide using electron paramagnetic resonance spectroscopy. *Cancer, 95*(6), 1191–1198. https://doi.org/10.1002/cncr.10817

Lopes-Coelho, F., Martins, F., Pereira, S. A., & Serpa, J. (2021). Anti-angiogenic therapy: Current challenges and future perspectives. *International Journal of Molecular Sciences, 22*(7). https://doi.org/10.3390/ijms22073765

Lugano, R., Ramachandran, M., & Dimberg, A. (2020). Tumor angiogenesis: Causes, consequences, challenges and opportunities. *Cellular and Molecular Life Sciences, 77*(9), 1745–1770. https://doi.org/10.1007/s00018-019-03351-7

Luo, Q., Wang, J., Zhao, W., Peng, Z., Liu, X., Li, B., Zhang, H., Shan, B., Zhang, C., & Duan, C. (2020). Vasculogenic mimicry in carcinogenesis and clinical applications. *Journal of Hematology & Oncology, 13*(1). https://doi.org/10.1186/s13045-020-00858-6

Mangoni, A. A., Rodionov, R. N., Mcevoy, M., Zinellu, A., Carru, C., & Sotgia, S. (2019). New horizons in arginine metabolism, ageing and chronic disease states. *Age and Ageing, 48*(6), 776–782. https://doi.org/10.1093/ageing/afz083

Mangoni, A. A., Tommasi, S., Sotgia, S., Zinellu, A., Paliogiannis, P., Piga, M., Cauli, A., Pintus, G., Carru, C., & Erre, G. L. (2021). Asymmetric dimethylarginine: A key player in the pathophysiology of endothelial dysfunction, vascular inflammation and atherosclerosis in rheumatoid arthritis? *Current Pharmaceutical Design, 27*(18), 2131–2140. https://doi.org/10.2174/1381612827666210106144247

Maniotis, A. J., Folberg, R., Hess, A., Seftor, E. A., Gardner, L. M. G., Pe'er, J., Trent, J. M., Meltzer, P. S., & Hendrix, M. J. C. (1999). Vascular channel formation by human melanoma cells in vivo and in vitro: Vasculogenic mimicry. *American Journal of Pathology, 155*(3), 739–752. https://doi.org/10.1016/S0002-9440(10)65173-5

Mao, X.-g., Xue, X.-y., Wang, L., Zhang, X., Yan, M., Tu, Y.-y., Lin, W., Jiang, X.-f., Ren, H.-g., Zhang, W., & Song, S.-j. (2013). CDH5 is specifically activated in glioblastoma stemlike cells and contributes to vasculogenic mimicry induced by hypoxia. *Neuro-Oncology, 15*(7), 865–879. https://doi.org/10.1093/neuonc/not029

Marotti, J. D., de Abreu, F. B., Wells, W. A., & Tsongalis, G. J. (2017). Triple-negative breast cancer: Next-generation sequencing for target identification. *American Journal of Pathology, 187*(10), 2133–2138. https://doi.org/10.1016/j.ajpath.2017.05.018. http://ajp.amjpathol.org/

Mitra, D., Bhattacharyya, S., Alam, N., Sen, S., Mitra, S., Mandal, S., Vignesh, S., Majumder, B., & Murmu, N. (2020). Phosphorylation of EphA2 receptor and vasculogenic mimicry is an indicator of poor prognosis in invasive carcinoma of the breast. *Breast Cancer Research and Treatment, 179*(2), 359–370. https://doi.org/10.1007/s10549-019-05482-8

Morales-Guadarrama, G., García-Becerra, R., Méndez-Pérez, E. A., García-Quiroz, J., Avila, E., & Díaz, L. (2021). Vasculogenic mimicry in breast cancer: Clinical relevance and drivers. *Cells, 10*(7). https://doi.org/10.3390/cells10071758

Morbidelli, L., Donnini, S., & Ziche, M. (2003). Role of nitric oxide in the modulation of angiogenesis. *Current Pharmaceutical Design, 9*(7), 521–530. https://doi.org/10.2174/1381612033391405

Murphy, R. B., Tommasi, S., Lewis, B. C., & Mangoni, A. A. (2016). Inhibitors of the hydrolytic enzyme dimethylarginine dimethylaminohydrolase (DDAH): Discovery, synthesis and development. *Molecules, 21*(5). https://doi.org/10.3390/molecules21050615

Nieto, M., Huang, R. Y. J., Jackson, R. A., & Thiery, J. P. (2016). EMT: 2016. *Cell, 166*(1), 21–45. https://doi.org/10.1016/j.cell.2016.06.028

Pérez, S. A., de Haro, C., Vicente, C., Donaire, A., Zamora, A., Zajac, J., Kostrhunova, H., Brabec, V., Bautista, D., & Ruiz, J. (2017). New acridine Thiourea Gold(I) anticancer agents: Targeting the nucleus and inhibiting vasculogenic mimicry. *ACS Chemical Biology, 12*(6), 1524–1537. https://doi.org/10.1021/acschembio.7b00090

Pacher, P., Beckman, J. S., & Liaudet, L. (2007). Nitric oxide and peroxynitrite in health and disease. *Physiological Reviews, 87*(1), 315–424. https://doi.org/10.1152/physrev.00029.2006

Papaevangelou, E., Boult, J. K. R., Whitley, G. S., Robinson, S. P., & Howe, F. A. (2018). Assessment of the direct effects of DDAH I on tumour angiogenesis in vivo. *Angiogenesis, 21*(4), 737–749. https://doi.org/10.1007/s10456-018-9617-6

Park, Y., & Kim, J. (2019). Regulation of IL-6 signaling by miR-125a and let-7e in endothelial cells controls vasculogenic mimicry formation of breast cancer cells. *BMB Reports, 52*(3), 214–219. https://doi.org/10.5483/bmbrep.2019.52.3.308

Pezzella, F., & Ribatti, D. (2022). Vascular co-option and vasculogenic mimicry mediate resistance to antiangiogenic strategies. *Cancer Reports, 5.*

Ragavan, V. N., Nair, P. C., Jarzebska, N., Angom, R. S., Ruta, L., Bianconi, E., Grottelli, S., Tararova, N. D., Ryazanskiy, D., Lentz, S. R., Tommasi, S., Martens-Lobenhoffer, J., Suzuki-Yamamoto, T., Kimoto, M., Rubets, E., Chau, S., Chen, Y., Hu, X., Bernhardt, N., … Rodionov, R. N. (2023). A multicentric consortium study demonstrates that dimethylarginine dimethylaminohydrolase 2 is not a dimethylarginine dimethylaminohydrolase. *Nature Communications, 14*(1). https://doi.org/10.1038/s41467-023-38467-9

Reddy, K. R. K., Dasari, C., Duscharla, D., Supriya, B., Ram, N. S., Surekha, M. V., Kumar, J. M., & Ummanni, R. (2018). Dimethylarginine dimethylaminohydrolase-1 (DDAH1) is frequently upregulated in prostate cancer, and its overexpression conveys tumor growth and angiogenesis by metabolizing asymmetric dimethylarginine (ADMA). *Angiogenesis, 21*(1), 79–94. https://doi.org/10.1007/s10456-017-9587-0

Rees, D. D., Palmer, R. M. J., Schulz, R., Hodson, H. F., & Moncada, S. (1990). Characterization of three inhibitors of endothelial nitric oxide synthase in vitro and in vivo. *British Journal of Pharmacology, 101*(3), 746–752. https://doi.org/10.1111/j.1476-5381.1990.tb14151.x

Ribatti, D., Solimando, A. G., & Pezzella, F. (2021). The anti-VEGF(R) drug discovery legacy: Improving attrition rates by breaking the vicious cycle of angiogenesis in cancer. *Cancers, 13*.

Salimian Rizi, B., Achreja, A., & Nagrath, D. (2017). Nitric oxide: The forgotten child of tumor metabolism. *Trends in Cancer, 3*(9), 659–672. https://doi.org/10.1016/j.trecan.2017.07.005

Salinas-Vera, Y. M., Marchat, L. A., García-Vázquez, R., González de la Rosa, C. H., Castañeda-Saucedo, E., Tito, N. N., Flores, C. P., Pérez-Plasencia, C., Cruz-Colin, J. L., Carlos-Reyes, Á., López-González, J. S., Álvarez-Sánchez, M. E., & López-Camarillo, C. (2018). Cooperative multi-targeting of signaling networks by angiomiR-204 inhibits vasculogenic mimicry in breast cancer cells. *Cancer Letters, 432*, 17–27. https://doi.org/10.1016/j.canlet.2018.06.003

Schnegg, C. I., Yang, M. H., Ghosh, S. K., & Hsu, M. Y. (2015). Induction of vasculogenic mimicry overrides VEGF-A silencing and enriches stem-like cancer cells in melanoma. *Cancer Research, 75*(8), 1682–1690. https://doi.org/10.1158/0008-5472.CAN-14-1855

Shah, R. J., Tommasi, S., Faull, R., Gleadle, J. M., Mangoni, A. A., & Selvanayagam, J. B. (2021). Arginine metabolites as biomarkers of myocardial ischaemia, assessed with cardiac magnetic resonance imaging in chronic kidney disease. *Biomolecules, 11*(3), 1–13. https://doi.org/10.3390/biom11030416

Shevde, L. A., Metge, B. J., Mitra, A., Xi, Y., Ju, J., King, J. A., & Samant, R. S. (2010). Spheroid-forming subpopulation of breast cancer cells demonstrates vasculogenic mimicry via hsa-miR-299-5p regulated de novo expression of osteopontin. *Journal of Cellular and Molecular Medicine, 14*(6 B), 1693–1706. https://doi.org/10.1111/j.1582-4934.2009.00821.x

Shiozawa, T., Iyama, S., Toshima, S., Sakata, A., Usui, S., Minami, Y., Sato, Y., Hizawa, N., & Noguchi, M. (2016). Dimethylarginine dimethylaminohydrolase 2 promotes tumor angiogenesis in lung adenocarcinoma. *Virchows Archiv, 468*(2), 179–190. https://doi.org/10.1007/s00428-015-1863-z

Shirakawa, K., Heike, Y., Wakasugi, H., Terada, M., Kobayashi, H., Kawamoto, S., Brechbiel, M. W., Kasumi, F., Iwanaga, T., Kasumi, F., & Konishi, F. (2002). Hemodynamics in vasculogenic mimicry and angiogenesis of inflammatory breast cancer xenograft. *Cancer Research, 62*(2), 560–566.

Shirakawa, K., Tsuda, H., Heike, Y., Kato, K., Asada, R., Inomata, M., Sasaki, H., Kasumi, F., Yoshimoto, M., Iwanaga, T., Konishi, F., Terada, M., & Wakasugi, H. (2001). Absence of endothelial cells, central necrosis, and fibrosis are associated with aggressive inflammatory breast cancer. *Cancer Research, 61*(2), 445–451.

Stuehr, D. J., Kwon, N. S., Nathan, C. F., Griffith, O. W., Feldman, P. L., & Wiseman, J. (1991). N omega-hydroxy-L-arginine is an intermediate in the biosynthesis of nitric oxide from L-arginine. *Journal of Biological Chemistry, 266*(10), 6259–6263. https://doi.org/10.1016/s0021-9258(18)38112-2

Sun, H., Zhang, D., Yao, Z., Lin, X., Liu, J., Gu, Q., Dong, X., Liu, F., Wang, Y., Yao, N., Cheng, S., Li, L., & Sun, S. (2017). Anti-angiogenic treatment promotes triple-negative breast cancer invasion via vasculogenic mimicry. *Cancer Biology & Therapy, 18*(4), 205–213. https://doi.org/10.1080/15384047.2017.1294288

Sung, H., Ferlay, J., Siegel, R. L., Laversanne, M., Soerjomataram, I., Jemal, A., & Bray, F. (2021). Global cancer statistics 2020: GLOBOCAN estimates of incidence and mortality worldwide for 36 cancers in 185 countries. *CA: A Cancer Journal for Clinicians, 71*(3), 209–249. https://doi.org/10.3322/caac.21660

Tommasi, S., Elliot, D. J., Hulin, J. A., Lewis, B. C., McEvoy, M., & Mangoni, A. A. (2017). Human dimethylarginine dimethylaminohydrolase 1 inhibition by proton pump inhibitors and the cardiovascular risk marker asymmetric dimethylarginine: In vitro and in vivo significance. *Scientific Reports, 7*(1). https://doi.org/10.1038/s41598-017-03069-1

Tommasi, S., Zanato, C., Lewis, B. C., Nair, P. C., Dall'Angelo, S., Zanda, M., & Mangoni, A. A. (2015). Arginine analogues incorporating carboxylate bioisosteric functions are micromolar inhibitors of human recombinant DDAH-1. *Organic and Biomolecular Chemistry, 13*(46), 11315–11330. https://doi.org/10.1039/c5ob01843a

Tran, C. T. L., Fox, M. F., Vallance, P., & Leiper, J. M. (2000). Chromosomal localization, gene structure, and expression pattern of DDAH1: Comparison with DDAH2 and implications for evolutionary origins. *Genomics, 68*(1), 101–105. https://doi.org/10.1006/geno.2000.6262

Tu, D. G., Yu, Y., Lee, C. H., Kuo, Y. L., Lu, Y. C., Tu, C. W., & Chang, W. W. (2016). Hinokitiol inhibits vasculogenic mimicry activity of breast cancer stem/progenitor cells through proteasome-mediated degradation of epidermal growth factor receptor. *Oncology Letters, 11*(4), 2934–2940. https://doi.org/10.3892/ol.2016.4300

Vallance, P., & Leiper, J. (2004). Cardiovascular biology of the asymmetric dimethylarginine:dimethylarginine dimethylaminohydrolase pathway. *Arteriosclerosis, Thrombosis, and Vascular Biology, 24*(6), 1023–1030. https://doi.org/10.1161/01.atv.0000128897.54893.26

van der Schaft, D. W. J., Seftor, R. E. B., Seftor, E. A., Hess, A. R., Gruman, L. M., Kirschmann, D. A., Yokoyama, Y., Griffioen, A. W., & Hendrix, M. J. C. (2004). Effects of angiogenesis inhibitors on vascular network formation by human endothelial and melanoma cells. *Journal of the National Cancer Institute, 96*(19), 1473–1477. https://doi.org/10.1093/jnci/djh267

Vanella, L., Di Giacomo, C., Acquaviva, R., Santangelo, R., Cardile, V., Barbagallo, I., Abraham, N. G., & Sorrenti, V. (2011). The DDAH/NOS pathway in human prostatic cancer cell lines: Antiangiogenic effect of L-NAME. *International Journal of Oncology, 39*(5), 1303–1310. https://doi.org/10.3892/ijo.2011.1107

Wadham, C., & Mangoni, A. A. (2009). Dimethylarginine dimethylaminohydrolase regulation: A novel therapeutic target in cardiovascular disease. *Expert Opinion on Drug Metabolism and Toxicology, 5*(3), 303–319. https://doi.org/10.1517/17425250902785172

Wagenblast, E., Soto, M., Gutiérrez-Ángel, S., Hartl, C. A., Gable, A. L., Maceli, A. R., Erard, N., Williams, A. M., Kim, S. Y., Dickopf, S., Harrell, J. C., Smith, A. D., Perou, C. M., Wilkinson, J. E., Hannon, G. J., & Knott, S. R. V. (2015). A model of breast cancer heterogeneity reveals vascular mimicry as a driver of metastasis. *Nature, 520*(7547), 358–362. https://doi.org/10.1038/nature14403

Wang, D., Gill, P. S., Chabrashvili, T., Onozato, M. L., Raggio, J., Mendonca, M., Dennehy, K., Li, M., Modlinger, P., Leiper, J., Vallance, P., Adler, O., Leone, A., Tojo, A., Welch, W. J., & Wilcox, C. S. (2007). Isoform-specific regulation by NG,NG-dimethylarginine dimethylaminohydrolase of rat serum asymmetric dimethylarginine and vascular endothelium-derived relaxing factor/NO. *Circulation Research, 101*(6), 627–635. https://doi.org/10.1161/CIRCRESAHA.107.158915

Wang, H., Huang, B., Li, B. M., Cao, K. Y., Mo, C. Q., Jiang, S. J., Pan, J. C., Wang, Z. R., Lin, H. Y., Wang, D. H., & Qiu, S. P. (2018). ZEB1-mediated vasculogenic mimicry formation associates with epithelial–mesenchymal transition and cancer stem cell phenotypes in prostate cancer. *Journal of Cellular and Molecular Medicine, 22*(8), 3768–3781. https://doi.org/10.1111/jcmm.13637

Wang, Y., Sun, H., Zhang, D., Fan, D., Zhang, Y., Dong, X., Liu, S., Yang, Z., Ni, C., Li, Y., Liu, F., & Zhao, X. (2018). TP 53 INP 1 inhibits hypoxia-induced vasculogenic mimicry formation via the ROS/snail signalling axis in breast cancer. *Journal of Cellular and Molecular Medicine, 22*(7), 3475–3488. https://doi.org/10.1111/jcmm.13625

Wei, X., Chen, Y., Jiang, X., Peng, M., Liu, Y., Mo, Y., Ren, D., Hua, Y., Yu, B., Zhou, Y., Liao, Q., Wang, H., Xiang, B., Zhou, M., Li, X., Li, G., Li, Y., Xiong, W., & Zeng, Z. (2021). Mechanisms of vasculogenic mimicry in hypoxic tumor microenvironments. *Molecular Cancer, 20*(1). https://doi.org/10.1186/s12943-020-01288-1

Xu, M. R., Wei, P. F., Suo, M. Z., Hu, Y., Ding, W., Su, L., Zhu, Y. D., Song, W. J., Tang, G. H., Zhang, M., & Li, P. (2019). Brucine suppresses vasculogenic mimicry in human triple-negative breast cancer cell line MDA-MB-231. *BioMed Research International, 2019*. https://doi.org/10.1155/2019/6543230

Yang, J. P., Liao, Y. D., Mai, D. M., Xie, P., Qiang, Y. Y., Zheng, L. S., Wang, M. Y., Mei, Y., Meng, D. F., Xu, L., Cao, L., Yang, Q., Yang, X. X., Wang, W. B., Peng, L. X., Huang, B. J., & Qian, C. N. (2016). Tumor vasculogenic mimicry predicts poor prognosis in cancer patients: A meta-analysis. *Angiogenesis, 19*(2), 191–200. https://doi.org/10.1007/s10456-016-9500-2

Yuan, Q., Hu, C. P., Gong, Z. C., Bai, Y. P., Liu, S. Y., Li, Y. J., & Jiang, J. L. (2015). Accelerated onset of senescence of endothelial progenitor cells in patients with type 2 diabetes mellitus: Role of dimethylarginine dimethylaminohydrolase 2 and asymmetric dimethylarginine. *Biochemical and Biophysical Research Communications, 458*(4), 869–876. https://doi.org/10.1016/j.bbrc.2015.02.050

Zhang, P., Xu, X., Hu, X., Wang, H., Fassett, J., Huo, Y., Chen, Y., Bache, R. J., & Li, J. (2013). DDAH1 deficiency attenuates endothelial cell cycle progression and angiogenesis. *PLoS One, 8*(11). https://doi.org/10.1371/journal.pone.0079444

Zhao, X., Sun, B., Liu, T., Shao, B., Sun, R., Zhu, D., Zhang, Y., Gu, Q., Dong, X., Liu, F., Zhao, N., Zhang, D., Li, Y., Meng, J., Gong, W., Zheng, Y., & Zheng, X. (2018). Long noncoding RNA n339260 promotes vasculogenic mimicry and cancer stem cell development in hepatocellular carcinoma. *Cancer Science, 109*(10), 3197–3208. https://doi.org/10.1111/cas.13740

PART III

Nitric oxide reactions in the cells and models to check its biological properties

CHAPTER

3

Decoding S-nitrosylation: Theoretical aspects and analytical approaches

Gianmarco Matrullo[1], *Chiara Pecorari*[2] and *Giuseppe Filomeni*[1,2]

[1]Department of Biology, University of Rome "Tor Vergata", Rome, Italy; [2]Redox Biology, Danish Cancer Institute, Copenhagen, Denmark

Abbreviations

AMPK AMP-Dependent Protein Kinase
AP Apurinic/Apyrimidinic site
APE Apurinic/Apyrimidinic Site Endonuclease
BSA Biotin-Switch Assay
cAMP Cyclic Adenosine Monophosphate
CH_2O_3 Performic Acid
CXCL12 C-X-C motif chemokine ligand 12
DAF Diaminofluorescein
DC Dendritic Cell
DTT Dithiothreitol
ECM Extracellular Matrix
EDRF Endothelium-Derived Relaxing Factor
EGF Epithelial Growth Factor
EGFR Epithelial Growth Factor Receptor
EMT Epithelial-to-mesenchymal transition
ESI Electrospray Ionization
FAK1 Focal Adhesion Kinase 1
FIH-1 factor-inhibiting HIF1a
GPCR G-Protein-Coupled Receptor
GPX4 Glutathione Peroxidase 4
GSH Glutathione
GSNO *S*-nitrosoglutathione
GSNOR *S*-nitrosoglutathione reductase
HGF Hepatocyte Growth factor

Therapeutic Applications of Nitric Oxide in Cancer and Inflammatory Disorders
https://doi.org/10.1016/B978-0-443-22354-9.00003-6

HIF1a Hypoxia-inducible Factor 1a
HSP90 Heat-shock Protein 90
IAM Iodoacetamide
IFN-g Interferon Gamma
IL-12R Interleukin 12 Receptor
IP Immunoprecipitation
L-NAME N^G-nitro-L-arginine methyl ester
LC-MS Liquid Chromatography Coupled to Mass Spectrometry
LDL Low density Lipoprotein
LTP Long-Term Potentiation
MALDI Matrix-Assisted Laser Desorption Ionization
MAPK Mitogen-Activated Protein Kinase
MLC Myosin Light Chain
MMTS Methyl-methane Thiosulfonate
MS Mass Spectrometry
NADPH Nicotinamide Adenine Dinucleotide Phosphate
NER Nucleotide Excision Repair
NF-kB Nuclear Factor kappa-light-chain-enhancer of activated B cells
NMDA *N*-methyl-D-Aspartate
NMDAR *N*-methyl-D-Aspartate Receptor
NO Nitric Oxide
NOS Nitric Oxide Synthase
ONOO Peroxynitrate
ORC Organomercury Resin Capture
OXPHOS Oxidative Phosphorylation
PBMC Peripheral Blood Mononucleated Cell
PDGF Platelet-Derived Growth Factor
PHD Prolyl Hydroxylase
PKG Protein Kinase G
PKM2 Pyruvate Kinase M2
PSNO *S*-nitrosylated protein
PTEN Phosphate Tensin Homologue
RKIP Raf Kinase Inhibitor Protein
ROS Reactive Oxygen Species
SCoR *S*-nitroso-coenzyme A Reductase
SDH Succinate Dehydrogenase
SDS Sodium Dodecyl Sulfate
SDS-PAGE SDS-Polyacrylamide Gel Electrophoresis
SERCA Sarco/Endoplasmic Reticulum Calcium ATPase
sGC Soluble Guanylyl Cyclase
SNO *S*-Nitrosothiol
SNO-CoA *S*-nitroso-coenzyme A
SNO-RAC SNO-Resin-Assisted Capture
SO_3H Sulfonic Acid
***t*BHP** *tert*-Butyl Hydroperoxide
TCA Tricarboxylic acid cycle
TCR T-Cell Receptor
TET ten-eleven translocation proteins
TNF-a Tumor Necrosis Factor Alpha
TRAP1 Tumor Necrosis Factor Receptor-Associated Protein 1
TRX Thioredoxin
TSP1 Thrombospondin 1
VEGF Vascular Endothelial Growth Factor
VSMC Vascular Smooth Muscle Cell
WB Western-Blot

Introduction

Nitric oxide (NO) is a gaseous molecule synthesized through the enzymatic oxidation of arginine to citrulline by a family of NADPH-dependent enzymes known as nitric oxide synthases (NOSs). These enzymes play crucial roles in various biological processes (Förstermann & Sessa, 2012). Originally recognized as a bioactive molecule binding to and modulating the activation of soluble guanylate cyclase (sGC), NO exerts diverse effects in different tissues (Zhao et al., 2015). While this signaling pathway is integral to cell physiology, the regulatory impact of NO extends beyond sGC activation. Principal NO-induced signaling mechanisms involve diverse chemical interactions with amino acid residues (e.g., tyrosine and cysteine) or prosthetic groups (e.g., heme groups and *Fe*–*S* clusters) of a broad spectrum of proteins, providing NO with the potential to influence various biological fundamentals (Hawkins & Davies, 2019).

NO signaling

The discovery of NO in 1987 by Ignarro et al. marked its initial identification as the endothelium-derived relaxing factor (EDRF) involved in the relaxation of vascular smooth muscle cells (VSMCs) and vasodilation through sGC activation (Ignarro et al., 1987). Subsequent identification of NOSs, distributed widely across different tissues and organisms, indicated additional NO-based signaling pathways. NO signaling primarily involves reactions with other free radicals, leading to the production of reactive oxidants that modify proteins through nitrosation or nitration. A clear example is the iron ion in the heme group of NOS itself, which is modified to *Fe*–NO in a complex multistep process (Tejero et al., 2019). Tyrosine nitration, that is, the addition of a nitro (NO_2) group to a tyrosine phenolic ring, is another mechanism of NO-based signaling, often stated as a nonreversible covalent modification, mainly acting as a cytotoxic effector against invading organisms (Radi, 2018). Although nitration often results in protein loss of function, it can affect protein function and intracellular signaling by interfering with other tyrosine posttranslational modifications, such as phosphorylation. Although still matter of debate, several studies have proposed a reversible and dynamic enzymatically driven nature of tyrosine (Griswold-Prenner et al., 2023; Radi, 2018). *S*-nitrosation (or *S*-nitrosylation) is, instead, a reversible posttranslational modification, which involves the covalent binding of an NO moiety to a reduced thiol (-SH), generating low-molecular-weight *S*-nitrosothiols (SNOs) and *S*-nitrosylated proteins (PSNOs) (Rizza & Filomeni, 2020). In recent years, *S*-nitrosylation has emerged as a fundamental modification influencing various protein classes, regulating diverse signaling pathways. Up to now, 4000 SNO sites and over 3000 experimentally verified PSNOs have been identified (Shi & Qiu, 2020), linking NO to processes ranging from the cardiovascular regulation to the immune response, neurological functions, metabolism, and human diseases (Jia et al., 2014; Pacher et al., 2007). Examples include: (i) caspases (Mannick et al., 2001) and phosphatases (Xian et al., 2000), whose *S*-nitrosylation has been demonstrated to inhibit their activity; (ii) kinases, such as the Src oncoprotein, in which *S*-nitrosylation leads to autoactivation and sustained kinase activity (Akhand et al., 1999); and several transcription factors most of which are inhibited in their DNA-binding ability through *S*-nitrosylation (Gao et al., 2004). Given *S*-nitrosylation reversibility, the mechanism responsible for NO detachment from target cysteines—so called denitrosylation—plays a crucial regulatory role. Several denitrosylating enzymes, including *S*-nitrosoglutathione (GSNO) reductase, *S*-nitrosoCoA (SNO-CoA) reductase (SCoR), and thioredoxin (TRX), have been discovered (Benhar et al., 2009). Nonenzymatic denitrosylation, dependent on the presence of reducing agents, nucleophiles, transition metals, or induced by light exposure, has also been described. Understanding the in vivo formation of *S*-nitrosothiols is essential to comprehending the biological functions of NO and its broad range of actions. Free NO, originating from the constitutive NOS isoforms 1 (neuronal, nNOS) and 3 (endothelial, eNOS), or from the inducible NOS2 (iNOS), can diffuse and induce different outcomes. Thiol-containing molecules, such as free cysteine or glutathione (GSH)—the most abundant low-molecular weight thiol in cells—typically capture the majority of NO. When GSH reacts with NO, it forms *S*-nitrosoglutathione (GSNO) that can be reduced back to GSH by GSNOR or transfer the NO moiety to another cysteine via a thiol-nitrosothiol exchange reaction called *trans*-nitrosylation. Recent advancements in the field sustain the hypothesis that this mechanism is the main one responsible for *S*-nitrosylation. Along this line, it has been reported that in peripheral blood mononucleated cells (PBMCs) exposed to proinflammatory conditions (i.e., interferon g and oxidized LDL), *S*-nitrosylation is enzymatically driven via *trans*-nitrosylation by a multimolecular complex called *S*-nitrosylase (Jia et al., 2014). This complex is formed by NOS2, which acts as the NO *generator*, and two proteins, that is, S100A8 and S100A9, which act as the *adaptor*. Mechanistically, S100A9 binds NOS2 and gets *S*-nitrosylated on a specific cysteine residue (Cys3). NO moiety is then transferred via *trans*-nitrosylation from S100A9 to the *target* protein via the structural support of S100A8 (Jia et al., 2014). The glycolytic enzyme gliceraldhehyde-3-phospate dehydrogenase is the first *S*-nitrosylation target characterized, but more than 50 putative targets have also been identified (Jia et al., 2014). It is now clear that the interaction between NOSs and target proteins is one of the main determinants for *S*-nitrosylation specificity; other factors include the presence of allosteric regulators like Ca^{2+}, Mg^{2+}, oxygen, and redox-active compounds (that can alter thiol reactivity and accessibility), electrostatic interactions, and subcellular compartmentalization (Hess et al., 2005).

NO chemical properties are crucial for evaluating its mode of action. Despite being a free radical and a reactive molecule, NO is indeed a weak oxidant, especially under physiological conditions; this impeding a direct rection with amino acids from taking place if not after the further oxygen- or reactive oxygen species (ROS)-mediated oxidation of NO to nitrogen dioxide (NO_2), dinitrogen trioxide (N_2O_3), or peroxynitrate ($ONOO^-$) (Broniowska & Hogg, 2012).

Nitric oxide in physiology and pathology

Vasodilation and muscle cell contraction

Endothelium-derived NO synthesized by NOS3 functions as a potent vasodilator by stimulating sGC activity in VSMCs. The precise mechanism of NO-induced sGC activation remains incompletely understood; however, it is well known that the heterodimeric sGC contains a heme group that binds NO during signaling, forming a *Fe*–NO complex that modulates sGC structure and cyclase activity. Subsequent cGMP production activates protein kinase G (PKG), promoting Ca^{2+} uptake in the sarcoplasmic reticulum, and opening of potassium channels (Carvajal et al., 2000). Reduced Ca^{2+} concentrations impair myosin light chain (MLC) kinase activity, preventing myosin phosphorylation and ultimately resulting in muscle relaxation (Webb, 2003).

While NO primarily operates through sGC activation in endothelial cells, additional sGC-independent mechanisms have been observed. For example, NO-derived $ONOO^-$ has been reported to activate sarco/endoplasmic reticulum calcium ATPase (SERCA) activity, reducing intracellular Ca^{2+} concentration (Adachi et al., 2004), and *S*-nitrosylation has been reported to modulate vasodilation by regulating the expression and activity of G protein-coupled receptors (GPCRs), critical in the tone modulation of VSMCs (Whalen et al., 2007).

In addition to its role in vasodilation control, NO is also implicated in muscle contraction. It indeed stimulates AMP-dependent protein kinase (AMPK) in skeletal muscle, crucial for regulating glucose uptake, mitochondrial biogenesis, and fatty acid oxidation, sustaining energy demands during aerobic exercise (Lira et al., 2007). In line, NOS1 expression increases during physical exercise (McConell et al., 2007/2007; Tatchum-Talom et al., 2000), and NO levels also rise during muscle fiber contraction (Pye et al., 2007). Furthermore, both NOS1 and GSNOR appear crucial during myogenesis and muscle differentiation (Montagna et al., 2019), as evidenced by numerous reports of NOS1-signaling defects in various muscular disorders, such as Duchenne dystrophy, emphasizing the significance of NO in both smooth and skeletal muscle cell physiology.

Neurotransmission

Garthwaite et al. were the first to implicate NO (at that time still known as EDRF) in neurotransmission (Garthwaite et al., 1988) and its diverse roles in the nervous system continue to emerge. NO is associated with the activation of *N*-methyl-D-aspartate (NMDA) receptors, crucial in synaptic plasticity, where it acts as a postsynaptic retrograde neurotransmitter (Picón-Pagès et al., 2019). Briefly, upon glutamate activation of the NMDA receptor, Ca^{2+} ions are released causing NO production from NOS1 (NOS1 and NOS3 are both calcium-dependent, unlike NOS2). The NO produced in the postsynaptic ending diffuses to the presynaptic ending, stimulating cGMP-dependent glutamate release in a long-lasting positive loop called long-term potentiation (LTP), a mechanism underlying learning and memory (Picón-Pagès et al., 2019). Additionally, NO acts presynaptically in nitrergic nerves to induce neurotransmitter release, influencing activities like gastrointestinal motility control (via the enteric nervous system) and pain transmission in the spinal cord (Freire et al., 2009; Sanders, 2016). Interestingly, various evidence suggests that NO can exert protective or detrimental roles for neurons. Prolonged NMDA receptor activation indeed leads to excitotoxic cell death, triggered by the prolonged action of neurotransmitters (Armada-Moreira et al., 2020). However, persistent NOS1 activity can result in NMDA receptor *S*-nitrosylation, which, instead, inhibits its activation as a negative (and protective) feedback loop control (Calabrese et al., 2007).

Metabolism and bioenergetics

Among other roles, NO has been linked to metabolic reprogramming. It has indeed been reported that NO affects metabolism in different ways, including direct—namely by competition with molecular oxygen—or *via* *S*-nitrosylation (Cleeter et al., 1994). Targets of NO are the respiratory chain complexes, mostly the complex IV (Cytochrome *c* oxidase). The production of ROS, due to the impairment of the electron transfer chain, can eventually result in the production of $ONOO^-$, which further inhibits and damages all the mitochondrial complexes in a positive detrimental feedback loop that strongly affects mitochondrial respiration (Brown, 2007). However, NO can also affect metabolism through *S*-nitrosylation, with effects on respiratory complexes comprehensively documented. For example, complex I has been reported to undergo *S*-nitrosylation at Cys39, this reducing mitochondrial respiration and, in turn, ROS production. Complex V, that is, ATP synthase, is also *S*-nitrosylated at Cys294, this inhibiting its activity and reducing the ATP net yield (Montagna et al., 2020). In the last few years, different enzymes involved in glycolysis and tricarboxylic acid (TCA) cycle have indeed been reported to undergo *S*-nitrosylation. *S*-nitrosylation of pyruvate kinase M2 (PKM2), for example, has been shown to inhibit its activity, thereby shunting glucose to the pentose phosphate pathway for NADPH production, with protective effects toward oxidative stress (Lowenstein, 2019; Zhou et al., 2019). Other typical targets are the TCA enzymes aconitase, a-ketoglutarate dehydrogenase, and succinate dehydrogenase (SDH) (Prime et al., 2009).

Tumor necrosis factor receptor-associated protein 1 (TRAP1) has recently emerged as another protein substrate of *S*-nitrosylation (Papaleo et al., 2023; Rizza et al., 2016) involved in cellular metabolic regulation by NO. TRAP1 is, indeed, a mitochondrial chaperon belonging to the heat-shock protein 90 (HSP90) family, which plays a pivotal role in orchestrating mitochondria homeostasis and metabolism. Notably, TRAP1 exerts its influence by downregulating OXPHOS, thereby favoring a metabolic shift toward glycolysis and conferring resistance to oxidative stress (Faienza, Lambrughi, et al., 2020; Faienza, Rizza, et al., 2020; Hoter et al., 2018). The regulatory process of TRAP1 takes place through its modulation of SDH, a constituent of the Complex II of the mitochondrial electron transport chain. Through intricate control mechanisms, TRAP1-mediated inhibition of SDH leads to a disbalance of fumarate to succinate ratio (Guzzo et al., 2014; Rizza et al., 2016; Sciacovelli et al., 2013) affecting mitochondrial general metabolic metabolism and impacting the activity of the a-ketoglutarate-dependent dioxygenases, including ten-eleven translocation proteins (TETs), which epigenetically influence overall transcriptional activity (Faienza, Lambrughi, et al., 2020; Faienza, Rizza, et al., 2020).

S-nitrosylation of TRAP1 at Cys501 is required for TRAP1 degradation via the ubiquitin-proteasome system (Rizza et al., 2016). Intriguingly, recent discoveries have shed light on an additional layer of regulation through an allosteric mechanism, wherein *S*-nitrosylation governs TRAP1 ATPase activity, adding a new dimension to its functional modulation (Faienza, Lambrughi, et al., 2020; Faienza, Rizza, et al., 2020; Papaleo et al., 2023).

Altogether, these lines of evidence confirm the effect of NO in cellular metabolism, for example, by affecting pathways, enzymes within the TCA cycle and/or respiratory chain complexes, as well as key proteins like TRAP1, emphasizing the intricate regulatory network orchestrated by NO in metabolic processes.

Immune response

NOS2 is the inducible NOS which plays a major role in the immune response. NO production from NOS2, mainly by macrophages and monocytes, is induced in response to inflammatory cytokines such as IFN-g, TNF-a, interleukin-1, or lipopolysaccharide, by NF-kB activation (Guzik et al., 2003). Consequently, NOS2 produces a large amount of NO, well above the physiological level, exerting cytotoxic and antimicrobial effects. These effects mainly depend on $ONOO^-$ production that takes place when NO is generated alongside a large amount of superoxide anion (O_2^-), which rapidly reacts with NO to form peroxynitrite. $ONOO^-$ has been exhaustively reported to induce DNA damage via nitro-oxidation; irreversible modification of protein function (such as the above mentioned inhibition of mitochondrial respiratory chain complexes) via tyrosine nitration (Brown, 1999; Guzik et al., 2003).

While the pivotal role of NO in innate immunity is well-established, recent years have witnessed a growing interest in unraveling its involvement in adaptive immunity, particularly its regulatory role on T-cell differentiation and activation. The debate initially centered around the capacity of T lymphocytes to produce NO, which was traditionally attributed to M1 macrophages. Despite initial controversies, accumulating evidence, such as the detection of endothelial NOS3 mRNA in activated T lymphocytes (Reiling et al., 1996) and the identification of inducible NOS2 mRNA in the thymus of mice (Tai et al., 1997), began to support the notion that T cells contributed to NO production. The evidence that NOS3 expression responded to canonical T-cell receptor (TCR) signaling (Ibiza et al., 2006) provided also a mechanism elucidating how NO was produced in T cells. More recently, alternative TCR-independent induction of NOS2, mediated by the NF-kB signaling pathway and the chemokine CXCL12, has also been reported (Navasardyan & Bonavida, 2021), extending the role of T cells as effective NO-producing immune cells. Recent insights have highlighted the regulatory role of NO and NOS2 in the physiology of T cells and other immune cell populations. It has been reported that NO inhibits the proliferation and immune activity of type 17 helper T cells (Th17) implicated in autoimmune diseases (Niedbala et al., 2011). In support to this, it was observed that mice lacking NOS2 developing severe autoimmune encephalomyelitis (Ibiza et al., 2006) and NOS2 expression in $CD4^+$ T lymphocytes inhibits the differentiation of Th17 lymphocytes and effector dendritic cells (DCs) (Navasardyan & Bonavida, 2021).

Conversely, other reports suggest a pivotal role of NO in activated $CD4^+$ T cells in the induction and stability of Th17 cells (García-Ortiz & Serrador, 2018). Molecular mechanisms underlying these regulatory effects are gradually coming to light. In particular, it has been shown that the expression of NOS2 by the NF-kB/CXCL12 pathway leads to the *S*-nitrosylation of caspases, which in turn inhibit cytokine-deprivation-induced apoptosis of CD8 T cells—an essential mechanism in immune response control (Navasardyan & Bonavida, 2021). Mouse models have provided additional molecular insights, demonstrating that low NO concentrations favor T helper 1 subtype differentiation by upregulating the interleukin 12 receptor (IL-12R) through a cGMP-dependent signaling pathway. By contrast, high NO concentrations promote T helper 2 differentiation by suppressing IL-12 production from macrophages (García-Ortiz & Serrador, 2018).

The intricate involvement of NO in both innate and adaptive immunity, particularly in determining the differentiation fate of immune cells, highlights its biological relevance. However, the contribution of endogenous or exogenous NO in these regulatory mechanisms remains still not fully understood. The conflicting evidence poses the need for comprehensive studies to elucidate the multifaceted roles of NO in the immune response.

Tumorigenesis

In the field of cancer biology, the role of NO remains a subject of ongoing debate, characterized by its controversial implications contingent upon cancer type and concentration levels. Elevated concentrations, similar to those observed in immune responses, can induce cytotoxic effects, leading to stable DNA, protein, and mitochondrial damage in both normal and cancer cells (Korde Choudhari et al., 2013). Conversely, low concentrations of NO have been demonstrated to exert cytoprotective effects, fostering carcinogenesis by deregulating various oncogenic or onco-suppressor pathways.

Mutagenicity and genome instability

The mutagenic impact of NO is based on its capacity to deaminate and oxidize DNA bases, resulting in strand breaks and cross-linking events (Thomas et al., 1998). Nitrosylation of amines of DNA bases, such as guanine, by oxidized NO-derivatives like N_2O_3, makes them more susceptible to deamination, leading to an abasic site and single-strand breaks or misrepair (Nguyen et al., 1992). A clear example of NO mutagenic potential is the deamination of cytosine in CpG rich sites of the p53 gene, resulting in a G-C to A-T transition. NO_2 and $ONOO^-$ represent additional NO-derived oxidant species responsible for DNA nitro-oxidation and the formation of 8-nitroguanine and 8-hydroxydeoxyguanosine, which rapidly depurinate, producing G-T transversion mutations. Deamination and oxidation thus stand as the principal mechanisms through which NO inflicts DNA damage (deRojas-Walker et al., 1995). Furthermore, NO is involved in the regulation of DNA damage repair, inhibiting the nucleotide excision repair (NER) system (Chien et al., 2004), and apurinic/apyrimidinic site (AP) endonuclease (APE), which cleaves DNA to form a gap subsequently filled with the correct nucleotide by polymerase-b. NO likely induces nuclear export of APE by *S*-nitrosylation of cysteine 93 and 310, thus disrupting its DNA repair function (Qu et al., 2007).

Cell death

NO plays a pivotal role in programmed cell death, known as apoptosis. Its involvement in this process is intricate and context-dependent, as it can exert both proapoptotic and antiapoptotic effects depending on NO concentration, cell type, and cellular environment. Low levels of NO typically confer protection by acting through sGC activation, up-regulating intracellular antioxidants (such as glutathione), or inactivating proapoptotic proteins such as caspases and Bax (Khan et al., 2020). Conversely, elevated levels of NO may promote apoptosis by activating proapoptotic signals or irreversibly damaging cellular components; one example is the activation of MAPK pathways, resulting in p53 accumulation and caspase activation (Khan et al., 2020).

Over the years, several mechanisms of induced cell death, beyond the established paradigm of apoptosis, have been elucidated. Anoikis, for instance, represents an alternative form of apoptosis critical for protecting the organism against cells that have acquired the ability to grow in suspension and metastasize, that is, colonize even distant tissues from those they originate from. Therefore, anoikis sensitizes cells to detachment from other cells and the extracellular matrix (ECM) (Taddei et al., 2012). Notably, recent findings indicate that NO is implicated in anoikis evasion through the *S*-nitrosylation of focal adhesion kinase 1 (FAK1) (Rizza et al., 2023). It is no coincidence that FAK1, a pivotal regulator of anoikis, plays a crucial role in cancer progression and metastatic capability. Specifically, *S*-nitrosylation at Cys658 has been proposed to sustain FAK1 phosphorylation, resulting in its hyperactivation and, consequently, driving resistance to anoikis (Rizza et al., 2023).

Another distinct cell death program is ferroptosis, characterized by its iron-dependent nature and linked to redox imbalance. This process is associated with an aberrant increase in iron ions, diminished activity of protective antioxidant enzymes—such as glutathione peroxidase 4 (GPX4)—and severe lipid peroxidation, culminating in membrane damage (Tang et al., 2020). The presence of ferrous ions (Fe^{2+}) facilitates the production of hydroxyl radicals, initiating lipid peroxidation, which is typically counteracted by GPX4, the primary defender against ferroptosis (Homma et al., 2021). Intriguingly, treatments with various NO donors have demonstrated to suppress ferroptosis by inhibiting the lipid peroxidation chain reaction, reasonably through interaction with *tert*-butyl hydroperoxide (*t*BHP), the primary initiator of this oxidative chain reaction (Homma et al., 2021). Several studies indicate the involvement of the p38/JNK MAPK pathway in ferroptosis regulation (Ye et al., 2019; Yu et al., 2015). Given that *S*-nitrosylation inhibits p38 and MAP3K activity, one can speculate that *S*-nitrosylation may also play a role in ferroptosis regulation either via activation of upstream regulators or directly via inhibition of p38/JNK (He et al., 2022). While the implication of both NO and *S*-nitrosylation in ferroptosis regulation can be hypothesized, further in-depth analyses are imperative for a comprehensive understanding of their roles in this mode of cell death.

Angiogenesis

NO emerges as a fundamental regulator of angiogenesis through its actions on endothelial cells lining the inner walls of blood arteries. NO affects various mechanisms promoting the development, growth, and remodeling of new blood vessels. The angiogenesis process is finely controlled by the balance between proangiogenic factors, such as vascular endothelial growth factor (VEGF) and prostaglandins, and antiangiogenic factors, such as thrombospondin-1 (TSP1) or endostatin. Typically, the rapidly proliferating cells of tumor masses face a reduced supply of oxygen (a condition known as hypoxia) and nutrients. This condition promotes vascularization by the stabilization of hypoxia-inducible factor 1a (HIF1a), leading to the production of VEGF (Forsythe et al., 1996), which mediates NOS3 activation via AKT-mediated phosphorylation at Ser1177 (Fulton et al., 1999). Such a sustained NO flux can also inhibit TSP1, a regulator of blood pressure and, in turn, blood vessel formation (Morbidelli et al., 2003).

On the other hand, NO is also recognized as an upstream regulator of VEGF, exerting its influence by enhancing HIF1a activity (Kimura et al., 2002; Movafagh et al., 2015). This establishes the groundwork for an NO/HIF1/VEGF axis characterized by a positive feedback, although evidence also exists for NO-mediated inhibition of HIF1a. The intricate interplay unfolds with distinct effects observed at different NO concentrations. Moderate-to-high NO concentrations (0.5–1 mM) promote HIF1a expression and activity, while lower concentrations inhibit HIF1a stabilization under hypoxic conditions.

Several molecular mechanisms have been reported to underlie the regulation of HIF1a by NO. HIF1a stabilization is proposed to depend on posttranslational modifications of HIF1a itself or the inhibition of prolyl hydroxylases (PHDs) under normoxic conditions (Movafagh et al., 2015). PHDs, responsible for hydroxylating HIF1a, mark it for ubiquitination and subsequent degradation by the proteasome. Another player in this regulatory network is factor-inhibiting HIF1a (FIH-1), a hydroxylase that represses HIF1 transcriptional activity. FIH-1 achieves this by targeting Asn803 of HIF1a, thus preventing binding with the coactivator p300.

Crucially, both PHDs and FIH-1 exhibit susceptibility to *S*-nitrosylation, a modification that occurs similarly in normoxia and hypoxia conditions. This common feature results in the stabilization of HIF1a and the enhancement of its transcriptional activity, thereby reinforcing the intricate dynamics of the NO/HIF/VEGF axis (Kurokawa et al., 2019; Park et al., 2008). This multifaceted interaction sheds light on the nuanced regulatory roles played by NO within this axis, contributing to the complexity of cellular responses in environments with variable oxygen tension, like those characterizing cancer.

Epithelial-to-mesenchymal transition

Epithelial-to-mesenchymal transition (EMT) is an essential process required for the transformation of a benign tumor into an aggressive and invasive malignancy. High fluxes of NO are often correlated with the inhibition of NF-kB through *S*-nitrosylation of the p50 subunit, which prevents NF-kB DNA binding ability (Baritaki et al., 2010). Consequently, downstream EMT markers whose expression is related to NF-kB, such as Snail, Raf kinase inhibitor protein (RKIP), and phosphatase tensin homologue (PTEN) are downregulated, suppressing EMT and metastasis, as demonstrated in breast cancer cells (Hickok et al., 2011) and other cancer models. NO has been shown to regulate several transcription factors like Snail family proteins SNAI1 and SNAI2, ZEB1 and ZEB2, growth factors (EGF, PDGF, and HGF), as well as Wnt/b-catenin and ECM components, suppressing tumor migration and invasiveness (Khan et al., 2020).

Despite a well-documented suppressing action, promigratory effects have also been reported. Indeed, NO has been reported to activate EGFR and Src kinase via *S*-nitrosylation in ER^- breast cancer, activating b-catenin signaling and switching E-cadherin toward N-cadherin and vimentin expression (Switzer et al., 2012). In line with these lines of evidence, NOS inhibition by N^G-nitro-L-arginine methyl ester (L-NAME) treatment reduces invasiveness of C3L5 and C10 mammary tumor cells in mice (Orucevic et al., 1999), and stably expression of NOS2 makes HRT-18 human colorectal adenocarcinoma cells more invasive (Siegert et al., 2002).

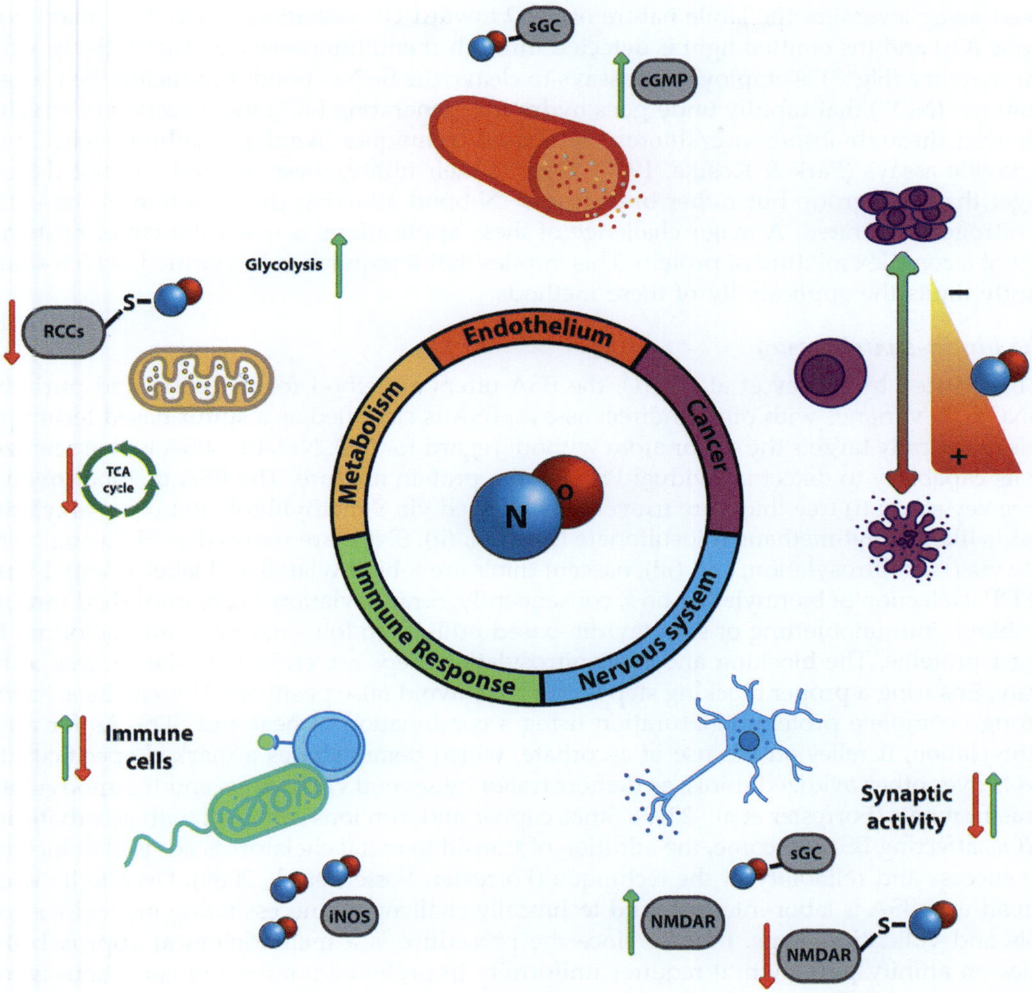

S-nitrosylation detection: Approaches and challenges

Indirect detection assays

In the pursuit of detecting *S*-nitrosylation (SNO), methods referred to as "NO based" or indirect detection strategies exploit the intrinsically weak nature of sulfur-nitrogen bond as primary target. Examples of such indirect approaches encompass ozone-based chemiluminescence, DAF-based fluorescence, Saville assay, and biotin-switch assay (BSA). The ozone-based assay leverages the labile nature of SNO toward UV radiation, where NO reacts with ozone (O_3) and the emitted light is detected through chemiluminescence. Alternatively, divalent mercury (Hg^{2+}) is employed in assays to cleave the S—NO bond, producing the nitrosonium ion (NO^+) that rapidly undergoes hydration, generating NO_2^-, the detection of which is achieved through absorbance/fluorescence-based techniques like diaminofluorescein (DAF) or Saville assays (Park & Kostka, 1997). Despite their utility, these methods do not directly target the SNO group but rather break the S—N bond allowing the detection of the sulfur or nitrogen generated. A major challenge of these applications is when the target protein is part of a complex mixture of protein. This implies that it requires to be purified, which significantly limits the applicability of these methods.

The biotin-switch assay

Introduced by Jaffrey et al. (2001), the BSA offers a method for detecting and purifying PSNOs. At variance with other indirect assays, BSA is classified as a sulfur-based technique, as it specifically targets the sulfur atom without regard for NO. Notably, BSA is characterized by its capability to detect individual PSNOs in a protein mixture. The BSA method involves three key steps: (i) free thiols are irreversibly blocked via *S*-methylthiolation using alkylating agents like methyl-methane thiosulfonate (MMTS); (ii), SNOs are reduced to SH using ascorbate via *trans*-nitrosylation; and (iii), nascent thiols are *S*-biotinylated and labeled with biotin-HPDP. Detection of biotinylation and, consequently, *S*-nitrosylation, is accomplished through antibiotin immunoblotting or streptavidin-based pull-down followed by immunoblotting for target proteins. The blocking and *trans*-nitrosylation steps are critical for the success of the assay. Ensuring a proper blocking step is crucial to avoid false positives. This can be achieved through complete protein denaturation using a combination of heat and SDS. As for *trans*-nitrosylation, it relies on the use of ascorbate, which demonstrates a marked specificity for SNOs over other oxidized thiols, as demonstrated by several validations and thermodynamic measurements (Forrester et al., 2007). Since copper and iron ions can react with ascorbate and SNOs, affecting BSA outcome, the addition of transition metal chelators is advised to increase the success and reliability of the technique (Forrester, Foster, et al., 2009). Despite its widespread use, BSA is labor-intensive and technically challenging, necessitating meticulous controls and validation steps. Indeed, since the procedure is a methodological approach that relies on affinity pulldown, it requires uniformity in protein amounts and conditions across samples (e.g., biotin, ascorbate, and beads quantity). The complexity of the protocol, which involves numerous washing and precipitation steps, make preliminary SDS-PAGE before pulldown necessary to ensure the reliability and reproducibility of subsequent analyses. Besides these loading controls or "input," to increase the robustness of BSA outcomes, the presence of negative and positive controls becomes crucial. A negative control, involving biotinylation in an ascorbate-free sample during the labeling phase, is deemed necessary to identify and mitigate potential sources of misleading results. The blocking phase is also equally critical, as certain protein thiols may exhibit resistance to blocking, leading to false-positive biotinylation outcomes. Nevertheless, with careful execution and the inclusion of negative controls, even in cases of inadequate blocking, the overall sensitivity of BSA may be only marginally compromised, leading to increased background noise without compromising specificity (Forrester, Foster, et al., 2009).

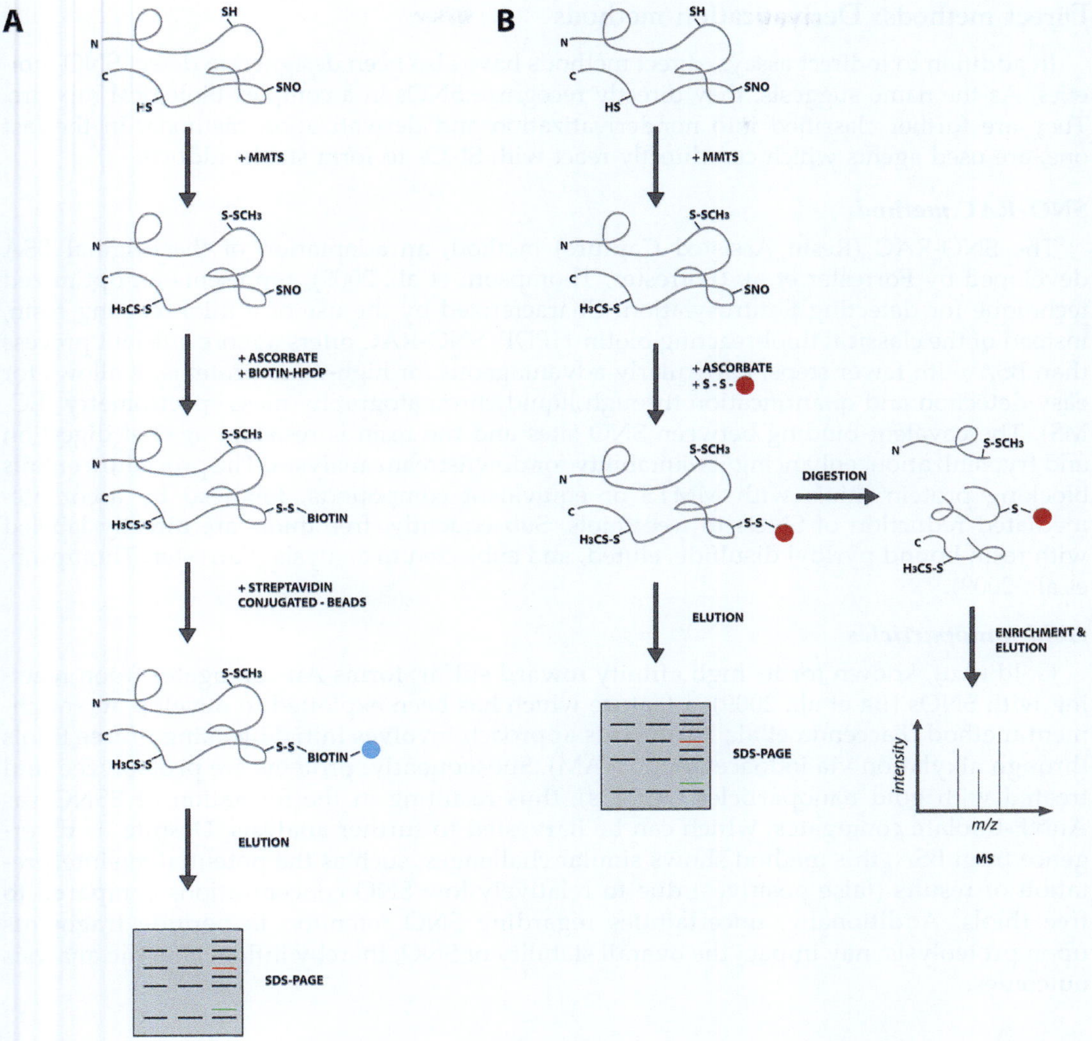
A
SH
N
C
SNO
HS
+ MMTS
S-SCH3
H3CS-S
+ ASCORBATE
+ BIOTIN-HPDP
S-S
BIOTIN
+ STREPTAVIDIN
CONJUGATED - BEADS
ELUTION
SDS-PAGE
B
+ S - S -
DIGESTION
ENRICHMENT &
ELUTION
intensity
m/z
MS

Direct methods: Derivatization methods

In addition to indirect assays, direct methods have also been designed to detect SNO moieties. As the name suggests, they directly recognize SNOs in a complex biological mixture. They are further classified into nonderivatization and derivatization methods. In the last one, are used agents which can directly react with SNOs to form stable adducts.

SNO-RAC method

The SNO-RAC (Resin Assisted Capture) method, an adaptation of the original BSA developed by Forrester et al. (Forrester, Thompson, et al., 2009), represents an optimized technique for detecting *S*-nitrosylation. Characterized by the use of a thiol-reacting resin, instead of the classical thiol-reacting biotin-HPDP, SNO-RAC offers a more efficient process than BSA with fewer steps. Particularly advantageous for high-mass proteins, it allows for easy detection and quantification through liquid chromatography-mass spectrometry (LC-MS). The covalent binding between SNO sites and the resin is resistant against digestion and trypsinization, enhancing its suitability for downstream analyses. The procedure entails blocking protein thiols with MMTS or equivalent compounds, followed by ascorbate-mediated reduction of SNOs to free thiols. Subsequently, free thiols are directly labeled with resin-bound pyridyl disulfide, eluted, and subjected to analysis (Forrester, Thompson, et al., 2009).

Gold nanoparticles

Gold (Au), known for its high affinity toward sulfur, forms Au-conjugates upon reacting with SNOs (Jia et al., 2009), a feature which has been exploited to develop an enrichment method (Faccenda et al., 2010). This approach involves initial blocking of free thiols through alkylation via iodoacetamide (IAM). Subsequently, proteins are proteolyzed and treated with gold nanoparticles (AuNPs), thus resulting in the formation of PSNOs as AuNP-thiolate conjugates, which can be harvested to further analysis. Despite its divergence from BSA, this method shows similar challenges, such as the potential misinterpretation of results (false positives) due to relatively low SNO concentrations compared to free thiols. Additionally, uncertainties regarding SNO retention in peptide fragments upon proteolysis may impact the overall stability of SNO, thereby influencing the analysis outcomes.

Organomercury resin capture

Organomercury compounds can form stable conjugates upon reacting with SNO (Bramanti et al., 2008). Also with this approach, the blocking phase emerges as a critical step. SNO proteins are captured by an organomercury compound (either conjugated with agarose resin or a biotin tag) for solid or solution-based enrichment. Elution is achieved with performic acid (CH_2O_3). It oxidizes the resulting thiols to sulfonic acid (SO_3H) derivatives providing a distinctive mass spectrometer peak signal which makes organomercury resin capture (ORC) an ideal method for SNO site-specific identification. Despite exhibiting greater selectivity for SNO over disulfides, the potential reactivity of organomercury compounds with free thiols requires blocking phase to make results reproducible and trusty. Moreover, not much is known regarding the reactivity of these compounds toward other cysteine PTMs, and given their high toxicity, this assay should be performed with extreme caution.

Organophosphine-based SNO labeling

Some reactions specifically target SNO groups and no other sulfur-based derivatives, such as thiols or disulfides. This is the feature exploited by a new technology developed by Devarie-Baez et al., which is based on organophosphine specific reactions with SNO, called "'bioorthogonal reactions'" (Devarie-Baez et al., 2013). This strategy utilizes triaryl-substituted phosphine, highly reactive against SNO, generating an azalide intermediate that can be converted into various products, such as sulfenamides, disulfide-phosphoimine adducts, dehydroalanine adducts, or disulfide conjugates. Upon treatment of with NO donor, cells are permeabilized and fixed, and then incubated with biotin-linked phosphine to bind PSNOs directly via a disulfide bond formation. The biotin label is then visualized by neutravidin/streptavidin-HRP or immunoblotting. Biotin-phosphine does not react with free thiols, avoiding one of major BSA pitfall. However, to avoid *trans*-nitrosylation reactions, a blocking phase with N-ethylmaleimide (NEM) is still required. Sample preparations that entail the use of reductants, like dithiothreitol (DTT) or b-mercaptoethanol able to reduce the disulfide bond formed between SNO and phosphine, are required as positive control. Vice versa, incubations which biotin that lacks the phosphine moiety should be used as negative control to assess the specificity of the approach (Zhang et al., 2010). Notwithstanding the high specificity to SNO groups, further validation and method refinements may be warranted.

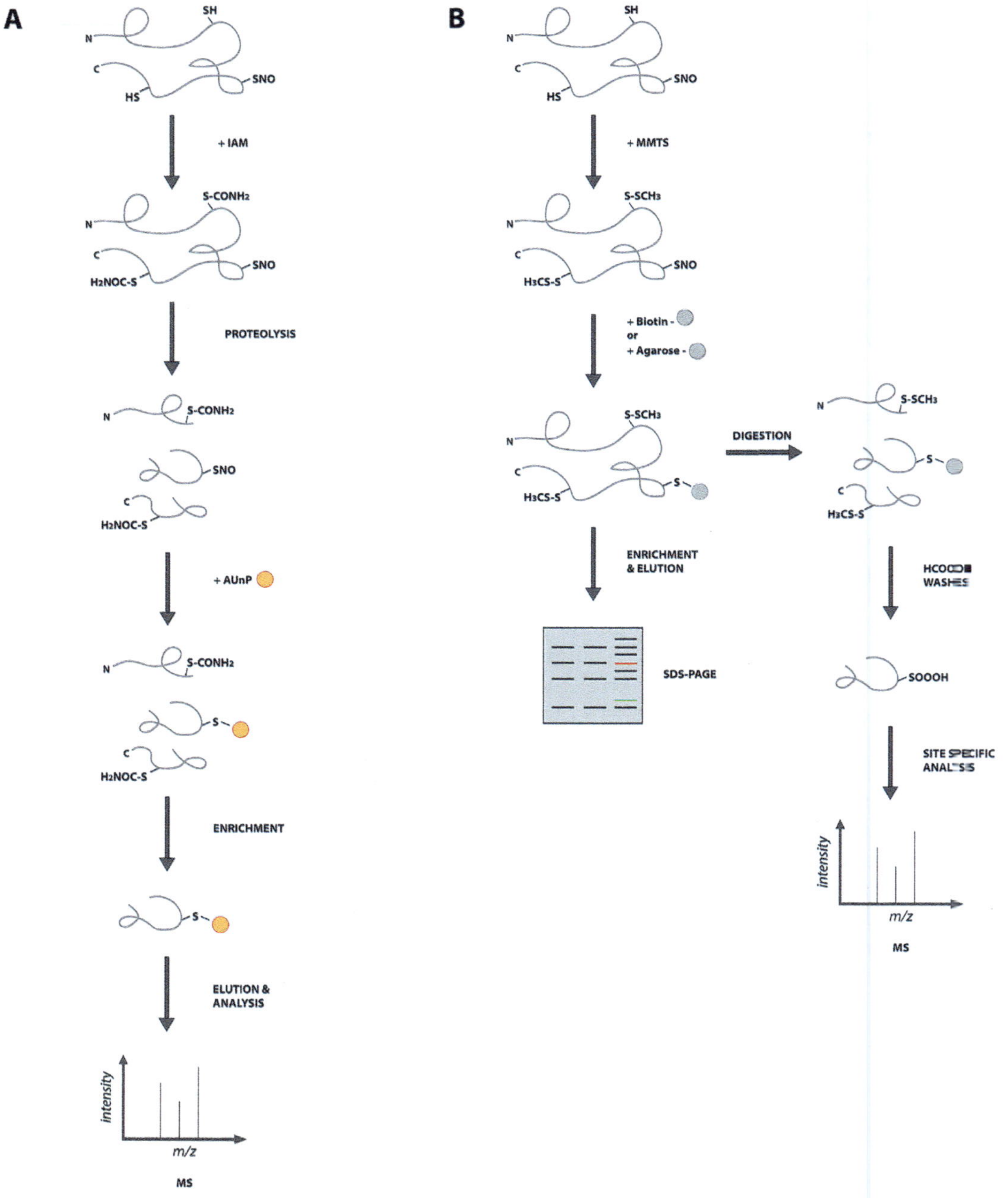
A
SH
N
C
SNO
HS
+ IAM
S-CONH2
H2NOC-S
PROTEOLYSIS
+ AUnP
ENRICHMENT
ELUTION &
ANALYSIS
intensity
m/z
MS
B
+ MMTS
S-SCH3
H3CS-S
+ Biotin -
or
+ Agarose -
DIGESTION
ENRICHMENT
& ELUTION
SDS-PAGE
SOOOH

Direct (nonderivative) methods

Mass spectrometry

Commonly employed mass spectrometry (MS) ionization methods, such as electrospray ionization (ESI) or matrix-assisted laser desorption ionization (MALDI), can dissociate the labile S—NO bond. However, the characterization of SNO-thioredoxin by ESI MS analysis suggests that a careful optimization of experimental conditions and parameters (e.g., voltage and collision energy) can result in a good and reliable signal (Wang et al., 2008). The application of MS to detect other PSNOs most likely requires an extensive setup for each PSNO of interest, making this technique extremely demanding and still challenging. Along this line, it has been reported that 3-nitrotyrosine might be identified as SNO moiety in the spectra (Lee et al., 2007), adding a significant incidence of false positive signals with this method.

Immunohistochemistry

Antibody-based techniques, such as immunoprecipitation (IP) and western-blot (WB), are normally not used to detect PSNOs as protein separation by SDS-PAGE easily disrupts S—NO bonds. However, these methods could be employed in nonreducing conditions. Several monoclonal and polyclonal antibodies for both IP and WB have been developed and reported, for example, in lung and vascular tissues (Greco et al., 2006; Munson et al., 2005). The specificity of these antibodies is not well characterized, arguing for positive and negative controls being always produced to test the reliability of the method.

Conclusion

Over the last decades, extensive investigations have shed light on the physiological role of NO, with a particular focus on its involvement in diverse cellular processes. The emergence of *S*-nitrosylation as a principal NO-mediated mechanism has expanded our understanding, yet intrinsic properties of NO still stimulate debates within the scientific community about the biological relevance of *S*-nitrosylation. Initially regarded as a typical posttranslational modification analogous to phosphorylation, the intrinsic weakness and reactivity of S—NO bonds, which can rapidly react with thiol-containing molecules, has raised concerns regarding its stability and specificity.

These controversies led to the idea of *S*-nitrosylation represents as a transient modification required for the formation of disulfide bonds, thus acting as a mediator to establish more stable end effectors (Wolhuter et al., 2018). Unlike phosphorylation, that entails kinases and phosphatases directly modifying target proteins, *S*-nitrosylation involves a complex interplay where nitrosylases and denitrosylases mostly act via *trans*-nitrosylation of intermediate low-molecular weight thiols, such as GSNO (Jia et al., 2014; Liu et al., 2001) and SNO-CoA (Hua-Lin Zhou et al., 2023; Zhou et al., 2019) to attach or remove NO moiety from reactive cysteines on target proteins. This new prototypic signal transduction has shifted the perspective from which to look at *S*-nitrosylation. Indeed, the evidence that most of the enzyme regulating *S*-nitrosylation are conserved along phylogeny reinforces the biological relevance of this redox-based posttranslational modification as a finely regulated mechanism underlying most of NO bioactivity.

NO orchestrates diverse roles in organisms, impacting systems from cardiovascular functions to cancer biology, intricately regulating both physiological and pathological processes. Despite persisting questions, the past 25 years have witnessed significant progress in unraveling NO signaling and homeostasis being mostly regulated by *S*-nitrosylation. Constantly emerging evidence on novel *S*-nitrosylase and denitrosylase proteins propels the field forward, aiming to decipher how disturbances in the NO and *S*-nitrosylation equilibrium contribute to pathologies and cancer biology. This evolving landscape promises novel therapeutic targets, fueling the quest for innovative therapies.

Conflict of interest

The authors declare no competing interest.

Acknowledgments

This work was supported by grants from the Danish Cancer Society (KBVU R146-A9414 and R231-A13855); the Italian Association for Cancer Research (IG2017-20719 and IG2023-29221), and the Novo Nordisk Foundation (NNF18OC0052550 and NNF22OC0079352) and PRIN 2022. Chiara Pecorari was a recipient of a PhD fellowship from the Danish Cancer Research Foundation (Dansk Kraeft-forskningsfond, DKF-0-0-532). Redox Biology Lab is part of the Center of Excellence for Autophagy, Recycling and Disease (CARD), funded by Danmarks Grundforskningsfond (DNRF125).

References

Adachi, T., Weisbrod, R. M., Pimentel, D. R., Ying, J., Sharov, V. S., Schöneich, C., & Cohen, R. A. (2004). S-Glutathiolation by peroxynitrite activates SERCA during arterial relaxation by nitric oxide. *Nature Medicine, 10*(11), 1200–1207. https://doi.org/10.1038/nm1119

Akhand, A. A., Pu, M., Senga, T., Kato, M., Suzuki, H., Miyata, T., Hamaguchi, M., & Nakashima, I. (1999). Nitric oxide controls Src kinase activity through a sulfhydryl group modification-mediated Tyr-527-independent and Tyr-416-linked mechanism. *Journal of Biological Chemistry, 274*(36), 25821–25826. https://doi.org/10.1074/jbc.274.36.25821

Armada-Moreira, A., Gomes, J. I., Pina, C. C., Savchak, O. K., Gonçalves-Ribeiro, J., Rei, N., Pinto, S., Morais, T. P., Martins, R. S., Ribeiro, F. F., Sebastião, A. M., Crunelli, V., & Vaz, S. H. (2020). Going the extra (synaptic) mile: Excitotoxicity as the road toward neurodegenerative diseases. *Frontiers in Cellular Neuroscience, 14*. https://doi.org/10.3389/fncel.2020.00090

Baritaki, S., Huerta-Yepez, S., Sahakyan, A., Karagiannides, I., Bakirtzi, K., Jazirehi, A. R., Bonavida, B., & Kugel, J. F. (2010). Mechanisms of nitric oxide-mediated inhibition of EMT in cancer: Inhibition of the metastasis-inducer snail and induction of the metastasis-suppressor RKIP. *Cell Cycle, 9*(24), 4931–4940. https://doi.org/10.4161/cc.9.24.14229

Benhar, M., Forrester, M. T., & Stamler, J. S. (2009). Protein denitrosylation: Enzymatic mechanisms and cellular functions. *Nature Reviews Molecular Cell Biology, 10*(10), 721–732. https://doi.org/10.1038/nrm2764

Bramanti, E., Jacovozzi, K., Dulivo, L., Vecoli, C., Zamboni, R., Mester, Z., & Dulivo, A. (2008). Determination of S-nitrosoglutathione and other nitrosothiols by p-hydroxymercurybenzoate derivatization and reverse phase chromatography coupled with chemical vapor generation atomic fluorescence detection. *Talanta, 77*(2), 684–694. https://doi.org/10.1016/j.talanta.2008.07.020

Broniowska, K. A., & Hogg, N. (2012). The chemical biology of S-nitrosothiols. *Antioxidants and Redox Signaling, 17*(7), 969–980. https://doi.org/10.1089/ars.2012.4590

Brown, G. C. (2007). Nitric oxide and mitochondria. *Frontiers in Bioscience, 12*(3), 1024–1033. https://doi.org/10.2741/2122

Brown, G. C. (1999). Nitric oxide and mitochondrial respiration. *Biochimica et Biophysica Acta (BBA) - Bioenergetics, 1411*(2–3), 351–369. https://doi.org/10.1016/S0005-2728(99)00025-0

Calabrese, V., Mancuso, C., Calvani, M., Rizzarelli, E., Butterfield, D. A., & Giuffrida Stella, A. M. (2007). Nitric oxide in the central nervous system: Neuroprotection versus neurotoxicity. *Nature Reviews Neuroscience, 8*(10), 766–775. https://doi.org/10.1038/nrn2214

Carvajal, J. A., Germain, A. M., Huidobro-Toro, J. P., & Weiner, C. P. (2000). Molecular mechanism of cGMP-mediated smooth muscle relaxation. *Journal of Cellular Physiology, 184*(3), 409–420. https://doi.org/10.1002/1097-4652(200009)184:3<409::AID-JCP16>3.0.CO;2-K

Chien, Y.-H., Bau, D.-T., & Jan, K.-Y. (2004). Nitric oxide inhibits DNA-adduct excision in nucleotide excision repair. *Free Radical Biology and Medicine, 36*(8), 1011–1017. https://doi.org/10.1016/j.freeradbiomed.2004.01.011

Cleeter, M. W. J., Cooper, J. M., Darley-Usmar, V. M., Moncada, S., & Schapira, A. H. V. (1994). Reversible inhibition of cytochrome c oxidase, the terminal enzyme of the mitochondrial respiratory chain, by nitric oxide. Implications for neurodegenerative diseases. *FEBS Letters, 345*(1), 50–54. https://doi.org/10.1016/0014-5793(94)00424-2

deRojas-Walker, T., Tamir, S., Ji, H., Wishnok, J. S., & Tannenbaum, S. R. (1995). Nitric oxide induces oxidative damage in addition to deamination in macrophage DNA. *Chemical Research in Toxicology, 8*(3), 473–477. https://doi.org/10.1021/tx00045a020

Devarie-Baez, N. O., Zhang, D., Li, S., Whorton, A. R., & Xian, M. (2013). Direct methods for detection of protein S-nitrosylation. *Methods, 62*(2), 171–176. https://doi.org/10.1016/j.ymeth.2013.04.018

Förstermann, U., & Sessa, W. C. (2012). Nitric oxide synthases: Regulation and function. *European Heart Journal, 33*(7), 829–837. https://doi.org/10.1093/eurheartj/ehr304

Faccenda, A., Bonham, C. A., Vacratsis, P. O., Zhang, X., & Mutus, B. (2010). Gold nanoparticle enrichment method for identifying S -nitrosylation and S -glutathionylation sites in proteins. *Journal of the American Chemical Society, 132*(33), 11392–11394. https://doi.org/10.1021/ja103591v

Faienza, F., Lambrughi, M., Rizza, S., Pecorari, C., Giglio, P., Salamanca Viloria, J., Allega, M. F., Chiappetta, G., Vinh, J., Pacello, F., Battistoni, A., Rasola, A., Papaleo, E., & Filomeni, G. (2020). S-nitrosylation affects TRAP1 structure and ATPase activity and modulates cell response to apoptotic stimuli. *Biochemical Pharmacology, 176*. https://doi.org/10.1016/j.bcp.2020.113869

Faienza, F., Rizza, S., Giglio, P., & Filomeni, G. (2020). TRAP1: A metabolic hub linking aging pathophysiology to mitochondrial S-nitrosylation. *Frontiers in Physiology, 11*. https://doi.org/10.3389/fphys.2020.00340

Forrester, M. T., Foster, M. W., & Stamler, J. S. (2007). Assessment and application of the biotin switch technique for examining protein S-nitrosylation under conditions of pharmacologically induced oxidative stress. *Journal of Biological Chemistry, 282*(19), 13977–13983. https://doi.org/10.1074/jbc.m609684200

Forrester, M. T., Foster, M. W., Benhar, M., & Stamler, J. S. (2009). Detection of protein S-nitrosylation with the biotin-switch technique. *Free Radical Biology and Medicine, 46*(2), 119–126. https://doi.org/10.1016/j.freeradbiomed.2008.09.034

Forrester, M. T., Thompson, J. W., Foster, M. W., Nogueira, L., Moseley, M. A., & Stamler, J. S. (2009). Proteomic analysis of S-nitrosylation and denitrosylation by resin-assisted capture. *Nature Biotechnology, 27*(6), 557–559. https://doi.org/10.1038/nbt.1545

Forsythe, J. A., Jiang, B. H., Iyer, N. V., Agani, F., Leung, S. W., Koos, R. D., & Semenza, G. L. (1996). Activation of vascular endothelial growth factor gene transcription by hypoxia-inducible factor 1. *Molecular and Cellular Biology, 16*(9), 4604–4613. https://doi.org/10.1128/MCB.16.9.4604

Freire, M. A. M., Guimarães, J. S., Leal, W. G., & Pereira, A. (2009). Pain modulation by nitric oxide in the spinal cord. *Frontiers in Neuroscience, 3*. https://doi.org/10.3389/neuro.01.024.2009

Fulton, D., Gratton, J. P., McCabe, T. J., Fontana, J., Fujio, Y., Walsh, K., Franke, T. F., Papapetropoulos, A., & Sessa, W. C. (1999). Regulation of endothelium-derived nitric oxide production by the protein kinase Akt. *Nature, 399*(6736), 597–601. https://doi.org/10.1038/21218

Gao, C., Guo, H., Wei, J., Mi, Z., Wai, P., & Kuo, P. C. (2004). S-nitrosylation of heterogeneous nuclear ribonucleoprotein A/B regulates osteopontin transcription in endotoxin-stimulated murine macrophages. *Journal of Biological Chemistry, 279*(12), 11236–11243. https://doi.org/10.1074/jbc.M313385200

García-Ortiz, A., & Serrador, J. M. (2018). Nitric oxide signaling in T cell-mediated immunity. *Trends in Molecular Medicine, 24*(4), 412–427. https://doi.org/10.1016/j.molmed.2018.02.002

Garthwaite, J., Charles, S. L., & Chess-Williams, R. (1988). Endothelium-derived relaxing factor release on activation of NMDA receptors suggests role as intercellular messenger in the brain. *Nature, 336*(6197), 385–388. https://doi.org/10.1038/336385a0

Greco, T. M., Hodara, R., Parastatidis, I., Heijnen, H. F. G., Dennehy, M. K., Liebler, D. C., & Ischiropoulos, H. (2006). Identification of S-nitrosylation motifs by site-specific mapping of the S-nitrosocysteine proteome in human vascular smooth muscle cells. *Proceedings of the National Academy of Sciences of the United States of America, 103*(19), 7420–7425. https://doi.org/10.1073/pnas.0600729103

Griswold-Prenner, I., Kashyap, A. K., Mazhar, S., Hall, Z. W., Fazelinia, H., & Ischiropoulos, H. (2023). Unveiling the human nitroproteome: Protein tyrosine nitration in cell signaling and cancer. *Journal of Biological Chemistry, 299*(8). https://doi.org/10.1016/j.jbc.2023.105038

Guzik, T. J., Korbut, R., & Adamek-Guzik, T. (2003). Nitric oxide and superoxide in inflammation and immune regulation. *Journal of Physiology & Pharmacology, 54*(4), 469–487.

Guzzo, G., Sciacovelli, M., Bernardi, P., & Rasola, A. (2014). Inhibition of succinate dehydrogenase by the mitochondrial chaperone TRAP1 has anti-oxidant and anti-apoptotic effects on tumor cells. *Oncotarget, 5*(23), 11897–11908. https://doi.org/10.18632/oncotarget.2472

Hawkins, C. L., & Davies, M. J. (2019). Detection, identification, and quantification of oxidative protein modifications. *Journal of Biological Chemistry, 294*(51), 19683–19708. https://doi.org/10.1074/jbc.REV119.006217

He, Q., Qu, M., Xu, C., Shi, W., Hussain, M., Jin, G., Zhu, H., Zeng, L.-H., & Wu, X. (2022). The emerging roles of nitric oxide in ferroptosis and pyroptosis of tumor cells. *Life Sciences, 290*. https://doi.org/10.1016/j.lfs.2021.120257

Hess, D. T., Matsumoto, A., Kim, S. O., Marshall, H. E., & Stamler, J. S. (2005). Protein S-nitrosylation: Purview and parameters. *Nature Reviews Molecular Cell Biology, 6*(2), 150–166. https://doi.org/10.1038/nrm1569

Hickok, J. R., Sahni, S., Mikhed, Y., Bonini, M. G., & Thomas, D. D. (2011). Nitric oxide suppresses tumor cell migration through N-Myc downstream-regulated gene-1 (NDRG1) expression: Role of chelatable iron. *Journal of Biological Chemistry, 286*(48), 41413–41424. https://doi.org/10.1074/jbc.M111.287052

Homma, T., Kobayashi, S., Conrad, M., Konno, H., Yokoyama, C., & Fujii, J. (2021). Nitric oxide protects against ferroptosis by aborting the lipid peroxidation chain reaction. *Nitric Oxide, 115*, 34–43. https://doi.org/10.1016/j.niox.2021.07.003

Hoter, A., El-Sabban, M., & Naim, H. (2018). The HSP90 family: Structure, regulation, function, and implications in health and disease. *International Journal of Molecular Sciences, 19*(9). https://doi.org/10.3390/ijms19092560

Ibiza, S., Víctor, V. M., Boscá, I., Ortega, A., Urzainqui, A., O'Connor, J. E., Sánchez-Madrid, F., Esplugues, J. V., & Serrador, J. M. (2006). Endothelial nitric oxide synthase regulates T cell receptor signaling at the immunological synapse. *Immunity, 24*(6), 753–765. https://doi.org/10.1016/j.immuni.2006.04.006

Ignarro, L. J., Buga, G. M., Wood, K. S., Byrns, R. E., & Chaudhuri, G. (1987). Endothelium-derived relaxing factor produced and released from artery and vein is nitric oxide. *Proceedings of the National Academy of Sciences of the United States of America, 84*(24), 9265–9269. https://doi.org/10.1073/pnas.84.24.9265

Jaffrey, S. R., Erdjument-Bromage, H., Ferris, C. D., Tempst, P., & Snyder, S. H. (2001). Protein S-nitrosylation: A physiological signal for neuronal nitric oxide. *Nature Cell Biology, 3*(2), 193–197. https://doi.org/10.1038/35055104

Jia, H. Y., Liu, Y., Zhang, X. J., Han, L., Du, L. B., Tian, Q., & Xu, Y. C. (2009). Potential oxidative stress of gold nanoparticles by induced-NO releasing in serum. *Journal of the American Chemical Society, 131*(1), 40–41. https://doi.org/10.1021/ja808033w

Jia, J., Arif, A., Terenzi, F., Willard, B., Plow, E. F., Hazen, S. L., & Fox, P. L. (2014). Target-selective protein S-nitrosylation by sequence motif recognition. *Cell, 159*(3), 623–634. https://doi.org/10.1016/j.cell.2014.09.032

Khan, F. H., Dervan, E., Bhattacharyya, D. D., McAuliffe, J. D., Miranda, K. M., & Glynn, S. A. (2020). The role of nitric oxide in cancer: Master regulator or not? *International Journal of Molecular Sciences, 21*(24), 1–30. https://doi.org/10.3390/ijms21249393

Kimura, H., Ogura, T., Kurashima, Y., Weisz, A., & Esumi, H. (2002). Effects of nitric oxide donors on vascular endothelial growth factor gene induction. *Biochemical and Biophysical Research Communications, 296*(4), 976–982. https://doi.org/10.1016/s0006-291x(02)02029-6

Korde Choudhari, S., Chaudhary, M., Bagde, S., Gadbail, A. R., & Joshi, V. (2013). Nitric oxide and cancer: A review. *World Journal of Surgical Oncology, 11*(1). https://doi.org/10.1186/1477-7819-11-118

Kurokawa, H., Ito, H., Terasaki, M., Matano, D., Taninaka, A., Shigekawa, H., Matsui, H., & Batra, S. K. (2019). Nitric oxide regulates the expression of heme carrier protein-1 via hypoxia inducible factor-1α stabilization. *PLoS One, 14*(9). https://doi.org/10.1371/journal.pone.0222074

Lee, S. J., Lee, J. R., Kim, Y. H., Park, Y. S., Park, S. I., Park, H. S., & Kim, K. P. (2007). Investigation of tyrosine nitration and nitrosylation of angiotensin II and bovine serum albumin with electrospray ionization mass spectrometry. *Rapid Communications in Mass Spectrometry, 21*(17), 2797–2804. https://doi.org/10.1002/rcm.3145

Lira, V. A., Soltow, Q. A., Long, J. H. D., Betters, J. L., Sellman, J. E., & Criswell, D. S. (2007). Nitric oxide increases GLUT4 expression and regulates AMPK signaling in skeletal muscle. *American Journal of Physiology - Endocrinology And Metabolism, 293*(4), E1062–E1068. https://doi.org/10.1152/ajpendo.00045.2007

Liu, L., Hausladen, A., Zeng, M., Que, L., Heitman, J., & Stamler, J. S. (2001). A metabolic enzyme for S-nitrosothiol conserved from bacteria to humans. *Nature, 410*(6827), 490–494. https://doi.org/10.1038/35068596

Lowenstein, C. J. (2019). Metabolism reprogrammed by the nitric oxide signalling molecule. *Nature, 565*(7737), 33–34. https://doi.org/10.1038/d41586-018-07457-z

Mannick, J. B., Schonhoff, C., Papeta, N., Ghafourifar, P., Szibor, M., Fang, K., & Gaston, B. (2001). S-nitrosylation of mitochondrial caspases. *Journal of Cell Biology, 154*(6), 1111–1116. https://doi.org/10.1083/jcb.200104008

McConell, G. K., Bradley, S. J., Stephens, T. J., Canny, B. J., Kingwell, B. A., & Rs, L.-Y. (2007). Skeletal muscle nNOS mu protein content is increased by exercise training in humans. *American Journal of Physiology - Regulatory, Integrative and Comparative Physiology, 293*(2), 821–828.

Montagna, C., Cirotti, C., Rizza, S., & Filomeni, G. (2020). When S -nitrosylation gets to mitochondria: From signaling to age-related diseases. *Antioxidants and Redox Signaling, 32*(12), 884–905. https://doi.org/10.1089/ars.2019.7872

Montagna, C., Rizza, S., Cirotti, C., Maiani, E., Muscaritoli, M., Musarò, A., Carrí, M. T., Ferraro, E., Cecconi, F., & Filomeni, G. (2019). nNOS/GSNOR interaction contributes to skeletal muscle differentiation and homeostasis. *Cell Death & Disease, 10*(5). https://doi.org/10.1038/s41419-019-1584-3

Morbidelli, L., Donnini, S., & Ziche, M. (2003). Role of nitric oxide in the modulation of angiogenesis. *Current Pharmaceutical Design, 9*(7), 521–530. https://doi.org/10.2174/1381612033391405

Movafagh, S., Crook, S., & Vo, K. (2015). Regulation of hypoxia-inducible factor-1a by reactive oxygen species: New developments in an old debate. *Journal of Cellular Biochemistry, 116*(5), 696–703. https://doi.org/10.1002/jcb.25074

Munson, D. A., Grubb, P. H., Kerecman, J. D., McCurnin, D. C., Yoder, B. A., Hazen, S. L., Shaul, P. W., & Ischiropoulos, H. (2005). Pulmonary and systemic nitric oxide metabolites in a baboon model of neonatal chronic lung disease. *American Journal of Respiratory Cell and Molecular Biology, 33*(6), 582–588. https://doi.org/10.1165/rcmb.2005-0182OC

Navasardyan, I., & Bonavida, B. (2021). Regulation of T Cells in cancer by nitric oxide. *Cells, 10*(10). https://doi.org/10.3390/cells10102655

Nguyen, T., Brunson, D., Crespi, C. L., Penman, B. W., Wishnok, J. S., & Tannenbaum, S. R. (1992). DNA damage and mutation in human cells exposed to nitric oxide in vitro. *Proceedings of the National Academy of Sciences of the United States of America, 89*(7), 3030–3034. https://doi.org/10.1073/pnas.89.7.3030

Niedbala, W., Alves-Filho, J. C., Fukada, S. Y., Vieira, S. M., Mitani, A., Sonego, F., Mirchandani, A., Nascimento, D. C., Cunha, F. Q., & Liew, F. Y. (2011). Regulation of type 17 helper T-cell function by nitric oxide during inflammation. *Proceedings of the National Academy of Sciences of the United States of America, 108*(22), 9220–9225. https://doi.org/10.1073/pnas.1100667108

Orucevic, A., Bechberger, J., Green, A. M., Shapiro, R. A., Billiar, T. R., & Lala, P. K. (1999). Nitric-oxide production by murine mammary adenocarcinoma cells promotes tumor-cell invasiveness. *International Journal of Cancer, 81*(6), 889–896. https://doi.org/10.1002/(SICI)1097-0215(19990611)81:6<889::AID-IJC9>3.0.CO;2-2

Pacher, P., Beckman, J. S., & Liaudet, L. (2007). Nitric oxide and peroxynitrite in health and disease. *Physiological Reviews, 87*(1), 315–424. https://doi.org/10.1152/physrev.00029.2006

Papaleo, E., Tiberti, M., Arnaudi, M., Pecorari, C., Faienza, F., Cantwell, L., Degn, K., Pacello, F., Battistoni, A., Lambrughi, M., & Filomeni, G. (2023). TRAP1 S-nitrosylation as a model of population-shift mechanism to study the effects of nitric oxide on redox-sensitive oncoproteins. *Cell Death & Disease, 14*(4). https://doi.org/10.1038/s41419-023-05780-6

Park, J. K. J., & Kostka, P. (1997). Fluorometric detection of BiologicalS-nitrosothiols. *Analytical Biochemistry, 249*(1), 61–66. https://doi.org/10.1006/abio.1997.2159

Park, Y.-K., Ahn, D.-R., Oh, M., Lee, T., Yang, E. G., Son, M., & Park, H. (2008). Nitric Oxide Donor, (±)- S -nitroso- N -acetylpenicillamine, stabilizes transactive hypoxia-inducible factor-1α by inhibiting von Hippel-Lindau recruitment and asparagine hydroxylation. *Molecular Pharmacology, 74*(1), 236–245. https://doi.org/10.1124/mol.108.045278

Picón-Pagès, P., Garcia-Buendia, J., & Muñoz, F. J. (2019). Functions and dysfunctions of nitric oxide in brain. *Biochimica et Biophysica Acta - Molecular Basis of Disease, 1865*(8), 1949–1967. https://doi.org/10.1016/j.bbadis.2018.11.007

Prime, T. A., Blaikie, F. H., Evans, C., Nadtochiy, S. M., James, A. M., Dahm, C. C., Vitturi, D. A., Patel, R. P., Hileye, C. R., Abakumova, I., Requejo, R., Chouchani, E. T., Hurd, T. R., Garvey, J. F., Taylor, C. T., Brookes, P. S., Smith, R. A. J., & Murphy, M. P. (2009). A mitochondria-targeted S-nitrosothiol modulates respiration, nitrosates thiols, and protects against ischemia-reperfusion injury. *Proceedings of the National Academy of Sciences of the United States of America, 106*(26), 10764–10769. https://doi.org/10.1073/pnas.0903250106

Pye, D., Palomero, J., Kabayo, T., & Jackson, M. J. (2007). Real-time measurement of nitric oxide in single mature mouse skeletal muscle fibres during contractions. *Journal of Physiology, 581*(1), 309–318. https://doi.org/10.1113/jphysiol.2006.125930

Qu, J., Liu, G.-H., Huang, B., & Chen, C. (2007). Nitric oxide controls nuclear export of APE1/Ref-1 through S-nitrosation of cysteines 93 and 310. *Nucleic Acids Research, 35*(8), 2522–2532. https://doi.org/10.1093/nar/gkl1163

Radi, R. (2018). Oxygen radicals, nitric oxide, and peroxynitrite: Redox pathways in molecular medicine. *Proceedings of the National Academy of Sciences, 115*(23), 5839–5848. https://doi.org/10.1073/pnas.1804932115

Reiling, N., Kröncke, R., Ulmer, A. J., Gerdes, J., Flad, H. D., & Hauschildt, S. (1996). Nitric oxide synthese: Expression of the endothelial, Ca^{2+}/calmodulin-dependent isoform in human B and T lymphocytes. *European Journal of Immunology, 26*(3), 511–516. https://doi.org/10.1002/eji.1830260302

Rizza, S., & Filomeni, G. (2020). Exploiting S- nitrosylation for cancer therapy: Facts and perspectives. *Biochemical Journal, 477*(19), 3649–3672. https://doi.org/10.1042/bcj20200064

Rizza, S., Montagna, C., Cardaci, S., Maiani, E., Di Giacomo, G., Sanchez-Quiles, V., Blagoev, B., Rasola, A., De Zio, D., Stamler, J. S., Cecconi, F., & Filomeni, G. (2016). S -nitrosylation of the mitochondrial chaperone TRAP1 sensitizes hepatocellular carcinoma cells to inhibitors of succinate dehydrogenase. *Cancer Research, 76*(14), 4170–4182. https://doi.org/10.1158/0008-5472.can-15-2637

Rizza, S., Di Leo, L., Pecorari, C., Giglio, P., Faienza, F., Montagna, C., Maiani, E., Puglia, M., Bosisio, F. M., Petersen, T. S., Lin, L., Rissler, V., Viloria, J. S., Luo, Y., Papaleo, E., De Zio, D., Blagoev, B., & Filomeni, G. (2023). GSNOR deficiency promotes tumor growth via FAK1 S-nitrosylation. *Cell Reports, 42*(1). https://doi.org/10.1016/j.celrep.2023.111997

Sanders, K. M. (2016). Enteric inhibitory neurotransmission, starting down under. *Advances in Experimental Medicine and Biology, 891*, 21–29. https://doi.org/10.1007/978-3-319-27592-5_3

Sciacovelli, M., Guzzo, G., Morello, V., Frezza, C., Zheng, L., Nannini, N., Calabrese, F., Laudiero, G., Esposito, F., Landriscina, M., Defilippi, P., Bernardi, P., & Rasola, A. (2013). The mitochondrial chaperone TRAP1 promotes neoplastic growth by inhibiting succinate dehydrogenase. *Cell Metabolism, 17*(6), 988–999. https://doi.org/10.1016/j.cmet.2013.04.019

Shi, X., & Qiu, H. (2020). Post-translational S-nitrosylation of proteins in regulating cardiac oxidative stress. *Antioxidants, 9*(11). https://doi.org/10.3390/antiox9111051

Siegert, A., Rosenberg, C., Schmitt, W. D., Denkert, C., & Hauptmann, S. (2002). Nitric oxide of human colorectal adenocarcinoma cell lines promotes tumour cell invasion. *British Journal of Cancer, 86*(8), 1310–1315. https://doi.org/10.1038/sj.bjc.6600224

Switzer, C. H., Glynn, S. A., Cheng, R. Y. S., Ridnour, L. A., Green, J. E., Ambs, S., & Wink, D. A. (2012). S-nitrosylation of EGFR and Src activates an oncogenic signaling network in human basal-like breast cancer. *Molecular Cancer Research, 10*(9), 1203–1215. https://doi.org/10.1158/1541-7786.MCR-12-0124

Taddei, M. L., Giannoni, E., Fiaschi, T., & Chiarugi, P. (2012). Anoikis: An emerging hallmark in health and diseases. *The Journal of Pathology, 226*(2), 380–393. https://doi.org/10.1002/path.3000

Tai, X. G., Toyo-oka, K., Yamamoto, N., Yashiro, Y., Mu, J., Hamaoka, T., & Fujiwara, H. (1997). Expression of an inducible type of nitric oxide (NO) synthase in the thymus and involvement of NO in deletion of TCR-stimulated double-positive thymocytes. *The Journal of Immunology, 158*(10), 4696–4703. https://doi.org/10.4049/jimmunol.158.10.4696

Tang, D., Kroemer, G., & Ferroptosis. (2020). *Current Biology, 30.*

Tatchum-Talom, R., Schulz, R., McNeill, J. R., & Khadour, F. H. (2000). Upregulation of neuronal nitric oxide synthase in skeletal muscle by swim training. *American Journal of Physiology - Heart and Circulatory Physiology, 279*(4), H1757–H1766. https://doi.org/10.1152/ajpheart.2000.279.4.h1757

Tejero, J., Hunt, A. P., Santolini, J., Lehnert, N., & Stuehr, D. J. (2019). Mechanism and regulation of ferrous heme-nitric oxide (NO) oxidation in NO synthases. *Journal of Biological Chemistry, 294*(19), 7904–7916. https://doi.org/10.1074/jbc.RA119.007810

Thomas, S., Lowe, J. E., Knowles, R. G., Green, I. C., & Green, M. H. L. (1998). Factors affecting the DNA damaging activity of superoxide and nitric oxide. *Mutation Research: Fundamental and Molecular Mechanisms of Mutagenesis, 402*(1–2), 77–84. https://doi.org/10.1016/S0027-5107(97)00284-4

Wang, Y., Liu, T., Wu, C., & Li, H. (2008). A strategy for direct identification of protein S-nitrosylation sites by quadrupole time-of-flight mass spectrometry. *Journal of the American Society for Mass Spectrometry, 19*(9), 1353–1360. https://doi.org/10.1016/j.jasms.2008.06.001

Webb, R. C. (2003). Smooth muscle contraction and relaxation. *American Journal of Physiology - Advances in Physiology Education, 27*(1–4), 201–206. https://doi.org/10.1152/advan.00025.2003

Whalen, E. J., Foster, M. W., Matsumoto, A., Ozawa, K., Violin, J. D., Que, L. G., Nelson, C. D., Benhar, M., Keys, J. R., Rockman, H. A., Koch, W. J., Daaka, Y., Lefkowitz, R. J., & Stamler, J. S. (2007). Regulation of β-adrenergic receptor signaling by S-nitrosylation of G-protein-coupled receptor kinase 2. *Cell, 129*(3), 511–522. https://doi.org/10.1016/j.cell.2007.02.046

Wolhuter, K., Whitwell, H. J., Switzer, C. H., Burgoyne, J. R., Timms, J. F., & Eaton, P. (2018). Evidence against stable protein S-nitrosylation as a widespread mechanism of post-translational regulation. *Molecular Cell, 69*(3), 438–450.e5. https://doi.org/10.1016/j.molcel.2017.12.019

Xian, M., Wang, K., Chen, X., Hou, Y., McGill, A., Chen, X., Zhou, B., Zhang, Z. Y., Cheng, J. P., & Wang, P. G. (2000). Inhibition of protein tyrosine phosphatases by low-molecular-weight S-nitrosothiols and S-nitrosylated human serum albumin. *Biochemical and Biophysical Research Communications, 268*(2), 310–314. https://doi.org/10.1006/bbrc.2000.2117

Ye, F., Chai, W., Xie, M., Yang, M., Yu, Y., Cao, L., & Yang. (2019). HMGB1 regulates erastin-induced ferroptosis via RAS-JNK/p38 signaling in HL-60/NRASQ61L cells. *American Journal of Cancer Research, 9*(4), 730–739.

Yu, Y., Xie, Y., Cao, L., Yang, L., Yang, M., Lotze, M. T., Zeh, H. J., Kang, R., & Tang, D. (2015). The ferroptosis inducer erastin enhances sensitivity of acute myeloid leukemia cells to chemotherapeutic agents. *Molecular & Cellular Oncology, 2*(4). https://doi.org/10.1080/23723556.2015.1054549

Zhang, J., Li, S., Zhang, D., Wang, H., Whorton, A. R., & Xian, M. (2010). Reductive ligation mediated one-step disulfide formation of S -nitrosothiols. *Organic Letters, 12*(18), 4208–4211. https://doi.org/10.1021/ol101863s

Zhao, Y., Vanhoutte, P. M., & Leung, S. W. S. (2015). Vascular nitric oxide: Beyond eNOS. *Journal of Pharmacological Sciences, 129*(2), 83–94. https://doi.org/10.1016/j.jphs.2015.09.002

Zhou, H. L., Zhang, R., Anand, P., Stomberski, C. T., Qian, Z., Hausladen, A., Wang, L., Rhee, E. P., Parikh, S. M., Karumanchi, S. A., & Stamler, J. S. (2019). Metabolic reprogramming by the S-nitroso-CoA reductase system protects against kidney injury. *Nature, 565*(7737), 96–100. https://doi.org/10.1038/s41586-018-0749-z

Zhou, H.-L., Grimmett, Z. W., Venetos, N. M., Stomberski, C. T., Qian, Z., McLaughlin, P. J., Bansal, P. K., Zhang, R., Reynolds, J. D., Premont, R. T., & Stamler, J. S. (2023). An enzyme that selectively S-nitrosylates proteins to regulate insulin signaling. *Cell, 186*(26), 5812–5825.e21. https://doi.org/10.1016/j.cell.2023.11.009

CHAPTER

4

Preclinical models to assess the pharmacological properties of NO derivatives

Lucia Morbidelli[1] *and Valerio Ciccone*[1,2]

[1]Department of Life Sciences, University of Siena, Siena, Italy; [2]Fondazione Veronesi, Milano, Italy

Abbreviations

ADME Absorption, distribution, metabolism, elimination
DOCA Deoxycorticosterone acetate
ECM Extracellular matrix
eNOS Endothelial nitric oxide synthase
FBS Fetal bovine serum
GLP Good laboratory practices
HUVECs Human umbilical vein endothelial cells
IL-1 Interleukin 1
IND Investigational new drug
iNOS Inducible nitric oxide synthase
MTT Thiazolyl blue tetrazolium bromide
NO Nitric oxide
NOS Nitric oxide synthase
PGE2 Prostaglandin E2
PPP Platelet-poor plasma
PRP Platelet-rich plasma
RT Room temperature
sGC Soluble guanylate cyclase
TME Tumor microenvironment
VEGF Vascular endothelial growth factor

No conflict statement: No potential conflicts of interest were disclosed.

Therapeutic Applications of Nitric Oxide in Cancer and Inflammatory Disorders
https://doi.org/10.1016/B978-0-443-22354-9.00004-8

From preclinical to clinical trials of drugs

The assessment of the role of nitric oxide (NO) and nitric oxide synthase (NOS) isoforms-related pathways in physiology and diseases continues to be a challenge to find novel therapeutic target(s) and develop new drugs (*see chapter by G. Varchi in this book*). Researchers are investing a lot of effort in synthetizing innovative NO-releasing or inhibiting drugs and their efficacy and safety, alone or in combination; these need to be assessed before any animal (and further human) use.

Once new molecules have been designed and synthesized, the long process of drug development starts by including preclinical and clinical studies. Preclinical studies are conducted using different protocols starting form isolated target moieties, cell cultures, and ultimately alive animals (Asati et al., 2022; Huang et al., 2020; McGonigle & Williams, 2017). Based on the computational design of the molecules to be selected, drug screening can be done through studies of receptor binding, on the activity of ion channels (by means of electrophysiological techniques), and on the enzymatic activity of specific biochemical pathways. The initial screening tests usually are performed on cultured primary or stabilized cells, isolated from the organs target of the drug action. Cells engineered to overexpress drug targets can be useful for the early screening of molecule libraries. Cells of different tissues can be further studied for their specific involved functions and are required in the tissue homeostatic regulation or physiopathologic changes typical of the pathology. Animal studies conducted according to internationally recognized legislation are then necessary to corroborate the efficacy of the novel drugs in experimental models of pathology and to assess their potential toxicity in normal animals (Huang et al., 2020; McGonigle & Williams, 2017). Pharmacokinetic characterization is mandatory for the proper drug formulation and administration criteria before any further human study.

This chapter focuses on the conventional methods to assess safety and efficacy of novel drugs (1) in different cells that are known to respond to NO, (2) in tissues from animals to evaluate ex vivo functions, and (3) functional responses in animal models of pathologies (mainly cardiovascular) linked to NO dysregulation.

Cell-based in vitro pharmacological screening and safety assessment

Cell components of the vascular system can be considered a gold-standard target of NO donors or derivatives. The main interest is on endothelial and smooth muscle cells, respectively, the inner part of the vessels with a variety of metabolic activities and the functional elements involved in the vessel tone control. Animal and human endothelial cells are commercially available and can be obtained from large vessels or microvasculature, or from specialized vascular areas such as the umbilical vein of newborn, blood vessels of specific organs or from solid tumors. Moreover, it is possible to get cells from healthy or ill subjects, for example, from diabetic patients. Endothelial cells from the human umbilical vein are the most used and usually derive from pooled donors. Microvascular endothelium can be studied in comparison with large vessels, and the responses from blood vessel endothelial cells can be compared with the lymphatic endothelium (Bachetti & Morbidelli, 2000; Ciccone et al., 2017).

Cell culture source and maintenance

All cell culture procedures are carried out at room temperature (RT) unless otherwise specified and in dedicated facilities under vertical hood and using sterile materials and reagents to avoid contaminations and guarantee sterility. For the study of NO derivatives, the mostly used cell types are endothelial and smooth muscle cells for cardiovascular therapeutic perspective. They can be freshly isolated from human tissue or acquired by specialized companies. For tumor therapy, human or animal tumor cell types are available from commercial sources or cell banks. Important issues are ethics clarification when using human or animal samplings and quality and safety certification from commercial providers.

The culture of human endothelial and other cells requires specialized culture media, which are indicated by the manufacturer instructions or published protocols. It is very important to follow the culture indications to ensure cell survival and maintenance of typical phenotypic patterns. If needed, the plastic is coated with gelatin or other proteins to facilitate cell adhesion. Primary endothelial and smooth muscle cells have a finite lifespan: they maintain a good proliferating rate for a maximum 8–10 passages, and then they become senescent and stop to proliferate (Bachetti & Morbidelli, 2000). Always, cell confluency and morphology have to be checked at the inverted microscope to monitor cell wellness. Mycoplasma contamination should be checked as well.

Cell survival through the MTT test

To evaluate the effect of NO donors or inhibitors on the survival of each cell line, the colorimetric quantitative assay of MTT (thiazolyl blue tetrazolium bromide) is performed (Genah et al., 2020; Koika et al., 2010; Ciccone et al., 2020). Cells are seeded at a cell density to give a semiconfluent monolayer in the appropriate medium supplemented with 10% fetal bovine serum (FBS). After 24 h of incubation to allow cell adherence, cells are stimulated for 24–48 h with drugs diluted in basal medium, supplemented with low serum. As positive control of cell viability, HUVECs are also stimulated with vascular endothelial growth factor (VEGF) used at 10 ng/mL or with medium additional with 10% serum. At the end of stimulation, medium is removed and replaced with a solution of 1.2 mM MTT (Sigma–Aldrich St. Louis, MO, the United States) in PBS (phosphate buffered saline, Euroclone, Milan, Italy), followed by 4 h incubation at 37°C. MTT solution is then removed and replaced with DMSO to solubilize the formazan crystals. The different intensity of the purple color is detected through a microplate reader (Infinite 200 Pro, Tecan Life Sciences, Switzerland) at 540 nm. Data are reported as 540 nm relative absorbance/well and normalized on the control and presented as fold change or absolute absorbance values.

Cell proliferation

Cell number can be evaluated following incubation with NO donors or other stimuli in the absence/presence of NOS or other metabolic inhibitors (as the ones depicted in Fig. 4.1) to assess the relevance of exogenous or endogenous NO in the promotion of cell proliferation (Ziche et al., 1994). Detached cells are counted using a hemocytometer. Cells are seeded at

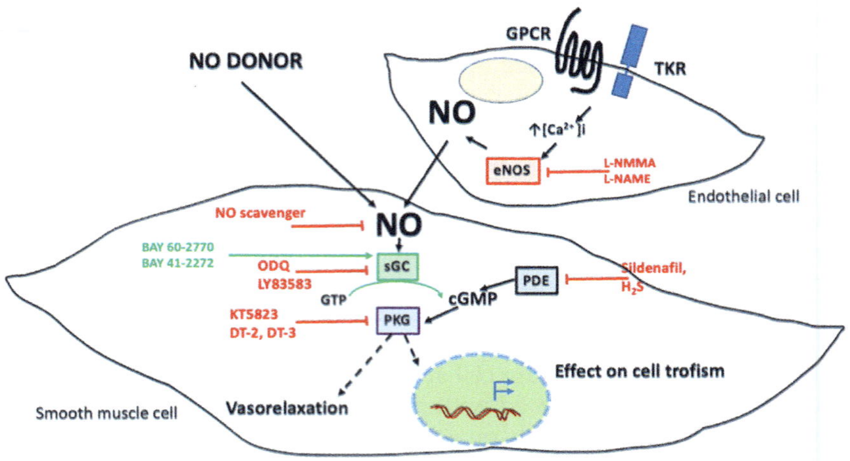

FIGURE 4.1 Drugs/molecules interfering with NOS/NO signaling pathway. After its production, NO diffuses out of the cells that produce it and enters the neighboring cells to generate a downstream response. NO synthase (NOS) catalyzes NO production by the deamination of the amino acid L-arginine. There are three isoforms of NOS. Endothelial cells have endothelial NOS (*eNOS*), nerve and muscle cells have neuronal NOS (*nNOS*), and macrophages produce inducible NOS (*iNOS*) upon exposure to various pathogens. eNOS is activated by intracellular calcium rise due to G protein–coupled (i.e., M3 muscarinic receptor) or tyrosine kinase (as VEGF-R2) receptors expressed by vascular endothelial cells. NO donors directly or through biotransformation release NO, which enters into the cells by simple diffusion. In the classical pathway, NO binds to the heme group of soluble guanylyl cyclase (sGC) and triggers it to convert GTP to cyclic GMP (cGMP). This leads to an increase in cGMP levels in the cytosol within seconds. cGMP then binds and activates protein kinase G (PKG), which is responsible for cell responses (relaxation of smooth muscle cells, inhibition of platelet aggregation, genomic changes in other cells). The quick rise in the cytosolic cGMP due to NO is compensated by its rapid degradation by the cGMP phosphodiesterase (PDE) enzyme, thereby balancing its levels. Some drugs, such as sildenafil, work by inhibiting this cGMP-PDE. Thus, the cGMP levels stay elevated for a prolonged period, causing relaxation of the smooth muscle of penile blood vessels, thereby increasing the blood flow. Other ways to increase downstream NO signaling are through sGC activation (BAY 60-2770, NO, and heme-independent sGC agonist) and stimulation (BAY 41-2272, binds to a regulatory site in the alpha subunit of sGC) (Hobbs & Moncada, 2003; Pankey et al., 2011). Inhibitors of the NOS/NO pathway include NOS enzymatic inhibitors (L-NMMA, L-NAME) (Morbidelli et al., 1996; Ziche et al., 1993, 1994), NO scavengers (Alsharabasy et al., 2023; Morbidelli et al., 2003), sGC inhibitors (ODQ, NS2028) (Morbidelli et al., 2010), and PKG inhibitors (KT5823; DT-2, DT-3) (Koika et al., 2010).

a subconfluent density in 96 multiplates in medium with 10% FBS. Cell adherence is facilitated in an incubator at 37°C for 4–5 h.

This figure reports the production of NO by synthase and the molecular mechanisms ignited by NO on target cells and how pharmacological or biochemical tools can interfere with the single steps of the metabolic cascades. This will allow to discover new drugs/strategies to control pathology occurrence and progression.

NO donors and inhibitors are appropriately dissolved and diluted and kept on ice.

After removing the media, cells are treated with NO derivatives in the absence or presence of pharmacological inhibitors of the NO or other signaling pathways (see Fig. 4.1 as examples). Treatment is carried out in medium with low serum concentration (0.1%) in a final volume of 100 μL. Each experimental point is repeated three times. VEGF (10–25 ng/mL) or 10%

serum can be added as positive control of cell proliferation. The multiplates are incubated at 37°C and 5% CO_2 for 2–5 consecutive days with NO derivatives freshly added every 2 days. Supernatants are gently removed, and the cells are fixed with methanol (4°C, for 2 h). Cell staining is obtained by adding Diff-Quik solution I (eosinophilic, orange dye) for 5 min; Diff-Quik solution II (basophilic, blue dye) is added for further 5 min. The multiplate is washed with tap water and let to dry. Cells are counted randomly at 20× original magnification in five fields with the aid of an ocular grid by an independent operator. Recently, the use of free image analysis software as Image J is recommended for cell counting on microscopic digital images. Data are reported as number of cells/well or as % of basal control (medium without stimuli). DNA synthesis and cell proliferation/apoptosis markers can be additionally evaluated to gain insights whether the treatment under study is affecting cell survival or promoting cell mitosis or, otherwise, apoptosis (Morbidelli et al., 1996; Ziche et al., 1993).

Adherent cell migration (scratch assay)

Cell migration, together with proliferation, is a typical feature of cell trophism and tissue remodeling or angiogenesis (the formation of new vessels starting from preexisting capillaries). Migration of adherent cells can be evaluated following incubation with NO donors or other stimuli in the absence/presence of metabolic inhibitors to assess the relevance of exogenous or endogenous NO (Fig. 4.1) in the promotion of cell migration (Monti, Ciccone, et al., 2016). Cells are harvested from Petri's dish and plated in 24 multiwells. Plates are incubated at 37°C in a humidified incubator at 5% CO_2 to allow cells to become 100% confluent. Appropriate solutions of NO donors and inhibitors are prepared and kept on ice. Cells in the central area of the well are scraped using the sterile tip of a pipette to create a wound of ±1 mm width. Detached cells are removed by gentle washing with PBS. The compounds previously prepared are added together with the antimitotic ARA-C (2.5 μg/mL) to block cell proliferation, thus ensuring that wound closure is due solely to migration. Duplicate images of the wound are acquired in each well under phase contrast microscope at 10× magnification from time 0 (to measure the initial wound width) for up to 18 h. The area of wound is calculated using Image J software or similar. The area value of time 0 is taken as 100% to calculate the percentage change in wound width of the area calculated for the image corresponding to the time selected. The lower the percentage, the higher the cell migration. Results are expressed as percentage of wound area (means ± SEM) respect to time 0.

Cell chemotaxis and chemoinvasion

The Boyden Chamber procedure is used to evaluate cell chemotaxis and chemoinvasion (Koika et al., 2010; Ziche et al., 1994). The method is based on the passage of single endothelial cells across porous filters against a concentration gradient of the migration effector. While, in chemotaxis, the filter is coated with extracellular matrix (ECM) proteins, during chemoinvasion, cells have to degrade a layer of ECM stratified on the filter (Ciccone, Monti, et al., 2018; Ciccone, Terzuoli et al., 2018). The Boyden Chamber consists of two compartments separated by a polyvinyl-pyrrolidone-free polycarbonate filter, 8 μm pore size, coated with type I collagen and fibronectin for chemotaxis and with gelatin or Matrigel solution for chemoinvasion. Filter coating protocol: (1) for chemotaxis: incubate the filter with type I collagen (100 μg/mL) for 1 min at RT; dry under hood; incubate with bovine serum fibronectin (10 μg/mL) for 2–3 min at RT; wash with medium to remove unspecific binding. (2) for chemoinvasion: incubate the filter in 1 mL of 1% gelatin solution or 0.5 mg/mL Matrigel for 2–3 min at 37°C; dry under hood (Taraboletti et al., 2000). Detached cells are counted using a hemocytometer, and cell density is adjusted as 2.5×10^5 cells/mL. Appropriate dilutions of NO donors and inhibitors are prepared and kept on ice. Cell suspension is eventually incubated in sterile tubes (500 μL for each experimental point) for 30 min with metabolic/signaling inhibitors or vehicle at 37°C. A chemoattractant solution (serum or angiogenic/growth factors dissolved in basal medium) is prepared and placed in the lower wells of Boyden Chamber (30 μL/well). At least one experimental point that does not contain any chemoattractant should be included as negative control. Each experimental point is performed in triplicate.

The lower chamber is covered with the precoated porous filter and rapidly fixed with the upper part of the chamber. 50 μL of eventually treated cell suspension is added to each upper well. The chamber is incubated at 37°C in a humidified atmosphere of 5% CO_2 for 4 h for chemotaxis and 18 h for chemoinvasion. The filter is removed and fixed in methanol overnight at 4°C in a Petri's dish. Nonmigrated cells on the upper surface of the filter are removed with a cotton swab. Migrated cells in the lower surface of the filter are stained with Diff-Quik as described in Cell proliferation section. The filter is mounted on a histological slide. Cells are randomly counted at 40× original magnification in five fields/well with the aid of an ocular grid. Data are reported as migrated cells/well or % of positive or negative controls.

Gelatin zymography

ECM degradation usually accompanies cell migration. The production of degradative enzymes is evaluated in media from subconfluent cell monolayers stimulated with NO donors or other stimuli in the absence/presence of metabolic inhibitors (Monti, Terzuoli, et al., 2016; Taraboletti et al., 2000). Cells secrete proteolytic enzymes in the supernatants, which are loaded in polyacrylamide gels containing enzymatic substrate as gelatin in case of collagenases. After electrophoresis, the proteolytic enzymes are allowed to degrade gelatin. Finally, the gel is stained with Coomassie Blue dye to detect the proteolytic activity as white bands in a blue background. The procedure is divided in three phases: supernatant collection from cells, preparation of polyacrylamide gel containing gelatin, detection of gelatinolytic activity. HT1080 cells are used as reference standard for the metalloproteinases MMP-2 and MMP-9 activity, typically present in endothelial cells (Taraboletti et al., 2000).

Cell preparation and supernatant collection. After detachment from Petri's dish, cells are counted using a hemocytometer and cell density adjusted at the value of 5×10^3 cells/100 μL/well to be plated in 96 multiwell plates. After adherence, growth medium is removed and replaced with basal medium without growth factors and with low serum (0.1% FBS) for 16 h.

NO donors and inhibitors are prepared and kept on ice. Supernatants are removed from cell monolayers, and fresh basal medium without serum is added. When foreseen, cells are pretreated with inhibitors (30 min) and then stimulated with NO donors or other growth factors for 24 h. Stimulation supernatants are sampled and immediately kept on ice or stored at −20°C if assessed later. Cell monolayers are fixed with methanol (2 h at 4°C) and stained with Diff-Quik and cells are counted (as described in Cell proliferation section).

Protocol for preparation of polyacrylamide gel containing gelatin and gel electrophoresis. An 8% polyacrylamide gel containing 0.1% gelatin is placed in a gel electrophoresis run system as described in the manufacturer's instructions. The tank is filled with cold running buffer and the system is put on ice. 50 μL of cell supernatants (prepared as described before) are mixed with 12.5 μL of 4× Laemmli Sample Buffer without heating the samples. Samples are added to the wells of the gel, including the molecular weight standard. The electrophoresis equipment is connected to power pack and run at 80 V for 10 min until the samples have entered in the separating gel. The running continues at 180 V till the dye front reaches the bottom of the gel. The gel plates then gently are separated with the use of a spatula, and the gel is rinsed with washing solution.

Detection of gelatinolytic activity. The gel is washed twice with gentle agitation for 15 min with 2.5% Triton X-100 to remove SDS. The developing solution is added and incubated for 48 h at 37°C. Coomassie Brilliant Blue R-250 solution 0.5% (w/v) is used to stain gels for 1 h with gentle agitation. After that, a destaining solution is replaced every 15 min with gentle agitation. The image of the gel is taken in Chemidoc (or similar systems) by the use of a digital camera. Quantification of gelatinolytic activity is done normalizing the optical density values of the white bands to the number of cells counted/well.

Matrigel assay to assess endothelial network formation

A method to analyze sprouting and network formation by endothelial cells in relation to angiogenesis response is based on the evaluation of network formation when grown on Matrigel 3D plugs (Monti, Terzuoli, et al., 2016; Terzuoli et al., 2015). Matrigel is a complex mixture of heterogenous matrix proteins present in the basal membrane such as collagen type IV, laminin, and heparan sulfate and is resistant to digestion by endogenous enzymes. The characteristic of Matrigel is the ability to solidify once kept at 37°C. Practically, Matrigel (store concentrated −20°C) is let to melt on ice at 4°C and left overnight to thaw completely, taking care not to let it warm at RT.

The top of any tips that will be used for Matrigel manipulation has to be cut to enlarge the pipette hole. Matrigel is mixed to homogeneity by gentle pipetting, and 250 μL of Matrigel is added to each well of a 24-well plate. The plate is incubated at 37°C for at least 30 min to let the Matrigel solidify. Appropriate concentrations of NO donors and/or signaling inhibitors are prepared and kept on ice.

Endothelial cells are harvested from Petri's dish, and 7×10^4 cells treated with NO donors and/or inhibitors are plated onto the layer of Matrigel. Medium with 0.1% FBS is included as negative control of the experiments, while medium with 10% FBS or angiogenic factors as positive controls. Plates are incubated for 8–18 h at 37°C in a humidified incubator at 5% CO_2. Images of the cells are captured using a brightfield inverted microscope at 4× and 10× magnification in 10 fields. To quantify the networks, the number of nodes is counted by image analysis software, evaluating the number of branch points and/or the number of complete circles. Results are finally reported as branch points/well or circles/well.

Measurement of endothelial permeability

Paracellular permeability is a feature of disruption of the endothelial barrier, typical of inflammation, and early phases of angiogenesis (Ciccone et al., 2017; Ciccone, Piragine, et al., 2022; Monti et al., 2014; Terzuoli et al., 2015).

Cells are seeded on collagen-coated insert membranes (Corning, Sigma–Aldrich, St. Louis, MO, the United States) containing a high density of 0.4 μm diameter pores, and the inserts are placed into a 12-multiwell plate. Cells are seeded at 8×10^4 cells/insert and cultured with complete medium for 72 h to reach complete confluency. Confluent monolayers are pre-treated with drugs (30 min) and then treated with IL-1β (100 ng/mL, 1–6 h) in the case of hypothetical protective effect by the molecule under investigation against an inflammatory mediator. 40-kDa FITC-Dextran (10 μM) is added on top of the cells, allowing the fluorescent molecules to pass through the endothelial cell monolayer in the lower compartment. The extent of permeability is determined by measuring the fluorescence of the medium in the lower compartments in a plate reader (Infinite 200 Pro SpectraFluor, Tecan, Mannedorf, Switzerland), at 485 and 535 nm excitation and emission, respectively. The paracellular dextran flux can be associated with immunofluorescence images of the cells labelled for VE-cadherin or zonula-occludens-1 (ZO-1), markers typical of endothelial layer integrity (Ciccone, Piragine, et al., 2022; Martelli et al., 2021).

PRP preparation and platelet aggregation

When using human platelets, the experiments are carried out in accordance with The Code of Ethics of the World Medical Association (Declaration of Helsinki). After a detailed informed consent is signed, blood (9 mL) is collected in the morning from resting and fasting subjects by a 19-G needle without stasis (Monti et al., 2014). The blood is then stored in acid/citrate/dextrose (ACD 1:9) for platelet studies. In particular, donors should not be treated in the previous 2 weeks with corticosteroids, nonsteroidal antiinflammatory drugs (except for acetaminophene) and antibacterial drugs, to avoid the influence of these drugs on platelet responses.

For platelet isolation, trisodium citrated blood is centrifuged, within 20 min from collection, at 100 x g for 20 min at 23°C to obtain platelet-rich plasma (PRP). This is separated from the top two-thirds of the supernatant to avoid contamination by other cells. The remaining blood is further centrifuged at 800 x g for 20 min to obtain platelet-poor plasma (PPP). PPP can be used to dilute PRP to the appropriate density.

Platelet aggregation studies are performed in an aggregometer (i.e., four-channel PAP-4 Bio Data Corp., Horsham, PA, the United States) and expressed as change in percentage of transmitted light (%TL). Platelet aggregation in PRP is induced by 5 μM ADP, while NO donors or stimulators of the NOS/sGC pathway are expected to inhibit thrombosis. All measurements are performed in duplicate and read 8 min after stimulation.

Clonogenic assay

A typical feature of tumor cells is the ability, when cultured very sparse, to survive and form colonies. The effect of drugs can be thus evaluated on cells before the dilution or during the colony formation. For clonogenic assay, cells are plated in 6-multiwell plates (at a density of 100 cells/well) in a medium containing 10% FBS. According to the protocol, cells can be pretreated for 30 min (at 37°C) with drugs/inhibitors and then grown in medium with 10% FBS for 1 week/10 days in the absence/presence of NO derivatives. Colonies (>50 cells/each) are fixed and stained as described in Cell proliferation section. Data are expressed as the number and size of colonies that arise after treatment of cells (Ciccone, Monti, et al., 2018).

2D and 3D cocultures

The interaction between two (or more) different cell types can be obtained (1) by exposing one cell type with the medium conditioned by the other cells, (2) by culturing one cell type on porous inserts, which allow the diffusion of the secretome (molecules, vesicles) toward the second cell type cultured on the bottom of multiwells or Petri's dishes, or (3) by culturing 3D multicellular spheroids.

2D cocultures can be established between endothelial and smooth muscle cells or endothelial cells and tumor cells, evaluating functions of the vascular endothelium in terms of angiogenic functional responses (migration, proliferation, permeability) (Ciccone, Terzuoli, et al., 2018) or endothelial to mesenchymal transition (Ciccone et al., 2022).

Multicellular 3D spheroids can be set up for different tumor types including in the mix tumor cells, endothelial cells, fibroblasts, or other cells of the tumor microenvironment (TME). The possibility to stably mark-specific cell types with fluorescent probes or protein allows to monitor their morphology and localization, and follow their destiny inside the spheroids (Ciccone, Terzuoli et al., 2018; Ciccone et al., 2022). The use of FACS sorting on the base of the probe can allow to apply innovative techniques on recovered singles cells, to gain information relevant to cell–cell communications within the TME and the response to drug treatment. Recently, the setup of spheroids has been seen as a great improvement in cancer research, starting to been recognized as an alternative approach to reduce the use of animals for further efficacy and pharmacokinetic studies (Habanjar et al., 2021; Pinto et al., 2020; Singh & Gautam, 2023).

In vitro tumorsphere formation

This assay tests the ability of single cells to form tumorspheres, the in vitro surrogate of stemness. Tumor cells (2×10^5 cells/well in 1.5 mL of medium) are distributed into ultralow attachment 6-well plates as reported (Ciccone, Terzuoli, et al., 2018). Tumorspheres are grown in complete medium supplemented with 10% FBS for 4 days. During tumorsphere formation, evaluated by daily observation under the inverted microscope, they can be treated with drugs. After 4 days, tumorspheres are collected and lysed for protein extraction. Additionally, tumorspheres can be disaggregated with trypsin/EDTA for expansion and analysis of further generation(s). Multicellular spheroids can be prepared with an appropriate proportion of the TME cellular components, mixed together on the ultralow adherence multiplates or other protocols as microfluidic devices (Singh & Gautam, 2023).

Ex vivo experiments

The evaluation of the efficacy of novel molecules on the vascular tone is conducted in fragments of perfused vessels, maintained in isolated organ systems, evaluating the vasorelaxation of precontracted vessels (Ziche et al., 2008). This is an example of ex vivo preparations. Involvement of the vascular endothelium and mediation by NOS/NO pathway can be verified by mechanical removal of the endothelium or the use of NOS/cGMP pathway inhibitors (Fig. 4.1). The measure of cGMP levels can substantiate the involvement of the cyclic nucleotide pathway.

The classic Langendorff perfused heart test, by perfusing hearts isolated from laboratory animals with nonoxygenated and/or sugar-free buffer (mimicking the ischemic condition), can be used to screen for molecules active on cardiomyocytes, the atrioventricular conduction system, and the coronary microcirculation in physiological and pathological settings. When studying the effect of NO derivatives on angiogenesis, aortic rings can be cultured in 3D gels and endothelial spouting can be followed during time.

Protocol examples are given in the following.

Aortic ring preparations for vessel tone measurements

Isolated thoracic aortic rings from normotensive or spontaneously hypertensive rats (SHRs), systemically treated with or without the drug under investigation, are tested (Monti, Terzuoli, et al., 2016). Clearance by the local and national ethical boards is necessary before any animal is purchased and treated. Please refer to the last version of the national legislation according to the European Directive 2010/63/UE. Following animal acclimation and treatment, if any, animals are scarified in a humane manner and aortae isolated avoiding vessel pulling, and kept in saline buffer. Adventitial tissue and fat are removed, and rings are produced by a microtome or a sharp surgical blade. Aortic rings are mounted in appropriate organ bath and let to equilibrate in the presence of specific buffer at 37°C gassed with 5% CO_2/95% O_2 mix. Changes in tension of aortic rings are processed by a pressure transducer and recorded by a computerized system (i.e., Biopac, Goleta, CA). The value of tension developed after phenylephrine (0.3 μM; Sigma–Aldrich) administration in the organ bath is taken as 100%, and the effects of increasing concentrations of the direct NO donor sodium nitroprusside (SNP) or acetylcholine (Ach) are referred to this value. The relevance of endogenous NO in inducing vasodilation is assayed by using aortic rings mechanically deprived of endothelium or following incubation of the vessel rings with an NOS inhibitor as L-NAME or others (Monti et al., 2014; Morbidelli et al., 2003, 2010; Ziche et al., 1994).

Isolated heart preparation and perfusion

Spontaneously beating hearts are rapidly explanted and mounted on a Langendorff apparatus for retrograde perfusion with physiological salt solution via the aorta at a constant flow rate of 10 mL/min (Pessina et al., 2020).

Heart contractility is measured as left ventricle developed pressure (LVDP = left ventricular systolic pressure minus left ventricular diastolic pressure) by means of a latex balloon, inserted into the left ventricle via the mitral valve and connected to a pressure transducer (BLPR, WPI, Berlin, Germany). Coronary perfusion pressure is recorded by a pressure transducer in the inflow line. Heart rate is calculated from LVDP. LVDP and coronary perfusion pressure are recorded with a digital PowerLab data acquisition system (PowerLab 8/30; ADInstruments, Australia) and analyzed with Chart Pro for Windows software (PowerLab; ADInstruments, Australia). At the end of the experiments or at fixed times, heart tissue and perfusate can be sampled for biochemical or molecular studies (Pessina et al., 2020).

Aortic ring angiogenesis assay

This protocol allows to measure pseudopapillary sprouting from vessel rings cultured in 3D gels (Morbidelli et al., 2010). Aortas from rats or mice are isolated in sterile conditions. Rings 1–2 mm long are produced and included in a fibrin gel obtained by adding 400 μL of a bovine fibrinogen solution (3 mg/mL in M199 medium; Sigma–Aldrich) into each well of 48-multiwell plates. Gelation of the fibrinogen is induced with bovine thrombin (1.5 units/mL; Sigma–Aldrich). After 20 min, 400 μL of complete medium is added to each well. After 24 h, the supernatant is removed, the gels are washed, and drug or vehicle (medium supplemented with 1% FBS and DMSO) is added for 30 min followed by treatment with VEGF or FGF-2 (10–20 ng/mL) used as positive control. Quantitative evaluation of newly formed structures (radial sprouts form the ring wall) is carried out on day 3. At a magnification of 200×, the area occupied by tubules is measured using an ocular grid. The result is expressed as the number of grid units/aortic ring (Macaluso et al., 2020; Morbidelli et al., 2010).

Measurement of cGMP levels

cGMP levels are measured on cell extracts from confluent cell monolayers, isolated aortic rings or platelets by commercially available kits as for example enzyme-immunoassay kit (Cayman, Ann Arbor (MI), the United States) as previously described (Monti et al., 2014; Ziche et al., 1993). Biological samples should be treated with 1 mM 3-isobutyl-5-methyl-xanthine (IBMX, an unspecific PDE inhibitor) (Sigma–Aldrich) for 15 min before stimulation to increase the measurement yield. cGMP levels are assayed in cell lysates according to the manufacturer instructions, while proteins to normalize the various experimental samples are measured in the pellet by Bradford's procedure.

In vivo experiments to assess efficacy and safety of NO derivatives

The next step concerns the use of living animals that mimic the human pathological conditions, such as spontaneously hypertensive rats (SHR, model of hypertension), strains of surgically manipulated rats (unephrectomized and exposed to deoxycorticosterone acetate [DOCA] and 1% NaCl in drinking water, called DOCA-salt rats). Blood pressure and ECG measurements can be made by the classic tail-cuff test on conscious animals, or by the sterile insertion of arterial catheters and plethysmography measurements (i.e., the recording of the volume variations of an organ, used above all for investigations on the heart) in anesthetized vital animals. Animal models of atherosclerosis are represented by selected (Watanabe rabbits) or engineered (e.g., $ApoE^{-/-}$, $LDLR^{-/-}$) rat or mice strains treated with high-cholesterol diets. Surgically, the events of atherosclerosis and restenosis can be reproduced in normal animals (from rats to pigs) by intraluminal balloon angioplasty or exposure to ameroid constrictor. This surgical procedure produces vascular damage with the development of atherosclerotic plaque and myocardial infarction, which can be followed in all their phases with the same instrumentation used in the clinic (echo-color-Doppler, angiography, circulating disease markers). Transient or permanent ligation of blood vessels can lead to ischemia and infarction in downstream tissue (e.g., left anterior descending artery in the heart; femoral artery in the lower limb) in both small animals and pigs. Colorimetric and histomorphological analyses on heart sections allow to measure the infarcted area and the cellular and molecular mechanisms involved in tissue damage and in the pharmacological effect of the substances under study.

A plethora of disease animal models is available for the study of drugs in different pathologies and approaches, i.e., preventive versus curative setting. Cancer growth and metastasis, ocular and skin disorders, neurodegeneration, and pain are other demanding examples, beside cardiovascular disorders.

Animal maintenance and treatment

The experiments can be performed on adult male C57BL/6 mice (6 weeks old), on male Wistar Kyoto rats (WKY, 8 weeks old) or spontaneously hypertensive rats (SHR, 8 weeks old) obtained from certified producer. All animals are housed in temperature- and humidity-controlled rooms (22°C, 50%) with lights on from 07:00 to 19:00, water and food available ad libitum, and left to acclimatize to these conditions for 1 week. All procedures should be carried out in accordance with national and internationals legislation as the European Directive 2010/63/UE. All efforts are made to minimize the number of animals used and their suffering, following the 3R practices.

Administration of the test compounds is carried out with the designed route of administration and drug posology. As an example, intraperitoneal injection (i.p.) of NO donors (doses of 1−10 mg/kg of body weight) can be scheduled at 10:00 a.m. every day for 2 or 4 weeks. Control animals are treated with the same volume of vehicle. Body weight, general behavior, food, and water consumption are weekly checked as safety assessment. Cardiovascular parameters can be followed during and after treatment. At the end of the experiment, animals are euthanized with an overdose of chloralum hydratum i.p. (Fluka, Sigma Aldrich, St. Louis, MO, the United States) and organs are collected (Monti, Terzuoli, et al., 2016). In particular, aorta is removed, immediately frozen in liquid nitrogen and stored at −80°C until use for further biochemical assessments, i.e., cGMP levels or western blotting for proteins of interest. Otherwise, the aorta can be sampled for aortic ring relaxation or angiogenesis experiments as described before in ex vivo studies.

In vivo measurement of arterial pressure and heart rate

Mean arterial pressure (MAP) and heart rate are monitored every 5 days in conscious restrained animals by tail-cuff plethysmography using the BP-2000 Blood Pressure Analysis System coupled with BP-2000 analysis software (Visitech Systems, Apex, NC, the United States) (Macaluso et al., 2020; Monti, Terzuoli, et al., 2016). Animals are conditioned at least four times to the procedure before beginning the experiments. Measurements are repeated every 30 min for 2 h, and data are reported as mean of four separate determinations. MAP is expressed as mmHg and heart rate as bpm (beat per minutes).

Effect of NO derivatives on tumor growth

The effect of NO donors or other derivatives can be studied in different protocols: xenograft implant of human tumor cells in immunodeficient mice both as heterotopic or orthotopic implants; implant of mouse tumor cells in mice, again as hereto- or orthotopic implant. The growth of the tumors is followed during time by caliper measurement (when feasible) or by in life imaging techniques, otherwise at fixed times following animal sacrifice and tumor isolation and weight. Protocols to assess antimetastatic potential by drugs can be applied in mice injected in the tail vein with suspensions of tumor cells and evaluating the tumor diffusion and take in distant organs. Tumor cells engineered with bioluminescence and fluorescence probes allow the easy monitoring by in-life imaging systems and recovery and sorting by ultimate FACS technologies. Importantly, investigations should be conducted in accordance with the ethical standards established in the national law, which for European countries acknowledge the European Directive 2010/63/UE.

In vivo tumor xenograft

To assess the involvement of NOS/NO pathway in tumor growth and angiogenesis, the xenograft protocol is largely used. As an example, immunodeficient mice (5-week-old female athymic mice, Envigo Huntingdon, the United Kingdom) are s.c. inoculated in the right flank with human tumor cells (10^7 MCF-7 cells/100 μL; 50 μL of cells and 50 μL of Matrigel). By using MCF-7 (mammary tumor cells), β-estradiol has to be administered i.m. (3 mg/kg), every 7 days (Ciccone, Terzuoli, et al., 2018). For other tumor cell types, please check literature reports for the number of cells to be inoculated, if alone or together with feeder cells as fibroblasts or matrix support as Matrigel. Depending on the experimental design, animals can be treated with drugs under investigation systemically or as local peritumor injection. Mice are daily observed, and caliper measurements of tumors are performed twice a week. Tumor dimensions are determined using the following formula: (shortest diameter × longest diameter × thickness of the tumor in mm). Data are reported as tumor volume in mm^3. After 2–3 weeks, mice are sacrificed. For each tumor, volume (mm^3) and weight (mg) are determined. During experiments, mouse weight is monitored for safety assessment. Each tumor is collected and split in two parts. One half is immediately frozen in liquid nitrogen for Western blot and RT-PCR analysis. The other part is embedded in Tissue-Tek O.C.T., cooled in isopentane and frozen in liquid nitrogen for histology. Seven-μm-thick cryostat sections from tissue samples are used for immunohistochemical staining with anti-Ki67 antibody (to evaluate tumor proliferation) and immunofluorescence with anti-CD31 and anti-NG2 antibodies (to evaluate angiogenesis and vessel normalization) or other markers of interest (Ciccone, Terzuoli, et al., 2018).

Three-dimensional ultrasound high-resolution tumor imaging can be performed at fixed times or at the end of experiments by using Vevo 2100 Imaging System (Visual Sonics Inc., Toronto, ON, Canada) to assess relative perfusion in a tumor area or a tumor volume in living animals (Ciccone, Terzuoli, et al., 2018). Mice are anesthetized with 2% isoflurane in oxygen and restrained on a heated stage (THM-100, Indus Instruments, Houston, TX) during imaging. Ultrasound coupling gel (Aquasonic 100, Parker Laboratories, Inc., Fairfield, NJ) is applied to the skin between the transducers and the mice, and images of the tumors are acquired through the ventral body wall in longitudinal orientation. Power Doppler Mode is used to visualize and measure flow in 3D. This mode allows to detect vascularity in and around subcutaneous tumors and produces a measure of relative quantification as percent vascularity, an index of relative vascular density. Positive areas appear in red. Power 3D-Mode, added to Power Doppler Mode, allows to reconstruct a volume that integrates the Power Doppler Mode color data with the surrounding B-Mode 3D volume (Ciccone, Terzuoli, et al., 2018).

Safety studies before clinical studies

During the early preclinical development process, also known as Go/No-Go decision, a drug candidate needs to pass through several steps, such as preliminary studies that aim to investigate the candidate safety including genotoxicity, mutagenicity, safety pharmacology and general toxicology, and determination of drug availability (studies on pharmacokinetics), and absorption, distribution, metabolism, and elimination (ADME) (Asati et al., 2022; Huang et al., 2020; McGonigle & Williams, 2017). These preliminary screening assays generally do not need to comply with Good Laboratory Practice (GLP) regulations. These studies aim at investigating the drug safety to obtain the first information about its tolerability in different systems that are relevant for further decisions. There are, however, other studies that should be performed according to GLP standards and are mandatory for the safe exposure to humans, such as repeated dose toxicity, genotoxicity, and safety pharmacology (Asati et al., 2022; Huang et al., 2020). These studies must be conducted before the investigational new drug (IND) application. The package of nonclinical studies should cover all information needed for the safe transposition of drugs from animals to humans, generally based on the nonobserved adverse effect level (NOAEL) obtained from general toxicity studies. After IND approval, other GLP experiments for the evaluation of chronic toxicity, reproductive and developmental toxicity, carcinogenicity, and genotoxicity are carried out during the clinical phase of development. However, the necessity of performing such studies depends on the clinical application purpose of the new drug.

Conclusions

Various experimental approaches have been presented in this chapter, allowing preclinical researchers to evaluate the efficacy and safety of the molecules or therapeutic strategies of interest. The multiplicity and redundancy of assays and protocols are based on the complexity of NO-related pathways and the multiple/divergent effects of NOS/NO pathways in various pathological conditions. The researchers should be aware of all the experimental procedures, both conventional and innovative, to prepare informative and properly written INDs for clinical trials. Efficacy studies on experimental models, together with safety assessment, when properly conducted, allow to obtain predictive information for the optimal design of safe human clinical trials.

Acknowledgements

The authors acknowledge funding from MUR-PRIN (project n. 2022RCRWE5) and EC HORIZON – MSCA – 2022 – DN (NO-CANCER-NET, Advanced Engineering of Nitric Oxide Based Therapeutics for Triple Negative Breast Cancer Training Network).

References

Alsharabasy, A. M., Aljaabary, A., Bohara, R., Farràs, P., Glynn, S. A., & Pandit, A. (2023). Nitric oxide-scavenging, anti-migration effects, and glycosylation changes after hemin treatment of human triple-negative breast cancer cells: A mechanistic study. *ACS Pharmacology & Translational Science, 6*(10), 1416–1432. https://doi.org/10.1021/acsptsci.3c00115

Asati, S., Pandey, V., Gour, V., Tiwari, R., Soni, V., Rajpoot, K., Tekade, M., Sharma, M. C., & Tekade, R. K. (2022). Toxicity and toxicokinetic considerations in product development and drug research. *Pharmacokinetics and Toxicokinetic Considerations - Vol II,* 401–424. https://doi.org/10.1016/B978-0-323-98367-9.00001-9

Bachetti, T., & Morbidelli, L. (2000). Endothelial cells in culture: A model for studying vascular functions. *Pharmacological Research, 42*(1), 9–19. https://doi.org/10.1006/phrs.1999.0655

Ciccone, V., Filippelli, A., Angeli, A., Supuran, C. T., & Morbidelli, L. (2020). Pharmacological inhibition of CA-IX impairs tumor cell proliferation, migration and invasiveness. *International Journal of Molecular Sciences, 21*(8), 2983. https://doi.org/10.3390/ijms21082983

Ciccone, V., Filippelli, A., Bacchella, C., Monzani, E., & Morbidelli, L. (2022). The nitric oxide donor [Zn(PipNONO)Cl] exhibits antitumor activity through inhibition of epithelial and endothelial mesenchymal transitions. *Cancers, 14*(17). https://doi.org/10.3390/cancers14174240

Ciccone, V., Monti, M., Monzani, E., Casella, L., & Morbidelli, L. (2018). The metal-nonoate Ni(SalPipNONO) inhibits in vitro tumor growth, invasiveness and angiogenesis. *Oncotarget, 9*(17), 13353–13365. https://doi.org/10.18632/oncotarget.24350

Ciccone, V., Piragine, E., Gorica, E., Citi, V., Testai, L., Pagnotta, E., Matteo, R., Pecchioni, N., Montanaro, R., Di Cesare Mannelli, L., Ghelardini, C., Brancaleone, V., Morbidelli, L., Calderone, V., & Martelli, A. (2022). Anti-inflammatory effect of the natural H2S-donor erucin in vascular endothelium. *International Journal of Molecular Sciences, 23*(24). https://doi.org/10.3390/ijms232415593

Ciccone, V., Monti, M., Antonini, G., Mattoli, L., Burico, M., Marini, F., Maidecchi, A., & Morbidelli, L. (2017). Efficacy of AdipoDren® in reducing interleukin-1-induced lymphatic endothelial hyperpermeability. *Journal of Vascular Research, 53*(5–6), 255–268. https://doi.org/10.1159/000452798

Ciccone, V., Terzuoli, E., Donnini, S., Giachetti, A., Morbidelli, L., & Ziche, M. (2018). Stemness marker ALDH1A1 promotes tumor angiogenesis via retinoic acid/HIF-1α/VEGF signalling in MCF-7 breast cancer cells\. *Journal of Experimental & Clinical Cancer Research, 37*(1). https://doi.org/10.1186/s13046-018-0975-0

Ciccone, V., Terzuoli, E., Ristori, E., Filippelli, A., Ziche, M., Morbidelli, L., & Donnini, S. (2022). ALDH1A1 overexpression in melanoma cells promotes tumor angiogenesis by activating the IL-8/Notch signaling cascade. *International Journal of Molecular Medicine, 50*(1). https://doi.org/10.3892/ijmm.2022.5155

Genah, S., Angeli, A., Supuran, C. T., & Morbidelli, L. (2020). Effect of Carbonic Anhydrase IX inhibitors on human endothelial cell survival. *Pharmacological Research, 159*. https://doi.org/10.1016/j.phrs.2020.104964

Habanjar, O., Diab-Assaf, M., Caldefie-Chezet, F., & Delort, L. (2021). 3D cell culture systems: Tumor application, advantages, and disadvantages. *International Journal of Molecular Sciences, 22*(22). https://doi.org/10.3390/ijms222212200

Hobbs, A. J., & Moncada, S. (2003). Antiplatelet properties of a novel, non-NO-based soluble guanylate cyclase activator, BAY 41-2272. *Vascular Pharmacology, 40*(3), 149–154. https://doi.org/10.1016/S1537-1891(03)00046-6

Huang, W., Percie du Sert, N., Vollert, J., & Rice, A. S. C. (2020). General principles of preclinical study design. *Handbook of Experimental Pharmacology, 257*, 55–69. https://doi.org/10.1007/164_2019_277

Koika, V., Zhou, Z., Vasileiadis, I., Roussos, C., Finetti, F., Monti, M., Morbidelli, L., & Papapetropoulos, A. (2010). PKG-I inhibition attenuates vascular endothelial growth factor-stimulated angiogenesis. *Vascular Pharmacology, 53*(5–6), 215–222. https://doi.org/10.1016/j.vph.2010.08.004

Macaluso, M., Bianchi, A., Sanmartin, C., Taglieri, I., Venturi, F., Testai, L., Flori, L., Calderone, V., De Leo, M., Braca, A., Ciccone, V., Donnini, S., Guidi, L., & Zinnai, A. (2020). By-products from winemaking and olive mill value chains for the enrichment of refined olive oil: Technological challenges and nutraceutical features. *Foods, 9*(10). https://doi.org/10.3390/foods9101390

Martelli, A., Piragine, E., Gorica, E., Citi, V., Testai, L., Pagnotta, E., Lazzeri, L., Pecchioni, N., Ciccone, V., Montanaro, R., Di Cesare Mannelli, L., Ghelardini, C., Brancaleone, V., Morbidelli, L., & Calderone, V. (2021). The H2S-donor erucin exhibits protective effects against vascular inflammation in human endothelial and smooth muscle cells. *Antioxidants, 10*(6). https://doi.org/10.3390/antiox10060961

McGonigle, P., & Williams, M. (2017). *Preclinical pharmacology and toxicology - contributions to the translational interface* ☆. Elsevier BV. https://doi.org/10.1016/b978-0-12-801238-3.05242-9

Monti, M., Solito, R., Puccetti, L., Pasotti, L., Roggeri, R., Monzani, E., Casella, L., & Morbidelli, L. (2014). Protective effects of novel metal-nonoates on the cellular components of the vascular system. *Journal of Pharmacology and Experimental Therapeutics, 351*(3), 500–509. https://doi.org/10.1124/jpet.114.218404

Monti, M., Terzuoli, E., Ziche, M., & Morbidelli, L. (2016). H2S dependent and independent anti-inflammatory activity of zofenoprilat in cells of the vascular wall. *Pharmacological Research, 113*, 426–437. https://doi.org/10.1016/j.phrs.2016.09.017

Monti, M., Ciccone, V., Pacini, A., Roggeri, R., Monzani, E., Casella, L., & Morbidelli, L. (2016). Anti-hypertensive property of a nickel-piperazine/NO donor in spontaneously hypertensive rats. *Pharmacological Research, 107*, 352–359. https://doi.org/10.1016/j.phrs.2016.03.033

Morbidelli, L., Donnini, S., Filippi, S., Messori, L., Piccioli, F., Orioli, P., Sava, G., & Ziche, M. (2003). Antiangiogenic properties of selected ruthenium(III) complexes that are nitric oxide scavengers. *British Journal of Cancer, 88*(9), 1484–1491. https://doi.org/10.1038/sj.bjc.6600906

Morbidelli, L., Chang, C. H., Douglas, J. G., Granger, H. J., Ledda, F., & Ziche, M. (1996). Nitric oxide mediates mitogenic effect of VEGF on coronary venular endothelium. *American Journal of Physiology - Heart and Circulatory Physiology, 270*(1), H411–H415. https://doi.org/10.1152/ajpheart.1996.270.1.h411

Morbidelli, L., Pyriochou, A., Filippi, S., Vasileiadis, I., Roussos, C., Zhou, Z., Loutrari, H., Waltenberger, J., Stössel, A., Giannis, A., Ziche, M., & Papapetropoulos, A. (2010). The soluble guanylyl cyclase inhibitor NS-2028 reduces vascular endothelial growth factor-induced angiogenesis and permeability. *American Journal of Physiology - Regulatory, Integrative and Comparative Physiology, 298*(3), R824–R832. https://doi.org/10.1152/ajpregu.00222.2009

Pankey, E. A., Bhartiya, M., Badejo, A. M., Haider, U., Stasch, J. P., Murthy, S. N., Nossaman, B. D., & Kadowitz, P. J. (2011). Pulmonary and systemic vasodilator responses to the soluble guanylyl cyclase activator, BAY 60-2770, are not dependent on endogenous nitric oxide or reduced heme. *American Journal of Physiology - Heart and Circulatory Physiology, 300*(3), H792–H802. https://doi.org/10.1152/ajpheart.00953.2010

Pessina, F., Frosini, M., Marcolongo, P., Fusi, F., Saponara, S., Gamberucci, A., Valoti, M., Giustarini, D., Fiorenzani, P., Gorelli, B., Francardi, V., Botta, M., Dreassi, E., & Quintas, L. E. M. (2020). Antihypertensive, cardio- and neuro-protective effects of *Tenebrio molitor* (Coleoptera: Tenebrionidae) defatted larvae in spontaneously hypertensive rats. *PLoS One, 15*(5). https://doi.org/10.1371/journal.pone.0233788

Pinto, B., Henriques, A. C., Silva, P. M. A., & Bousbaa, H. (2020). Three-dimensional spheroids as in vitro preclinical models for cancer research. *Pharmaceutics, 12*(12), 1–38. https://doi.org/10.3390/pharmaceutics12121186

Singh, K., & Gautam, P. K. (2023). Macrophage infiltration in 3D cancer spheroids to recapitulate the TME and unveil interactions within cancer cells and macrophages to modulate chemotherapeutic drug efficacy. *BMC Cancer, 23*(1). https://doi.org/10.1186/s12885-023-11674-9

Taraboletti, G., Morbidelli, L., Donnini, S., Parenti, A., Granger, H. J., Giavazzi, R., & Ziche, M. (2000). The heparin binding 25 kDa fragment of thrombospondin-1 promotes angiogenesis and modulates gelatinase and TIMP-2 production in endothelial cells. *Federation of American Societies for Experimental Biology Journal, 14*(12), 1674–1676. https://doi.org/10.1096/fj.99-0931fje

Terzuoli, E., Monti, M., Vellecco, V., Bucci, M., Cirino, G., Ziche, M., & Morbidelli, L. (2015). Characterization of zofenoprilat as an inducer of functional angiogenesis through increased H 2 S availability. *British Journal of Pharmacology, 172*(12), 2961–2973. https://doi.org/10.1111/bph.13101

Ziche, M., Morbidelli, L., Masini, E., Amerini, S., Granger, H. J., Maggi, C. A., Geppetti, P., & Ledda, F. (1994). Nitric oxide mediates angiogenesis in vivo and endothelial cell growth and migration in vitro promoted by substance P. *Journal of Clinical Investigation, 94*(5), 2036–2044. https://doi.org/10.1172/JCI117557

Ziche, M., Donnini, S., Morbidelli, L., Monzani, E., Roncone, R., Gabbini, R., & Casella, L. (2008). Nitric oxide releasing metal–diazeniumdiolate complexes strongly induce vasorelaxation and endothelial cell proliferation. *ChemMedChem, 3*(7), 1039–1047. https://doi.org/10.1002/cmdc.200700354

Ziche, M., Morbidelli, L., Masini, E., Granger, H., Geppetti, P., & Ledda, F. (1993). Nitric oxide promotes DNA synthesis and cyclic GMP formation in endothelial cells from postcapillary venules. *Biochemical and Biophysical Research Communications, 192*(3), 1198–1203. https://doi.org/10.1006/bbrc.1993.1543

PART IV

Nitric oxide and neurodegeneration

CHAPTER

5

Nitric oxide-mediated neuroinflammatory pathways as treatment targets in neurodegeneration

Jennifer Cale, Tracy D. Farr, Sébastien Serres and Joern R. Steinert

Division of Physiology, Pharmacology and Neuroscience, School of Life Sciences, University of Nottingham, Nottingham, United Kingdom

Abbreviations

AD Alzheimer's disease
APP Amyloid precursor protein
ATP Adenosine triphosphate
Aβ Amyloid beta
BACE1 β-site APP-cleaving enzyme 1
BAP Biological antioxidant potential
BBB Blood-brain barrier
CBF Cerebral blood flow
cGMP Cyclic guanosine monophosphate
CNS Central nervous system
CSF1R Colony-stimulating factor-1 receptor
dROM Diacron reactive oxygen metabolite
EC Endothelial cells
EDHF Endothelium-derived hyperpolarizing factor
eNOS Endothelial NOS
ET-1 Endothelin-1
H_2O_2 Hydrogen peroxide
HMG-CoA 3-hydroxy-3-methylglutaryl coenzyme A
iNOS Inducible NOS
MCAO Middle cerebral artery occlusion

Therapeutic Applications of Nitric Oxide in Cancer and Inflammatory Disorders
https://doi.org/10.1016/B978-0-443-22354-9.00005-X

MCI Mild cognitive impairment
NADPH Nicotinamide adenine dinucleotide phosphate oxidase
NF-κB Nuclear factor kappa B
NLRP3 NLR Family pyrin domain containing 3
NMDAR N-methyl-D-aspartate receptors
nNOS Neuronal NOS
NO Nitric oxide
NOS Nitric oxide synthase
NVU Neurovascular unit
O_2 Superoxide
RNS Reactive nitrogen species
ROS Reactive oxygen species
sGC Soluble guanylyl cyclase
TBI Traumatic brain injury
VCI Vascular cognitive impairment
VSMC Vascular smooth cells

Introduction

Redox stress is associated with neuroinflammation, can be caused by ischemia/reperfusion injury, mitochondrial dysfunction or protein aggregation and is linked to several neurological diseases (Aguilera et al., 2018; Allen & Bayraktutan, 2009; Barnham et al., 2004; Picca et al., 2020) including dementia which affects memory, psychological responses, and behavior thus interfering with day-to-day life. There are several subtypes of dementia, each with their own respective causes, symptoms, and treatments. Vascular cognitive impairment (VCI) is the second most common type of dementia and refers to cognitive disorders with a vascular origin (Burns et al., 2022). The prevalence of vascular dementia ranges from 1.2% to 4.2% of people aged over 65 years old and appears to increase linearly with age (Hébert & Brayne, 2004). The main vascular risk factors include age, hypertension, diabetes, stroke, and smoking (Ritchie et al., 2001; Román et al., 2002) with executive dysfunction being the typical manifestation of VCI among a variety of different pathologies and symptoms, diagnosis, which makes VCI classification challenging.

Current consensus suggests that multiinfarct dementia, which is a result of several mini strokes, is a prominent subtype along with subcortical ischemic vascular dementia, including Binswanger's disease, which is the result of microvascular pathology (Skrobot et al., 2017). This condition includes thickening of blood vessel walls and basement membranes causing them to become stiff and enriched with collagen deposition which ultimately damages the brain white matter. Mixed dementia also represents a major subtype of VCI in which vascular pathologies coexist with other forms of dementia, such as Alzheimer's and Lewy body's dementia and a final subtype referring to cognitive decline following a stroke. There is an urgent unmet clinical need to identify therapeutic and treatment interventions for VCI (Kalaria, 2016; Langa et al., 2004; Uwagbai & Kalish, 2022) which presents an ongoing challange caused by the complexity of this multifaceted and heterogeneous condition. It is therefore important to identify vital targets of neuropathological mechanisms impacting on the neurovascular unit (NVU) function.

The NVU is a tightly controlled and well-connected structural and functional complex composed of microglia, astrocytes, pericytes, oligodendrocytes, neurons, interneurons, vascular smooth cells (VSMCs), brain microvascular endothelial cells (ECs), basement membrane, and extracellular matrix (Fig. 5.1). Importantly, ECs, VSMCs, pericytes, and perivascular astrocytic end-feet make up the blood–brain barrier (BBB) which selectively controls the signaling between the central nervous system (CNS) and peripheral blood. Moreover, the

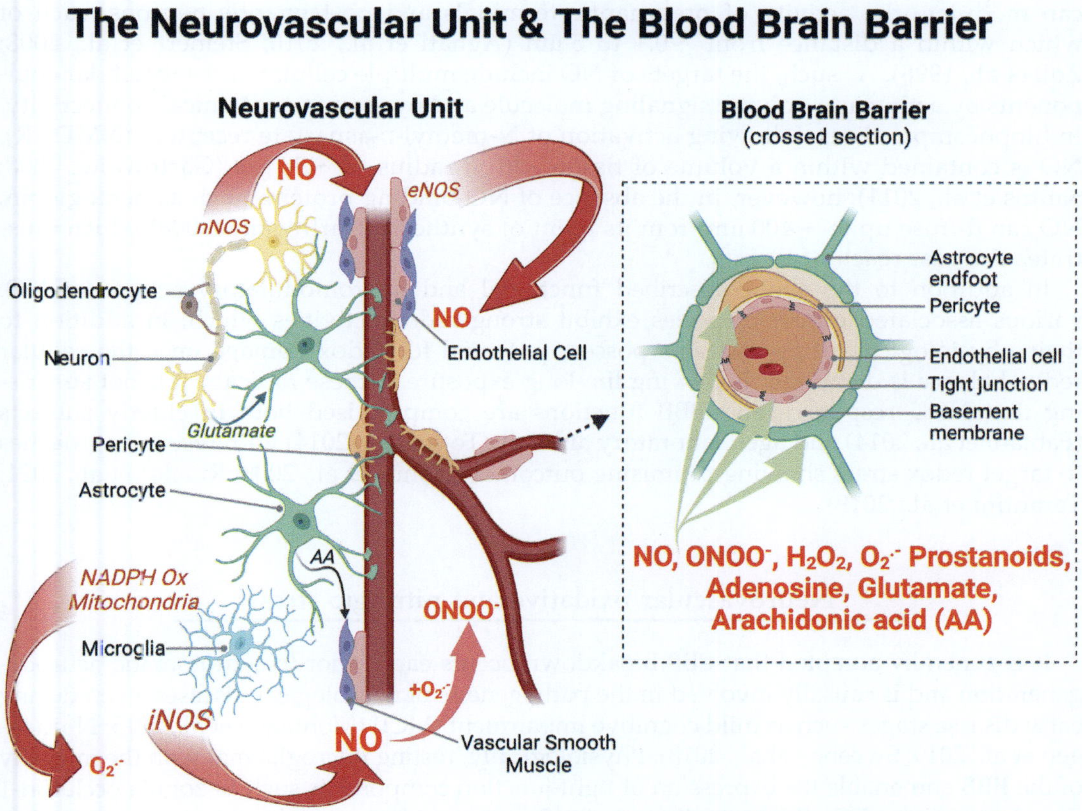

FIGURE 5.1 Composition and redox signaling of the neurovascular unit (NVU) and blood–brain barrier (BBB). The NVU consists of interacting vascular and neuronal compartments. Neurons, glia, astrocytes, and oligodendrocytes reside within the CNS and continuously communicate with vascular partners like pericytes, endothelial cell, and smooth muscle cells to adjust blood flow and energy and oxygen supply to the brain. Multiple communication routes are important in neurovascular coupling processes, with main vasoactive molecules including NO, $ONOO^-$, H_2O_2, $O_2^{\bullet -}$, prostaglandins, adenosine, glutamate, and arachidonic acid (AA). Activated microglia are a major source of iNOS-generated NO, endothelial cells, and neurons express Ca^{2+}-dependent eNOS and nNOS, respectively, to allow nitrergic communication across and within the NVU. The overwhelming production of reactive oxygen and nitrogen species will affect all components of the NVU which leads to dysfunction of pericyte, smooth muscle, and endothelial signaling followed by breakdown of the BBB in pathologies. The main sources for oxidative stress are glial NADPH oxidase and mitochondria. The assembly and function of above cell types creates the BBB to bridge communication between the vascular and central nervous systems and dysfunction of its components compromises this barrier function.

NVU maintains the balance of metabolites in the CNS by providing sufficient nutrients to the brain and disposing metabolic by-products in the cerebral blood fluid (CBF), both in response to alterations in neuronal activity and neuronal requirements. This unit is under a tight and constant control of signaling molecules such as nitric oxide (NO) but also oxygen-free radicals which makes it susceptible to age-dependent dysfunction linked to risks of developing neurodegeneration (Fig. 5.1).

The unique properties of NO to diffuse in both aqueous and lipid environments and thus within and between cells make this signaling molecule highly essential in the communication within the NVU. NO is typically produced in the soma of neurons and can modulate the activity of presynaptic terminals and postsynaptic neurons, each of which within a distance from ~0.1 to 5 μm (Agnati et al., 2010; Steinert et al., 2008; Zoli et al., 1998). As such, the targets of NO include multiple cellular and subcellular components by acting as a volume signaling molecule and bypassing anatomical connectivity. In hippocampal slices, following activation of N-methyl-D-aspartate receptors (NMDAR), NO is contained within a volume of tissue with a radius of ~50 μm (Garthwaite, 2008; Santos et al., 2011); however, in the absence of NO-binding proteins such as hemoglobin, NO can diffuse up to ~400 μm from its point of synthesis (Garthwaite, 2008) which illustrates the vast reach of targets.

In addition to the above-described functional and neuromodulatory actions of NO, various associated nitrergic species exhibit strong redox activities which, in addition to their physiological properties, also possess potential for redox damage once the cellular redox balance is disturbed. Following life-long exposure to these radicals, it is not surprising that NVU responses and BBB functions are compromised both in elderly subjects (Fabiani et al., 2014) and aged laboratory animals (Toth et al., 2014) with rescue approaches to target redox stress showing promising outcomes (Hamel et al., 2008; Rinaldi et al., 2021; Tarantini et al., 2019).

Neurovascular oxidative and nitrergic stress

It is generally accepted that BBB breakdown occurs early prior to symptomatic neurodegeneration and is causally involved in the pathogenesis of neurological diseases, even during early disease stages such as mild cognitive impairment (MCI) (Montagne et al., 2015; Nordengen et al., 2019; Sweeney et al., 2018). Physiologically, resting microglia maintain the integrity of the BBB and enable the expression of tight-junction components such as zonula occludin-1 on microvascular ECs (Mehrabadi et al., 2017), or provide essential tight-junction proteins, such as claudin-5 (Li et al., 2023). However, a BBB dysfunction leads to uncontrolled exposure of the brain to inflammatory molecules, contributing to circulatory neuroinflammation which subsequently can compromise the NVU. Specifically, microglia-mediated dysfunction of different components in the NVU, such as ECs, astrocytes, and pericytes, can compromise the integrity of the NVU resulting in its impairment thereby creating a positive feedback loop (Haruwaka et al., 2019). This has been shown during sustained inflammation, whereby microglia can phagocytose astrocytic end-feet (Haruwaka et al., 2019) (Fig. 5.1). Predominant factors in microglial-derived neuroinflammatory signaling are redox-active molecules like oxygen-free radicals and various nitrergic species.

Oxygen-mediated signaling

The brain consumes significant levels of oxygen and is therefore exposed to excessive amounts of oxygen-free radicals. It is rich in polyunsaturated fatty acids (Bazinet & Layé, 2014) which are sensitive to peroxidation *via* oxygen-free radicals and lacks a powerful antioxidant capacity (Halliwell, 1992; Luca et al., 2015; Uttara et al., 2009) leaving the brain vulnerable to pathological disturbances caused by redox signaling (Halliwell, 2006). An increase in reactive oxidants, predominantly superoxide ($O_2 \bullet^-$) in the vascular system, a concurrent reduction of NO bioavailability, and endothelial dysfunction in vascular disease implicates oxidative stress as a crucial contributor to vascular dementia (Bennett et al., 2009).

The damage in the endothelium and the NVU appears to play a key role in the early pathogenesis of VCI (Martins-Filho et al., 2021). Dysfunctional mitochondria which are a major source of oxygen-free radicals have been found in cerebral ECs in Alzheimer's disease (AD) patients. The loss of mitochondrial membrane potential, increased production of mitochondrial reactive oxygen species (ROS), and an increased mitochondrial membrane permeability are responsible for the release of cytochrome C and mitochondrial DNA into the cytoplasm of ECs (Parodi-Rullán et al., 2021). Collectively, this reduces the coverage of pericytes of the endothelial layer in AD (Ding et al., 2020; Sengillo et al., 2013) suggesting that in early disease stages of AD there is close link between NVU disruption and BBB damage. This idea is supported by findings showing a limited capillary flow in AD patients due to capillary constrictions caused by pericytes *via* aberrant ROS-endothelin-1 (ET-1) signaling (Nortley et al., 2019).

Although the pathogeneses of different types of dementia may be linked, the exact pathophysiologies related to oxidative stress signalling seem to differ. When comparing serum levels of diacron reactive oxygen metabolites (dROMs) reflecting the amount of organic hydroperoxide, biological antioxidant potential (BAP), endogenous plasma antioxidants, such as albumin, total bilirubin and uric acid in AD, cerebrovascular dementia and vascular dementia patients, all showed white matter hyperintensity and multiple cavities (Hatanaka et al., 2015). However, dROM levels are increased in vascular dementia but not in AD patients, whereas BAP levels were only reduced in the mixed dementia patient group.

Further findings show that redox-active components in the serum from patients with AD and vascular dementia differ, although they might share a common oxidative stress-related pathogenesis. Vascular dementia and late-onset AD might maintain some distinctive features with individuals with elevated homocysteine and uric acid having a high probability of being affected by vascular dementia, but not those with late-onset AD (Cervellati et al., 2014). These findings are supported by animal studies showing a redox-sensitive component of neurovascular dysfunction in AD (Li et al., 2021) which links to amyloid beta (Aβ)-activated microglia secretion of proinflammatory factors to induce pericyte loss (Jansson et al., 2016; Takata et al., 2011).

Nitric oxide-mediated signaling

An additional potential key component of vascular dementia involves the highly reactive radical NO produced by nitric oxide synthases (NOSs). The NOS enzyme exists in three isoforms: neuronal NOS (nNOS/NOS1), inducible NOS (iNOS/NOS2), and endothelial NOS (eNOS/NOS3). The calcium-independent iNOS is commonly expressed in activated microglia in response to invading microorganisms and chemokines, while calcium-dependent eNOS and nNOS are expressed in different tissues including ECs and neurons, respectively (Hecker et al., 1999; Villanueva & Giulivi, 2010).

NO plays a critical role in maintaining the homeostasis of neurophysiological functions including the cerebrovascular circulation, immune response, neurotransmission, and astrocyte activity. Although physiologically produced NO is neuroprotective, it will exert neurotoxic functions when produced excessively over longer periods following iNOS expression, and subsequent reactions with superoxide anions form peroxynitrite and other reactive nitrogen species. Established evidence shows that an excessive production of NO and the presence of ROS presents a pathological feature that contributes to the onset of several diseases such as Alzheimer's, Huntington's, and Parkinson's disease (Padovan-Neto et al., 2019; Zhang et al., 2006; Heneka & O'Banion, 2007; Malinski, 2007).

Due to their high reactivities, these reactive nitrogen and oxygen species may react with thiols and tyrosine groups within multiple proteins resulting in NO-mediated posttranslational modifications which can modify protein functions leading to potential malfunction or misfolding as in the case of Aβ (Bradley & Steinert, 2016; Kummer et al., 2011; Chen & Liu, 2017; Rinaldi et al., 2021; Sobrevia et al., 2016; Steinert & Amal, 2023; Steinert et al., 2010). The accrual of peroxynitrite, in particular, can lead to nitrergic stress, ultimately damaging lipids, proteins, and DNA (Cyr et al., 2020; Picón-Pagès et al., 2019). Direct evidence for the NO−pericyte interactions comes from a study showing that iNOS activation correlates with disruption of the cerebral microcirculation (Zheng et al., 2023).

Physiologically, NO produced by eNOS or nNOS attenuates L-type Ca^{2+} channel activities suggesting the role of endothelial and neuronal NO/NOS pathway in regulating vascular and neuronal function (Rakhit et al., 2001; Tozer et al., 2012). Vascular smooth muscle cells and microvascular ECs, both exposed to NO, retain vascular function, in which NO contributes to maintaining the homeostasis by acting as a vasodilatory and antiinflammatory signaling molecule. An NO imbalance produces proinflammatory and prothrombotic environments, ultimately leading to vascular impairment, thus affecting the regulation of blood flow (Cyr et al., 2020; Incalza et al., 2018). A general endothelial dysfunction associated with an NO deficiency has been linked to systemic hypertension, pre-eclampsia and diabetes, all of which are risk factors for vascular dementia (Carey et al., 2024; Fleissner & Thum, 2011).

Furthermore, a reduced eNOS expression is associated with an increase in cerebrovascular concentration of Aβ resulting in an increased expression of amyloid precursor protein (APP) and β-site APP-cleaving enzyme 1 (BACE1) (Austin & Katusic, 2020; Austin et al., 2010; Tan et al., 2015) suggesting a reduced bioavailability of NO being a precursor for cerebral amyloid angiopathy in developing AD (Ledo et al., 2021). This link has also been reported in a model of early cognitive dysfunction in a rat model of diabetes (Gonçalves et al., 2022).

Redox stress in the vascular, neuronal, and astrocytic compartments

Smooth muscle and endothelial cells

Vascular smooth muscle cells and ECs are vital for the modulation of vascular tone, supply of nutrients and oxygen to the brain, and removal of harmful metabolites. The vascular endothelium releases both vasoconstrictor and vasodilator substances which include ET-1 and NO, prostacyclin, and endothelium-derived hyperpolarizing factor (EDHF), respectively. EC production of NO provides both vasodilatory and antiatherogenic effects, which can inhibit VMSC proliferation and migration, leukocyte adhesion and migration, and platelet activation (Herouvi et al., 2013). Endothelial-derived, but also neuronally produced NO diffuses into nearby VSMC within the brain to activate soluble guanylyl cyclase (sGC) (Garthwaite, 2010). The activation of sGC then produces cyclic guanosine monophosphate (cGMP) and activates kinases that trigger vascular relaxation (Bender et al., 2013). Aging has been strongly associated with the dysfunction of EC, and subsequently VSMC vasoactive functions, due to the reduction of NO bioavailability (Lourenço et al., 2017a,b, 2018). The important interplay between NO and oxygen radicals is further amplified following the uncoupling of eNOS, resulting from particular high levels of oxidative stress or the lack of availability of the NOS cofactor tetrahydrobiopterin, which as a result leads to the production of superoxide in the vascular ECs (Kinoshita et al., 1997; Kuzkaya et al., 2003; Takaishi et al., 2021). Treatment of ECs with H_2O_2 in vitro disrupts tight junctions, resulting in BBB breakdown at concentrations that do not cause cell death. The associated decrease of NO impedes vascular relaxation as well as induces EC apoptosis, leading to restricted blood flow and disruption of the fine balance of vasoconstrictors and vasodilators (Papapetropoulos et al., 1997).

Neurons

Due to its high-energy requirement but limited energy reserves, the brain is highly dependent on continuous supply of oxygen and glucose which is under activity-dependent control through neurovascular coupling to meet the demand in glucose oxidation within neurons. Despite ongoing research into the underlying mechanisms of neurovascular coupling, much of it remains elusive due to the sheer diversity of potential signaling pathways and different brain regions. The degree of neuronally derived NO strongly depends on NMDAR- and calcium-dependent nNOS activation (Rameau et al., 2007; Sattler et al., 1999). Following overstimulation and limited scavenging capacity, a potential overproduction of NO during high levels of activity or excitotoxicity, oxidative and nitrergic stress may modify neuronal proteins and negatively impact on neuronal function (Girouard et al., 2009; Sattler et al., 1999). This can be aggravated under conditions of neuroinflammation and neurodegeneration whereby iNOS-derived NO further shifts the balance toward enhanced NO toxicity levels (Wijasa et al., 2020; Steinert et al., 2010).

Astrocytes and pericytes

Astrocytes involved in neurovascular coupling ensure optimal energy supply to neurons. In this mechanism, glutamate released by neurons signals to astrocytes *via* metabotropic G-protein-coupled receptor activation (mGluR) which leads to the activation of phospholipase A_2 and promotes the release of arachidonic acid (AA). AA is then converted into active metabolites by cyclooxygenases, lipoxygenases, cytochrome P450 to produce various vasoactive compounds like 20-hydroxyeicosatetraenoic acid (20-HETE), epoxyeicosatrienoic acids (EETs, i.e., 5,6-, 8,9-, 11,12- and 14,15-EET), prostaglandins, thromboxane, and leukotrienes (Dalkara & Alarcon-Martinez, 2015; Gordon et al., 2008; Harder et al., 1998). Importantly, the nonenzymatic metabolism of AA involves ROS production by vascular cells (endothelium, smooth muscle) and perivascular astrocytes. ROS generated by ECs include $O_2^{\cdot-}$, hydrogen peroxide (H_2O_2), and hydroxyl radicals ($^{\cdot}OH$). ROS can induce peroxidation of AA, which gives rise to the isoprostanes. Several scavenger systems are counteracting ROS activities, including enzymes such as superoxide dismutase (SOD), catalase (CAT) and glutathione peroxidase (GPx), and nonenzymatic antioxidants such as vitamin E, vitamin C, β-carotene, and heme-binding proteins (Fridovich, 1997).

The function of pericytes in neurovascular coupling may further be compromised due to their ability to produce significant amounts of mitochondrial ROS under pathological conditions (Mathiisen et al., 2010). Nicotinamide adenine dinucleotide phosphate (NADPH) oxidase, a major superoxide-producing enzyme is highly expressed in brain pericytes (Kuroda et al., 2014; Manea et al., 2005). Indeed, ROS has been shown to cause a sustained increase in intracellular Ca^{2+} levels in cultured human brain microvascular pericytes (Nakamura et al., 2009) and both ROS and NO metabolites, especially peroxynitrite, have been reported to induce pericyte contraction during focal ischemia/reperfusion in the intact mouse brain (Yemisci et al., 2009). Traumatic brain injury (TBI) induces neuroinflammation and oxidative stress and may act as an initiating factor for pathological damage which in turn impairs cerebral microcirculatory function. Microglia are activated following TBI to promote the production of inflammatory factors, upregulate NF-κB and iNOS expression which is linked to pericyte loss (Zheng et al., 2023).

Pericytes dysfunction and loss can be prevented by antioxidant treatments (Mito-TEMPO) which results in a protection of BBB function and improving cognition in a hyperglycemic mouse model (Lin et al., 2023). This reinforces the importance of ROS signalling and pericytes function in regulating the brain microvascular functions. Furthermore, inhibition of TNF-alpha by infliximab which downregulates iNOS expression, attenuated inflammation and oxidative damage and recovered loss of pericytes (Zheng et al., 2023). Additional evidence for ROS involvement in astrocytic dysfunction has been reported in AD mice in which an augmented calcium-induced ROS production was reduced by Tempol application (Li et al., 2021). The ROS-mediated changes could potentially perturb neuron-astrocyte communication, which in turn could result in aberrant astrocytic calcium signals, and thereby affecting astrocyte modulation of local blood flow (Zonta et al., 2003).

The process by which astrocytes are involved in neurovascular coupling is still debated and it is suggested that a pathway may also operate through ATP release into the extracellular space (Iadecola & Gottesman, 2019). This suggests that ATP and adenosine formed from its hydrolysis activate purinergic receptors in VSMC and EC, promoting vasodilator effects (Pelligrino et al., 2011). This has been shown *via* P2Y1 activation in ECs with subsequent production of NO by the endothelial isoform of NOS (Toth et al., 2015). Although a full understanding of the mechanism of the neurovascular coupling between neurons and astrocytes remains elusive and divided, the connection between NVU dysfunction and neurodegenerative diseases such as AD and vascular dementia is well established (Iadecola, 2004, 2017).

Therapeutic targets of NO signaling

Based on the reported interactions between neuroinflammation, neurovascular coupling, and neurodegeneration, it is important to target a neuroinflammatory component early during disease to maintain a functional NVU. Suppressing neuroinflammatory and associated redox signaling has been shown to slowdown several types of neurodegeneration (Askew et al., 2023; Bourgognon et al., 2021; Steinert & Amal, 2023). Evidence from one study in a mouse model shows that inhibition of the colony-stimulating factor-1 receptor (CSF1R), a key regulator of microglial proliferation, alleviates cerebrovascular white matter pathology and cognitive deficits (Askew et al., 2023). Moreover, inhibition of the NLR family pyrin domain containing 3 (NLRP3) inflammasome activation possesses an attractive target for clinical investigation of neuroinflammatory disorders (Harrison et al., 2023). Suppression of neuroinflammation also reduces Iba-1 coverage (marker of activated microglia), lowered the number of reactive microglia, and improved memory in AD mouse model (Balu et al., 2023).

As a consequence of the overwhelming evidence for neuroinflammatory involvement in several neurodegenerative diseases, selective suppression of neuroinflammatory nitrergic and oxidative stress shows beneficial effects in slowing various pathologies (Askew et al., 2023; Bourgognon et al., 2021; Chiu et al., 2023). Conversely, enhancing NO signaling predominantly through the activation of eNOS also shows positive effects on the neurovascular function and alleviates pathologies. Further beneficial effect of stimulating the NO/cMGP pathways has been shown through sensitizing sGC to NO and preclinical studies have established that CNS-penetrant sGC stimulators exert beneficial effects in experimental models of neuroinflammation and neurodegeneration (Correia et al., 2021).

Inhibition of NO signaling

Early evidence suggests that the inhibition of iNOS and NADPH oxidase improves behavior and attenuates redox stress in animal models of hypertension-induced vascular dementia or Parkinson's disease (Broom et al., 2011; Sharma & Singh, 2013). The potentially cytotoxic effects of aberrant NO signaling are established in numerous neurodegenerative diseases. Consequently, several approaches to minimize or block NOS activities have attracted attention for many years (Yang et al., 2015), with treatment strategies proven to be beneficial in various disease conditions (Bourgognon et al., 2021; Broom et al., 2011; Sil et al., 2017). A well-established mechanism of inhibition of NMDARs by memantine reduces excitotoxicity and posttranslational protein modifications mediated by NO in numerous neurodegenerative conditions (Gu et al., 2010; Xia et al., 2010). This protective mechanism is also applicable to neurovascular conditions where an allosteric approach to tune NMDAR activity by nitromemantine may hold therapeutic potential for cerebrovascular disorders (Takahashi et al., 2015). In addition to specific NMDAR-dependent actions, memantine also seems to negatively affect nNOS phosphorylation thereby reducing its activity and alleviating NO-mediated intracerebral hemorrhage outcomes (Chen et al., 2021). Similarly, inhibition of either nNOS or iNOS by troxerutin and cerebroprotein hydrolysate application in mice following middle cerebral artery occlusion (MCAO) shows neuroprotective effects and minimizes the damage to the BBB (Zhào et al., 2021). In the case of rat retinal neurovascular dysfunction as a result of diabetic retinopathy, agmatine, an endogenous decarboxylated L-arginine, provides protective effects predominantly *via* reducing NOS/NO pathways to restore retinal vascular permeability, mostly through antioxidant, antiapoptotic capacity (Abo El Gheit et al., 2021), illustrating the benefits of targeting the cytotoxic nature of nitrergic stress.

Activation of NO signaling

In addition to reducing nitrergic stress, activation of NO signaling has also been reported to have beneficial outcomes in neurovascular deficiencies. Numerous studies have shown that AD patients have reduced CBF and decreased eNOS activity or expression resulting in a markedly atrophic microvascular endothelium (Inoue et al., 2023; Katusic et al., 2009), suggesting that failing endothelial function plays a critical role in the pathogenesis of AD. These findings suggest that enhancing eNOS activity to support the structural and functional integrity of the microvascular endothelium may increase cerebral perfusion and improve cognition in AD.

Simvastatin, a 3-hydroxy-3-methylglutaryl coenzyme A (HMG-CoA) reductase synthesis inhibitor, dramatically restores cerebrovascular function by increasing eNOS activity through activation of the PI3K/Akt pathway resulting in greater bioavailability of NO (Tong & Hamel, 2015). Similarly, when AD or MCI patients with mild disease traits were treated sequentially with simvastatin (weeks 0–16), L-arginine (weeks 4–16), and tetrahydrobiopterin (weeks 8–16), they showed a significant increase in CBF at week eight in the limbic and in the cerebral cortex (Degrush et al., 2022). Additional animal studies revealed that supplementation with NO donors (sodium nitroprusside) can alleviate high glucose-induced dysfunction of vascular components of the NVU (Chhabria et al., 2019). In an *in vivo* mouse model of ischemia (MCAO), activation of endothelial NO production following inhibition of serine racemase also alleviates symptoms of stroke illustrating the beneficial effects of eNOS-derived NO within the NVU (Watanabe et al., 2016). Further support for benefits of boosting eNOS-produced NO in an MCAO model has been shown following coapplication of troxerutin, naturally occurring flavonoid, and cerebroprotein hydrolysate (Zhào et al., 2021). The combination of both compounds resulted in reduced microglia activation and peroxynitrite formation while enhancing endothelial NO production.

Conclusions

Building on the current knowledge with focus on the redox-sensitive changes within the NVU, it emerges that neuroinflammatory signaling has a profound effect on the functioning of neurovascular coupling. The origin of enhanced neuroinflammatory activity stems from underlying neurodegenerative diseases associated with misfolded protein expression, systemic vascular effects and injury like stroke or TBI, respectively, and diabetes with a confounding contribution within an aging tissue (see reviews (Lourenço et al., 2017b; Garry et al., 2015; Gonçalves et al., 2022)). The general theme that enhanced levels of reactive oxygen and nitrogen species compromise several functions of the NVU components like pericytes, endothelial and smooth muscle cells, and neurons has been shown in preclinical and human studies (Lourenço & Laranjinha, 2021; Lourenço et al., 2017a). Based on these findings, it emerges that reducing neuroinflammatory ROS levels and iNOS-derived NO has beneficial effects on the NVU leading to improved vascular and cognitive function. Conversely, endothelial-derived NO, as a result of eNOS activity, is fundamentally important for a physiological NVU function and suppression of eNOS renders ECs and vascular responses dysfunctional. Therefore, reinstating an endothelial NO production has been shown to improve vascular and brain cognitive function. The balance between an overproduction of reactive nitrogen and oxygen species and cellular scavenging capacities *versus* physiological production of above reactive species determines the functionality of the NVU and ultimately neuronal and cognitive functions.

Acknowledgments

This work was supported by the University of Nottingham, UK, and the BBSRC (JC).

No conflict statement

No potential conflicts of interests were disclosed.

References

Abo El Gheit, R. E., Soliman, N. A., Badawi, G. A., Madi, N. M., El-Saka, M. H., Badr, S. M., & Emam, M. N. (2021). Retinoprotective effect of agmatine in streptozotocin-induced diabetic rat model: Avenues for vascular and neuronal protection: Agmatine in diabetic retinopathy. *Journal of Physiology and Biochemistry, 77*(2), 305–320. https://doi.org/10.1007/s13105-021-00799-9

Agnati, L. F., Guidolin, D., Guescini, M., Genedani, S., & Fuxe, K. (2010). Understanding wiring and volume transmission. *Brain Research Reviews, 64*(1), 137–159. https://doi.org/10.1016/j.brainresrev.2010.03.003

Aguilera, G., Colín-González, A. L., Rangel-López, E., Chavarría, A., & Santamaría, A. (2018). Redox signaling, neuroinflammation, and neurodegeneration. *Antioxidants and Redox Signaling, 28*(18), 1626–1651. https://doi.org/10.1089/ars.2017.7099

Allen, C. L., & Bayraktutan, U. (2009). Oxidative stress and its role in the pathogenesis of ischaemic stroke. *International Journal of Stroke, 4*(6), 461–470. https://doi.org/10.1111/j.1747-4949.2009.00387.x

Askew, K. E., Beverley, J., Sigfridsson, E., Szymkowiak, S., Emelianova, K., Dando, O., Hardingham, G. E., Duncombe, J., Hennessy, E., Koudelka, J., Samarasekera, N., Salman, R. A. S., Smith, C., Tavares, A. A. S., Gomez-Nicola, D., Kalaria, R. N., McColl, B. W., & Horsburgh, K. (2023). Inhibiting CSF1R alleviates cerebrovascular white matter disease and cognitive impairment. *Glia*. https://doi.org/10.1002/glia.24481

Austin, S. A., & Katusic, Z. S. (2020). Partial loss of endothelial nitric oxide leads to increased cerebrovascular beta amyloid. *Journal of Cerebral Blood Flow and Metabolism, 40*(2), 392–403. https://doi.org/10.1177/0271678X18822474

Austin, S. A., Santhanam, A. V., & Katusic, Z. S. (2010). Endothelial nitric oxide modulates expression and processing of amyloid precursor protein. *Circulation Research, 107*(12), 1498–1502. https://doi.org/10.1161/circresaha.110.233080

Balu, D., Valencia-Olvera, A. C., Nguyen, A., Patnam, M., York, J., Peri, F., Neumann, F., LaDu, M. J., & Tai, L. M. (2023). A small-molecule TLR4 antagonist reduced neuroinflammation in female E4FAD mice. *Alzheimer's Research and Therapy, 15*(1). https://doi.org/10.1186/s13195-023-01330-6

Barnham, K. J., Masters, C. L., & Bush, A. I. (2004). Neurodegenerative diseases and oxidatives stress. *Nature Reviews Drug Discovery, 3*(3), 205–214. https://doi.org/10.1038/nrd1330

Bazinet, R. P., & Layé, S. (2014). Polyunsaturated fatty acids and their metabolites in brain function and disease. *Nature Reviews Neuroscience, 15*(12), 771–785. https://doi.org/10.1038/nrn3820

Bender, S. B., McGraw, A. P., Jaffe, I. Z., & Sowers, J. R. (2013). Mineralocorticoid receptor-mediated vascular insulin resistance: An early contributor to diabetes-related vascular disease? *Diabetes, 62*(2), 313–319. https://doi.org/10.2337/db12-0905

Bennett, S., Grant, M. M., & Aldred, S. (2009). Oxidative stress in vascular dementia and Alzheimer's disease: A common pathology. *Journal of Alzheimer's Disease, 17*(2), 245–257. https://doi.org/10.3233/JAD-2009-1041

Bourgognon, J. M., Spiers, J. G., Robinson, S. W., Scheiblich, H., Glynn, P., Ortori, C., Bradley, S. J., Tobin, A. B., & Steinert, J. R. (2021). Inhibition of neuroinflammatory nitric oxide signaling suppresses glycation and prevents neuronal dysfunction in mouse prion disease. *Proceedings of the National Academy of Sciences of the United States of America, 118*(10). https://doi.org/10.1073/pnas.2009579118

Bradley, S. A., & Steinert, J. R. (2016). Nitric oxide-mediated posttranslational modifications: Impacts at the synapse. *Oxidative Medicine and Cellular Longevity, 2016*. https://doi.org/10.1155/2016/5681036

Broom, L., Marinova-Mutafchieva, L., Sadeghian, M., Davis, J. B., Medhurst, A. D., & Dexter, D. T. (2011). Neuroprotection by the selective iNOS inhibitor GW274150 in a model of Parkinson disease. *Free Radical Biology and Medicine, 50*(5), 633–640. https://doi.org/10.1016/j.freeradbiomed.2010.12.026

Burns, S., Selman, A., Sehar, U., Rawat, P., Reddy, A. P., & Reddy, P. H. (2022). Therapeutics of Alzheimer's disease: Recent developments. *Antioxidants*, (12).

Carey, C., Mulcahy, E., McCarthy, F. P., Jennings, E., Kublickiene, K., et al. (2024). Hypertensive disorders of pregnancy and the risk of maternal dementia: a systematic review and meta-analysis. *American Journal of Obstetrics and Gynecology*, 196–210. https://doi.org/10.1016/j.ajog.2024.01.013

Cervellati, C., Romani, A., Seripa, D., Cremonini, E., Bosi, C., Magon, S., Passaro, A., Bergamini, C. M., Pilotto, A., & Zuliani, G. (2014). Oxidative balance, homocysteine, and uric acid levels in older patients with late onset Alzheimer's Disease or vascular dementia. *Journal of the Neurological Sciences, 337*(1–2), 156–161. https://doi.org/10.1016/j.jns.2013.11.041

Chen, L., & Liu, B. (2017). Relationships between stress granules, oxidative stress, and neurodegenerative diseases. *Oxidative Medicine and Cellular Longevity, 2017*, 1–10. https://doi.org/10.1155/2017/1809592

Chen, X., Xiang, X., Xie, T., Chen, Z., Mou, Y., Gao, Z., Xie, X., Song, M., Huang, H., Gao, Z., & Chen, M. (2021). Memantine protects blood–brain barrier integrity and attenuates neurological deficits through inhibiting nitric oxide synthase ser1412 phosphorylation in intracerebral hemorrhage rats: Involvement of peroxynitrite-related matrix metalloproteinase-9/NLRP3 inflammasome activation. *NeuroReport, 32*(3), 228–237. https://doi.org/10.1097/wnr.0000000000001577

Chhabria, K., Vouros, A., Gray, C., MacDonald, R. B., Jiang, Z., Wilkinson, R. N., Plant, K., Vasilaki, E., Howarth, C., & Chico, T. J. A. (2019). Sodium nitroprusside prevents the detrimental effects of glucose on the neurovascular unit and behaviour in zebrafish. *Disease Models and Mechanisms, 12*(9). https://doi.org/10.1242/dmm.039867

Chiu, Y. J., Lin, C. H., Lin, C. Y., Yang, P. N., Lo, Y. S., Chen, Y. C., Chen, C. M., Wu, Y. R., Yao, C. F., Chang, K. H., & Lee-Chen, G. J. (2023). Investigating therapeutic effects of indole derivatives targeting inflammation and oxidative stress in neurotoxin-induced cell and mouse models of Parkinson's disease. *International Journal of Molecular Sciences, 24*(3). https://doi.org/10.3390/ijms24032642

Correia, S. S., Liu, G., Jacobson, S., Bernier, S. G., Tobin, J. V., Schwartzkopf, C. D., Atwater, E., Lonie, E., Rivers, S., Carvalho, A., Germano, P., Tang, K., Iyengar, R. R., Currie, M. G., Hadcock, J. R., Winrow, C. J., & Jones, J. E. (2021). The CNS-penetrant soluble guanylate cyclase stimulator CYR119 attenuates markers of inflammation in the central nervous system. *Journal of Neuroinflammation, 18*(1). https://doi.org/10.1186/s12974-021-02275-z

Cyr, A. R., Huckaby, L. V., Shiva, S. S., & Zuckerbraun, B. S. (2020). Nitric oxide and endothelial dysfunction. *Critical Care Clinics, 36*(2), 307–321. https://doi.org/10.1016/j.ccc.2019.12.009

Dalkara, T., & Alarcon-Martinez, L. (2015). Cerebral microvascular pericytes and neurogliovascular signaling in health and disease. *Brain Research, 1623*, 3–17. https://doi.org/10.1016/j.brainres.2015.03.047

Degrush, E., Shazeeb, M. S., Drachman, D., Vardar, Z., Lindsay, C., Gounis, M. J., & Henninger, N. (2022). Cumulative effect of simvastatin, l-arginine, and tetrahydrobiopterin on cerebral blood flow and cognitive function in Alzheimer's disease. *Alzheimer's Research and Therapy, 14*(1). https://doi.org/10.1186/s13195-022-01076-7

Ding, R., Hase, Y., Ameen-Ali, K. E., Ndung'u, M., Stevenson, W., Barsby, J., Gourlay, R., Akinyemi, T., Akinyemi, R., Uemura, M. T., Polvikoski, T., Mukaetova-Ladinska, E., Ihara, M., & Kalaria, R. N. (2020). Loss of capillary pericytes and the blood−brain barrier in white matter in poststroke and vascular dementias and Alzheimer's disease. *Brain Pathology, 30*(6), 1087−1101. https://doi.org/10.1111/bpa.12888

Fabiani, M., Gordon, B. A., Maclin, E. L., Pearson, M. A., Brumback-Peltz, C. R., Low, K. A., McAuley, E., Sutton, B. P., Kramer, A. F., & Gratton, G. (2014). Neurovascular coupling in normal aging: A combined optical, erp and fMRI study. *NeuroImage, 85*, 592−607. https://doi.org/10.1016/j.neuroimage.2013.04.113

Fleissner, F., & Thum, T. (2011). Critical role of the nitric oxide/reactive oxygen species balance in endothelial progenitor dysfunction. *Antioxidants and Redox Signaling, 15*(4), 933−948. https://doi.org/10.1089/ars.2010.3502

Fridovich, I. (1997). Superoxide anion radical ($O^{-}\cdot^{2}$), superoxide dismutases, and related matters. *Journal of Biological Chemistry, 272*(30), 18515−18517. https://doi.org/10.1074/jbc.272.30.18515

Garry, P. S., Ezra, M., Rowland, M. J., Westbrook, J., & Pattinson, K. T. S. (2015). The role of the nitric oxide pathway in brain injury and its treatment — from bench to bedside. *Experimental Neurology, 263*, 235−243. https://doi.org/10.1016/j.expneurol.2014.10.017

Garthwaite, J. (2008). Concepts of neural nitric oxide-mediated transmission. *European Journal of Neuroscience, 27*(11), 2783−2802. https://doi.org/10.1111/j.1460-9568.2008.06285.x

Garthwaite, J. (2010). New insight into the functioning of nitric oxide-receptive guanylyl cyclase: Physiological and pharmacological implications. *Molecular and Cellular Biochemistry, 334*(1−2), 221−232. https://doi.org/10.1007/s11010-009-0318-8

Girouard, H., Wang, G., Gallo, E. F., Anrather, J., Zhou, P., Pickel, V. M., & Iadecola, C. (2009). NMDA receptor activation increases free radical production through Nitric Oxide and NOX2. *Journal of Neuroscience, 29*(8), 2545−2552. https://doi.org/10.1523/JNEUROSCI.0133-09.2009

Gonçalves, J. S., Seiça, R. M., Laranjinha, J., & Lourenço, C. F. (2022). Impairment of neurovascular coupling in the hippocampus due to decreased nitric oxide bioavailability supports early cognitive dysfunction in type 2 diabetic rats. *Free Radical Biology and Medicine, 193*, 669−675. https://doi.org/10.1016/j.freeradbiomed.2022.11.009

Gordon, G. R. J., Choi, H. B., Rungta, R. L., Ellis-Davies, G. C. R., & MacVicar, B. A. (2008). Brain metabolism dictates the polarity of astrocyte control over arterioles. *Nature, 456*(7223), 745−750. https://doi.org/10.1038/nature07525

Gu, Z., Nakamura, T., & Lipton, S. A. (2010). Redox reactions induced by nitrosative stress mediate protein misfolding and mitochondrial dysfunction in neurodegenerative diseases. *Molecular Neurobiology, 41*(2−3), 55−72. https://doi.org/10.1007/s12035-010-8113-9

Hébert, R., & Brayne, C. (2004). Epidemiology of vascular dementia. *Neuroepidemiology, 14*(5), 240−257. https://doi.org/10.1159/000109800

Halliwell, B. (2006). Oxidative stress and neurodegeneration: Where are we now? *Journal of Neurochemistry, 97*(6), 1634−1658. https://doi.org/10.1111/j.1471-4159.2006.03907.x

Halliwell, B. (1992). Reactive oxygen species and the central nervous system. *Journal of Neurochemistry, 59*(5), 1609−1623. https://doi.org/10.1111/j.1471-4159.1992.tb10990.x

Hamel, E., Nicolakakis, N., Aboulkassim, T., Ongali, B., & Tong, X.-K. (2008). Oxidative stress and cerebrovascular dysfunction in mouse models of Alzheimer's disease. *Experimental Physiology, 93*(1), 116−120. https://doi.org/10.1113/expphysiol.2007.038729

Harder, D. R., Alkayed, N. J., Lange, A. R., Gebremedhin, D., & Roman, R. J. (1998). Functional hyperemia in the brain: Hypothesis for astrocyte-derived vasodilator metabolites. *Stroke, 29*(1), 229−234. https://doi.org/10.1161/01.STR.29.1.229

Harrison, D., Billinton, A., Bock, M. G., Doedens, J. R., Gabel, C. A., Holloway, M. K., Porter, R. A., Reader, V., Scanlon, J., Schooley, K., & Watt, A. P. (2023). Discovery of clinical candidate NT-0796, a brain-penetrant and highly potent NLRP3 inflammasome inhibitor for neuroinflammatory disorders. *Journal of Medicinal Chemistry, 66*(21), 14897−14911. https://doi.org/10.1021/acs.jmedchem.3c01398

Haruwaka, K., Ikegami, A., Tachibana, Y., Ohno, N., Konishi, H., Hashimoto, A., Matsumoto, M., Kato, D., Ono, R., Kiyama, H., Moorhouse, A. J., Nabekura, J., & Wake, H. (2019). Dual microglia effects on blood brain barrier permeability induced by systemic inflammation. *Nature Communications, 10*(1). https://doi.org/10.1038/s41467-019-13812-z

Hatanaka, H., Hanyu, H., Fukasawa, R., Hirao, K., Shimizu, S., Kanetaka, H., & Iwamoto, T. (2015). Differences in peripheral oxidative stress markers in A lzheimer's disease, vascular dementia and mixed dementia patients. *Geriatrics and Gerontology International, 15*(S1), 53–58. https://doi.org/10.1111/ggi.12659

Hecker, M., Cattaruzza, M., & Wagner, A. H. (1999). Regulation of inducible nitric oxide synthase gene expression in vascular smooth muscle cells. *General Pharmacology: The Vascular System, 32*(1), 9–16. https://doi.org/10.1016/s0306-3623(98)00082-2

Heneka, M. T., & O'Banion, M. K. (2007). Inflammatory processes in Alzheimer's disease. *Journal of Neuroimmunology*, 69–91. https://doi.org/10.1016/j.jneuroim.2006.11.017

Herouvi, D., Karanasios, E., Karayianni, C., & Karavanaki, K. (2013). Cardiovascular disease in childhood: The role of obesity. *European Journal of Pediatrics, 172*(6), 721–732. https://doi.org/10.1007/s00431-013-1932-8

Iadecola, C., & Gottesman, R. F. (2019). Neurovascular and cognitive dysfunction in hypertension. *Circulation Research, 124*(7), 1025–1044. https://doi.org/10.1161/circresaha.118.313260

Iadecola, C. (2004). Neurovascular regulation in the normal brain and in Alzheimer's disease. *Nature Reviews Neuroscience, 5*(5), 347–360. https://doi.org/10.1038/nrn1387

Iadecola, C. (2017). The neurovascular unit coming of age: A journey through neurovascular coupling in health and disease. *Neuron, 96*(1), 17–42. https://doi.org/10.1016/j.neuron.2017.07.030

Incalza, M. A., D'Oria, R., Natalicchio, A., Perrini, S., Laviola, L., & Giorgino, F. (2018). Oxidative stress and reactive oxygen species in endothelial dysfunction associated with cardiovascular and metabolic diseases. *Vascular Pharmacology, 100*, 1–19. https://doi.org/10.1016/j.vph.2017.05.005

Inoue, Y., Shue, F., Bu, G., & Kanekiyo, T. (2023). Pathophysiology and probable etiology of cerebral small vessel disease in vascular dementia and Alzheimer's disease. *Molecular Neurodegeneration, 18*(1). https://doi.org/10.1186/s13024-023-00640-5

Jansson, D., Scotter, E. L., Rustenhoven, J., Coppieters, N., Smyth, L. C., Oldfield, R. L., et al. (2016). Interferon-gamma blocks signalling through PDGFRbeta in human brain pericytes. *Journal of Neuroinflammation, 13*(1).

Kalaria, R. N. (2016). Neuropathological diagnosis of vascular cognitive impairment and vascular dementia with implications for Alzheimer's disease. *Acta Neuropathologica, 131*(5), 659–685. https://doi.org/10.1007/s00401-016-1571-z

Katusic, Z. S., d'Uscio, L. V., & Nath, K. A. (2009). Vascular protection by tetrahydrobiopterin: Progress and therapeutic prospects. *Trends in Pharmacological Sciences, 30*(1), 48–54. https://doi.org/10.1016/j.tips.2008.10.003

Kinoshita, H., Milstien, S., Wambi, C., & Katusic, Z. S. (1997). Inhibition of tetrahydrobiopterin biosynthesis impairs endothelium-dependent relaxations in canine basilar artery. *American Journal of Physiology - Heart and Circulatory Physiology, 273*(2), H718–H724. https://doi.org/10.1152/ajpheart.1997.273.2.h718

Kummer, M. P., Hermes, M., Delekarte, A., Hammerschmidt, T., Kumar, S., Terwel, D., Walter, J., Pape, H. C., König, S., Roeber, S., Jessen, F., Klockgether, T., Korte, M., & Heneka, M. T. (2011). Nitration of tyrosine 10 critically enhances amyloid β aggregation and plaque formation. *Neuron, 71*(5), 833–844. https://doi.org/10.1016/j.neuron.2011.07.001

Kuroda, J., Ago, T., Nishimura, A., Nakamura, K., Matsuo, R., Wakisaka, Y., Kamouchi, M., & Kitazono, T. (2014). Nox4 is a major source of superoxide production in human brain pericytes. *Journal of Vascular Research, 51*(6), 429–438. https://doi.org/10.1159/000369930

Kuzkaya, N., Weissmann, N., Harrison, D. G., & Dikalov, S. (2003). Interactions of peroxynitrite, tetrahydrobiopterin, ascorbic acid, and thiols: Implications for uncoupling endothelial nitric-oxide synthase. *Journal of Biological Chemistry, 278*(25), 22546–22554. https://doi.org/10.1074/jbc.M302227200

Langa, K. M., Foster, N. L., & Larson, E. B. (2004). Mixed dementia: Emerging concepts and therapeutic implications. *JAMA, 292*(23), 2901–2908. https://doi.org/10.1001/jama.292.23.2901

Ledo, A., Lourenço, C. F., Cadenas, E., Barbosa, R. M., & Laranjinha, J. (2021). The bioactivity of neuronal-derived nitric oxide in aging and neurodegeneration: Switching signaling to degeneration. *Free Radical Biology and Medicine, 162*, 500–513. https://doi.org/10.1016/j.freeradbiomed.2020.11.005

Li, L., Tong, X. K., Hosseini Kahnouei, M., Vallerand, D., Hamel, E., & Girouard, H. (2021). Impaired hippocampal neurovascular coupling in a mouse model of Alzheimer's disease. *Frontiers in Physiology, 12*. https://doi.org/10.3389/fphys.2021.715446

Li, W., Niu, X., Dai, Y., Wu, X., Li, J., & Sheng, W. (2023). Rnf-213 knockout induces pericyte reduction and blood-brain barrier impairment in mouse. *Molecular Neurobiology, 60*(11), 6188–6200. https://doi.org/10.1007/s12035-023-03480-y

Lin, L., Chen, Z., Huang, C., Wu, Y., Huang, L., Wang, L., Ke, S., & Liu, L. (2023). Mito-TEMPO, a mitochondria-targeted antioxidant, improves cognitive dysfunction due to hypoglycemia: An association with reduced pericyte loss and blood-brain barrier leakage. *Molecular Neurobiology, 60*(2), 672–686. https://doi.org/10.1007/s12035-022-03101-0

Lourenço, C. F., & Laranjinha, J. (2021). Nitric oxide pathways in neurovascular coupling under normal and stress conditions in the brain: Strategies to rescue aberrant coupling and improve cerebral blood flow. *Frontiers in Physiology, 12*. https://doi.org/10.3389/fphys.2021.729201

Lourenço, C. F., Ledo, A., Barbosa, R. M., & Laranjinha, J. (2017a). Neurovascular uncoupling in the triple transgenic model of Alzheimer's disease: Impaired cerebral blood flow response to neuronal-derived nitric oxide signaling. *Experimental Neurology, 291*, 36–43. https://doi.org/10.1016/j.expneurol.2017.01.013

Lourenço, C. F., Ledo, A., Barbosa, R. M., & Laranjinha, J. (2017b). Neurovascular-neuroenergetic coupling axis in the brain: Master regulation by nitric oxide and consequences in aging and neurodegeneration. *Free Radical Biology and Medicine, 108*, 668–682. https://doi.org/10.1016/j.freeradbiomed.2017.04.026

Lourenço, C. F., Ledo, A., Caetano, M., Barbosa, R. M., & Laranjinha, J. (2018). Age-dependent impairment of neurovascular and neurometabolic coupling in the hippocampus. *Frontiers in Physiology, 9*. https://doi.org/10.3389/fphys.2018.00913

Luca, M., Luca, A., & Calandra, C. (2015). The role of oxidative damage in the pathogenesis and progression of Alzheimer's disease and vascular dementia. *Oxidative Medicine and Cellular Longevity, 2015*, 1–8. https://doi.org/10.1155/2015/504678

Malinski, T. (2007). Nitric oxide and nitroxidative stress in Alzheimer's disease. *Journal of Alzheimer's Disease*, 207–218. https://doi.org/10.3233/jad-2007-11208

Manea, A., Raicu, M., & Simionescu, M. (2005). Expression of functionally phagocyte-type NAD(P)H oxidase in pericytes: Effect of angiotensin II and high glucose. *Biology of the Cell, 97*(9), 723–734. https://doi.org/10.1042/bc20040107

Martins-Filho, R. K., Zotin, M. C., Rodrigues, G., & Pontes-Neto, O. (2021). Biomarkers related to endothelial dysfunction and vascular cognitive impairment: A systematic review. *Dementia and Geriatric Cognitive Disorders, 49*(4), 365–374. https://doi.org/10.1159/000510053

Mathiisen, T. M., Lehre, K. P., Danbolt, N. C., & Ottersen, O. P. (2010). The perivascular astroglial sheath provides a complete covering of the brain microvessels: An electron microscopic 3D reconstruction. *Glia, 58*(9), 1094–1103. https://doi.org/10.1002/glia.20990

Mehrabadi, A. R., Korolainen, M. A., Odero, G., Miller, D. W., & Kauppinen, T. M. (2017). Poly(ADP-ribose) polymerase-1 regulates microglia mediated decrease of endothelial tight junction integrity. *Neurochemistry International, 108*, 266–271. https://doi.org/10.1016/j.neuint.2017.04.014

Montagne, A., Barnes, S. R., Sweeney, M. D., Halliday, M. R., Sagare, A. P., Zhao, Z., Toga, A. W., Jacobs, R. E., Liu, C. Y., Amezcua, L., Harrington, M. G., Chui, H. C., Law, M., & Zlokovic, B. V. (2015). Blood-Brain barrier breakdown in the aging human hippocampus. *Neuron, 85*(2), 296–302. https://doi.org/10.1016/j.neuron.2014.12.032

Nakamura, K., Kamouchi, M., Kitazono, T., Kuroda, J., Shono, Y., Hagiwara, N., Ago, T., Ooboshi, H., Ibayashi, S., & Iida, M. (2009). Amiloride inhibits hydrogen peroxide-induced Ca2+ responses in human CNS pericytes. *Microvascular Research, 77*(3), 327–334. https://doi.org/10.1016/j.mvr.2008.12.001

Nordengen, K., Kirsebom, B. E., Henjum, K., Selnes, P., Gísladóttir, B., Wettergreen, M., Torsetnes, S. B., Grøntvedt, G. R., Waterloo, K. K., Aarsland, D., Nilsson, L. N. G., & Fladby, T. (2019). Glial activation and inflammation along the Alzheimer's disease continuum. *Journal of Neuroinflammation, 16*(1). https://doi.org/10.1186/s12974-019-1399-2

Nortley, R., Korte, N., Izquierdo, P., Hirunpattarasilp, C., Mishra, A., Jaunmuktane, Z., & et al.. (2019). Amyloid beta oligomers constrict human capillaries in Alzheimer's disease via signaling to pericytes. *Science, 365*(6450).

Padovan-Neto, F. E., Jurkowski, L., Murray, C., Stutzmann, G. E., Kwan, M., et al. (2019). Age- and sex-related changes in cortical and striatal nitric oxide synthase in the Q175 mouse model of Huntington's disease. *Nitric Oxide*, 40–50. https://doi.org/10.1016/j.niox.2018.12.002

Papapetropoulos, A., García-Cardeña, G., Madri, J. A., & Sessa, W. C. (1997). Nitric oxide production contributes to the angiogenic properties of vascular endothelial growth factor in human endothelial cells. *Journal of Clinical Investigation, 100*(12), 3131–3139. https://doi.org/10.1172/JCI119868

Parodi-Rullán, R. M., Javadov, S., & Fossati, S. (2021). Dissecting the crosstalk between endothelial mitochondrial damage, vascular inflammation, and neurodegeneration in cerebral amyloid angiopathy and Alzheimer's disease. *Cells, 10*(11). https://doi.org/10.3390/cells10112903

Pelligrino, D. A., Vetri, F., & Xu, H. L. (2011). Purinergic mechanisms in gliovascular coupling. *Seminars in Cell and Developmental Biology, 22*(2), 229–236. https://doi.org/10.1016/j.semcdb.2011.02.010

Picón-Pagès, P., Garcia-Buendia, J., & Muñoz, F. J. (2019). Functions and dysfunctions of nitric oxide in brain. *Biochimica et Biophysica Acta - Molecular Basis of Disease, 1865*(8), 1949–1967. https://doi.org/10.1016/j.bbadis.2018.11.007

Picca, A., Calvani, R., Coelho-Junior, H. J., Landi, F., Bernabei, R., & Marzetti, E. (2020). Mitochondrial dysfunction, oxidative stress, and neuroinflammation: Intertwined roads to neurodegeneration. *Antioxidants, 9*(8). https://doi.org/10.3390/antiox9080647

Rakhit, A., Maguire, C. T., Wakimoto, H., Gehrmann, J., Li, G. K., Kelly, R. A., Michel, T., & Berul, C. I. (2001). In vivo electrophysiologic studies in endothelial nitric oxide synthase (eNOS)-deficient mice. *Journal of Cardiovascular Electrophysiology, 12*(11), 1295–1301. https://doi.org/10.1046/j.1540-8167.2001.01295.x

Rameau, G. A., Tukey, D. S., Garcin-Hosfield, E. D., Titcombe, R. F., Misra, C., Khatri, L., Getzoff, E. D., & Ziff, E. B. (2007). Biphasic coupling of neuronal nitric oxide synthase phosphorylation to the NMDA receptor regulates AMPA receptor trafficking and neuronal cell death. *Journal of Neuroscience, 27*(13), 3445–3455. https://doi.org/10.1523/JNEUROSCI.4799-06.2007

Rinaldi, C., Donato, L., Alibrandi, S., Scimone, C., D'Angelo, R., & Sidoti, A. (2021). Oxidative stress and the neurovascular unit. *Life, 11*(8). https://doi.org/10.3390/life11080767

Ritchie, K., Artero, S., & Touchon, J. (2001). Classification criteria for mild cognitive impairment: A population-based validation study. *Neurology, 56*(1), 37–42. https://doi.org/10.1212/WNL.56.1.37

Román, G. C., Erkinjuntti, T., Wallin, A., Pantoni, L., & Chui, H. C. (2002). Subcortical ischaemic vascular dementia. *The Lancet Neurology, 1*(7), 426–436. https://doi.org/10.1016/s1474-4422(02)00190-4

Santos, R. M., Lourenço, C. F., Pomerleau, F., Huettl, P., Gerhardt, G. A., Laranjinha, J., & Barbosa, R. M. (2011). Brain nitric oxide inactivation is governed by the vasculature. *Antioxidants and Redox Signaling, 14*(6), 1011–1021. https://doi.org/10.1089/ars.2010.3297

Sattler, R., Xiong, Z., Lu, W. Y., Hafner, M., MacDonald, J. F., & Tymianski, M. (1999). Specific coupling of NMDA receptor activation to nitric oxide neurotoxicity by PSD-95 protein. *Science, 284*(5421), 1845–1848. https://doi.org/10.1126/science.284.5421.1845

Sengillo, J. D., Winkler, E. A., Walker, C. T., Sullivan, J. S., Johnson, M., & Zlokovic, B. V. (2013). Deficiency in mural vascular cells coincides with blood-brain barrier disruption in Alzheimer's disease. *Brain Pathology, 23*(3), 303–310. https://doi.org/10.1111/bpa.12004

Sharma, B., & Singh, N. (2013). Pharmacological inhibition of inducible nitric oxide synthase (iNOS) and nicotinamide adenine dinucleotide phosphate (NADPH) oxidase, convalesce behavior and biochemistry of hypertension induced vascular dementia in rats. *Pharmacology Biochemistry and Behavior, 103*(4), 821–830. https://doi.org/10.1016/j.pbb.2012.11.011

Sil, S., Ghosh, T., Ghosh, R., & Gupta, P. (2017). Nitric oxide synthase inhibitor, aminoguanidine reduces intracerebroventricular colchicine induced neurodegeneration, memory impairments and changes of systemic immune responses in rats. *Journal of Neuroimmunology, 303*, 51–61. https://doi.org/10.1016/j.jneuroim.2016.12.007

Skrobot, O. A., O'Brien, J., Black, S., Chen, C., DeCarli, C., Erkinjuntti, T., Ford, G. A., Kalaria, R. N., Pantoni, L., Pasquier, F., Roman, G. C., Wallin, A., Sachdev, P., Skoog, I., Taragano, F. E., Kril, J., Cavalieri, M., Jellinger, K. A., Kovacs, G. G., ... Kehoe, P. G. (2017). The vascular impairment of cognition classification consensus study. *Alzheimer's and Dementia, 13*(6), 624–633. https://doi.org/10.1016/j.jalz.2016.10.007

Sobrevia, L., Ooi, L., Ryan, S., & Steinert, J. R. (2016). Nitric oxide: A regulator of cellular function in health and disease. *Oxidative Medicine and Cellular Longevity, 2016*. https://doi.org/10.1155/2016/9782346

Steinert, J. R., & Amal, H. (2023). The contribution of an imbalanced redox signalling to neurological and neurodegenerative conditions. *Free Radical Biology and Medicine, 194*, 71–83. https://doi.org/10.1016/j.freeradbiomed.2022.11.035

Steinert, J. R., Chernova, T., & Forsythe, I. D. (2010). Nitric oxide signaling in brain function, dysfunction, and dementia. *The Neuroscientist, 16*(4), 435–452. https://doi.org/10.1177/1073858410366481

Steinert, J. R., Kopp-Scheinpflug, C., Baker, C., Challiss, R. A. J., Mistry, R., Haustein, M. D., Griffin, S. J., Tong, H., Graham, B. P., & Forsythe, I. D. (2008). Nitric oxide is a volume transmitter regulating postsynaptic excitability at a glutamatergic synapse. *Neuron, 60*(4), 642–656. https://doi.org/10.1016/j.neuron.2008.08.025

Sweeney, M. D., Sagare, A. P., & Zlokovic, B. V. (2018). Blood-brain barrier breakdown in Alzheimer disease and other neurodegenerative disorders. *Nature Reviews Neurology, 14*(3), 133–150. https://doi.org/10.1038/nrneurol.2017.188

Takahashi, H., Xia, P., Cui, J., Talantova, M., Bodhinathan, K., Li, W., Saleem, S., Holland, E. A., Tong, G., Piña-Crespo, J., Zhang, D., Nakanishi, N., Larrick, J. W., McKercher, S. R., Nakamura, T., Wang, Y., & Lipton, S. A. (2015). Pharmacologically targeted NMDA receptor antagonism by NitroMemantine for cerebrovascular disease. *Scientific Reports, 5*(1). https://doi.org/10.1038/srep14781

Takaishi, K., Kinoshita, H., Kawashima, S., & Kawahito, S. (2021). Human vascular smooth muscle function and oxidative stress induced by NADPH oxidase with the clinical implications. *Cells, 10*(8). https://doi.org/10.3390/cells10081947

Takata, F., Dohgu, S., Matsumoto, J., Takahashi, H., Machida, T., Wakigawa, T., Harada, E., Miyaji, H., Koga, M., Nishioku, T., Yamauchi, A., & Kataoka, Y. (2011). Brain pericytes among cells constituting the blood-brain barrier are highly sensitive to tumor necrosis factor-α, releasing matrix metalloproteinase-9 and migrating in vitro. *Journal of Neuroinflammation, 8*(1). https://doi.org/10.1186/1742-2094-8-106

Tan, X. L., Xue, Y. Q., Ma, T., Wang, X., Li, J. J., Lan, L., Malik, K. U., McDonald, M. P., Dopico, A. M., & Liao, F. F. (2015). Partial eNOS deficiency causes spontaneous thrombotic cerebral infarction, amyloid angiopathy and cognitive impairment. *Molecular Neurodegeneration, 10*(1). https://doi.org/10.1186/s13024-015-0020-0

Tarantini, S., Valcarcel-Ares, M. N., Toth, P., Yabluchanskiy, A., Tucsek, Z., Kiss, T., Hertelendy, P., Kinter, M., Ballabh, P., Süle, Z., Farkas, E., Baur, J. A., Sinclair, D. A., Csiszar, A., & Ungvari, Z. (2019). Nicotinamide mononucleotide (NMN) supplementation rescues cerebromicrovascular endothelial function and neurovascular coupling responses and improves cognitive function in aged mice. *Redox Biology, 24*. https://doi.org/10.1016/j.redox.2019.101192

Tong, X. K., & Hamel, E. (2015). Simvastatin restored vascular reactivity, endothelial function and reduced string vessel pathology in a mouse model of cerebrovascular disease. *Journal of Perinatology, 35*(3), 512–520. https://doi.org/10.1038/jcbfm.2014.226

Toth, P., Tarantini, S., Davila, A., Valcarcel-Ares, M. N., Tucsek, Z., Varamini, B., Ballabh, P., Sonntag, W. E., Baur, J. A., Csiszar, A., & Ungvari, Z. (2015). Purinergic glio-endothelial coupling during neuronal activity: Role of P2Y 1 receptors and eNOS in functional hyperemia in the mouse somatosensory cortex. *American Journal of Physiology - Heart and Circulatory Physiology, 309*(11), H1837–H1845. https://doi.org/10.1152/ajpheart.00463.2015

Toth, P., Tarantini, S., Tucsek, Z., Ashpole, N. M., Sosnowska, D., Gautam, T., Ballabh, P., Koller, A., Sonntag, W. E., Csiszar, A., & Ungvari, Z. (2014). Resveratrol treatment rescues neurovascular coupling in aged mice: Role of improved cerebromicrovascular endothelial function and downregulation of NADPH oxidase. *American Journal of Physiology - Heart and Circulatory Physiology, 306*(3), H299–H308. https://doi.org/10.1152/ajpheart.00744.2013

Tozer, A. J. B., Forsythe, I. D., Steinert, J. R., & Barnes, S. (2012). Nitric oxide signalling augments neuronal voltage-gated L-type (CaV1) and P/Q-Type (CaV2.1) channels in the mouse medial nucleus of the trapezoid body. *PLoS One, 7*(2). https://doi.org/10.1371/journal.pone.0032256

Uttara, B., Singh, A. V., Zamboni, P., & Mahajan, R. T. (2009). Oxidative stress and neurodegenerative diseases: A review of upstream and downstream antioxidant therapeutic options. *Current Neuropharmacology, 7*(1), 65–74. https://doi.org/10.2174/157015909787602823

Uwagbai, O., & Kalish, V. B. (2022). Vascular Dementia in StatPearls. Treasure Island (FL). *StatPearls Publishing.*

Villanueva, C., & Giulivi, C. (2010). Subcellular and cellular locations of nitric oxide synthase isoforms as determinants of health and disease. *Free Radical Biology and Medicine, 49*(3), 307–316. https://doi.org/10.1016/j.freeradbiomed.2010.04.004

Watanabe, A., Sasaki, T., Yukami, T., Kanki, H., Sakaguchi, M., Takemori, H., Kitagawa, K., & Mochizuki, H. (2016). Serine racemase inhibition induces nitric oxide-mediated neurovascular protection during cerebral ischemia. *Neuroscience, 339*, 139–149. https://doi.org/10.1016/j.neuroscience.2016.09.036

Wijasa, T. S., Sylvester, M., Brocke-Ahmadinejad, N., Schwartz, S., Santarelli, F., et al. (2020). Quantitative proteomics of synaptosome S-nitrosylation in Alzheimer's disease. *Journal of Neurochemistry*, 710–726. https://doi.org/10.1111/jnc.14870

Xia, P., Chen, H. S. V., Zhang, D., & Lipton, S. A. (2010). Memantine preferentially blocks extrasynaptic over synaptic NMDA receptor currents in hippocampal autapses. *Journal of Neuroscience, 30*(33), 11246–11250. https://doi.org/10.1523/JNEUROSCI.2488-10.2010

Yang, Y., Yu, T., Lian, Y. J., Ma, R., Yang, S., & Cho, J. Y. (2015). Nitric oxide synthase inhibitors: A review of patents from 2011 to the present. *Expert Opinion on Therapeutic Patents, 25*(1), 49–68. https://doi.org/10.1517/13543776.2014.979154

Yemisci, M., Gursoy-Ozdemir, Y., Vural, A., Can, A., Topalkara, K., & Dalkara, T. (2009). Pericyte contraction induced by oxidative-nitrative stress impairs capillary reflow despite successful opening of an occluded cerebral artery. *Nature Medicine, 15*(9), 1031–1037. https://doi.org/10.1038/nm.2022

Zhang, L., Dawson, V. L., & Dawson, T. M. (2006). Role of nitric oxide in Parkinson's disease. *Pharmacology & Therapeutics*, 33–41. https://doi.org/10.1016/j.pharmthera.2005.05.007

Zhào, H., Wang, R., Zhang, Y., Liu, Y., & Huang, Y. (2021). Neuroprotective effects of troxerutin and cerebroprotein hydrolysate injection on the neurovascular unit in a rat model of Middle cerebral artery occlusion. *International Journal of Neuroscience, 131*(3), 264–278. https://doi.org/10.1080/00207454.2020.1738431

Zheng, S., Wang, C., Lin, L., Mu, S., Liu, H., Hu, X., Chen, X., & Wang, S. (2023). TNF-α impairs pericyte-mediated cerebral microcirculation via the NF-κB/iNOS Axis after experimental traumatic brain injury. *Journal of Neurotrauma, 40*(3–4), 349–364. https://doi.org/10.1089/neu.2022.0016

Zoli, M., Torri, C., Ferrari, R., Jansson, A., Zini, I., Fuxe, K., & Agnati, L. F. (1998). The emergence of the volume transmission concept. *Brain Research Reviews, 26*(2–3), 136–147. https://doi.org/10.1016/S0165-0173(97)00048-9

Zonta, M., Angulo, M. C., Gobbo, S., Rosengarten, B., Hossmann, K. A., Pozzan, T., & Carmignoto, G. (2003). Neuron-to-astrocyte signaling is central to the dynamic control of brain microcirculation. *Nature Neuroscience, 6*(1), 43–50. https://doi.org/10.1038/nn980

PART V

Nitric oxide and therapeutics

CHAPTER

6

The role of nitric oxide in gemcitabine resistance of pancreatic cancer cells

Mariachiara Gani, Eros Di Giorgio, Raffaella Picco, Luigi E. Xodo and Valentina Rapozzi

Department of Medicine, University of Udine, Udine, Italy

Abbreviations

1400 W N-{[3- (aminomethyl)phenyl]methyl}ethenimidamide
5-FU 5-fluorouracil
ACC acetyl-CoA carboxylase
ACLY ATP citrate lyase
ADH5 alcohol dehydrogenase class-3
AHR aryl hydrocarbon receptor
AKT protein kinase B
APE1/Ref1 apurinic/apyrimidinic endonuclease1/redox factor-1
ATP adenosine triphosphate
Bcl-2 B-cell lymphoma-2
CA9 carbonic anhydrase 9
CAV-1 caveolin-1
CDP cytidine diphosphate
CRC colorectal cancer
CXCL12 C-X-C motif chemokine 12
CXCR4 C-X-C motif chemokine receptor 4
dCDP deoxycytidine diphosphate
dCK deoxycytidine kinase
DNL, de novo lipogenesis
Dnm2 dynamin 2
Drp1 dynamin-related protein 1
EGFR epidermal growth factor receptor
EMT epithelial–mesenchymal transition
eNOS or NOS1 endothelial nitric oxide synthase

Therapeutic Applications of Nitric Oxide in Cancer and Inflammatory Disorders
https://doi.org/10.1016/B978-0-443-22354-9.00006-1

ERK extracellular signal–regulated kinase
FAK focal adhesion kinase
FASN fatty acid synthases
FOXO forkhead box-O
GEM gemcitabine
GLS glutaminase
GLUTs glucose transporters
GMP guanosine monophosphate
GSNOR s-nitrosoglutathione reductase
hCNT1 human concentrative nucleoside transporter 1
hENT1 human equilibrative nucleoside transporter 1
HIF-1 hypoxia-inducible factor
HK2 hesokinase 2
HMGA1 high mobility group A1
Hsp27 heat shock protein 27
IDH1 isocitrate dehydrogenase 1
IDO1 indoleamine 2,3-dioxygenase 1
IL-1 β interleukin-1 beta
IL-2 interleukin-2
IL-6 interleukin-6
IL-8 interleukin-8
iNOS or NOS2 inducible nitric oxide synthase
JAG1 jagged-1
KRAS Kristen rat sarcoma virus
L-NAME L-NG-nitroarginine methyl ester
L-NMMA N^G-monomethyl-L-arginine
LABC locally advanced breast cancer
LDHA lactate dehydrogenase A
MAPK mitogen-activated protein kinase
MDSCs myeloid-derived suppressor cells
Mfn1-Mfn2 mitofusins 1 and 2
MPK-1 protein kinase phosphatases-1
MRP5 (or ABCC5) multidrug resistance associated protein 5
mTORC1 mammalian target of rapamycin complex 1
NF-κB nuclear factor kappa-light-chain-enhancer of activated B cells
nNOS or NOS3 neuronal nitric oxide synthase
NO nitric oxide
NOS nitric oxide synthase
Notch2 neurogenic locus notch homolog protein 2
NRF2 nuclear factor erythroid 2-related factor 2
OPA1 optic atrophy 1
PDAC pancreatic ductal adenocarcinoma
PI3K phosphoinositide 3-kinase
PKM2 pyruvate kinase isoform M2
PPNO propylamine propylamine nonoate
PTEN phosphatase and tensin homolog
Ras rat sarcoma virus
RNS reactive nitrogen species
ROS reactive oxygen species
RR ribonucleotide reductase
SerpinB2 plasminogen activator inhibitor-2
SNAP s-nitroso-N-acetylpenicillamine
SNO s-nitrosothiols

SRC proto-oncogene tyrosine-protein kinase Src
STAT5 signal transducer and activator of transcription 5
TCA tricarboxylic acid cycle
TIMP1 TIMP metallopeptidase inhibitor 1
TME tumor microenvironment
TNBC triple-negative breast cancer
TRAP1 tumor necrosis factor receptor-associated protein 1
YY1 Ying Yang 1

Introduction

Chemotherapy is one of the most important treatments for cancer, but its effectiveness can be compromised by resistance. There are two types of resistance to chemotherapy: intrinsic and acquired (Wang et al., 2019). Intrinsic resistance is due to factors present in the tumor that make the therapy ineffective, while acquired resistance develops during treatment. Since tumors exhibit a high degree of molecular heterogeneity (Holohan et al., 2013), drug resistance can originate from a small tumor cell population selected by the therapy (Wang et al., 2019).

Resistance to anticancer drugs depends on various factors, such as genetic mutations, epigenetic modifications, upregulation of drug efflux, and various other cellular and molecular mechanisms (Zheng, 2017) (Fig. 6.1). It is clear that a single mechanism alone cannot explain the emergence of resistance.

In this mini-review, we focus on the potential role of nitric oxide (NO) in the development of tumor progression and chemoresistance. BxPC-3 human pancreatic cancer cells treated with gemcitabine (GEM) served as an experimental model.

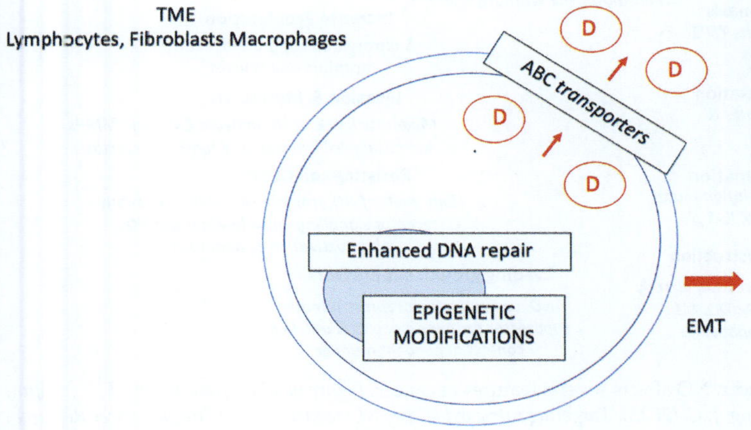

FIGURE 6.1 Mechanisms of chemoresistance in cancer. These mechanisms include changes in the tumor microenvironment (TME), increased DNA repair, epigenetic changes, epithelial to mesenchymal transition (EMT), and increased drug efflux (D).

Nitric oxide and tumor progression

NO is a gaseous free radical with signaling effects in biological systems. It reacts strongly with various molecules including Fe, O_2, and O_2^- and exerts its direct effects mainly through interactions with the heme moiety of soluble guanyl cyclase, producing cyclic guanosine monophosphate (cGMP) (Murad, 1998, 2006). An indirect effect of NO depends on other NO products, such as nitrogen oxides ($NO+$, N_2O_3, $ONOO-$) (Vasudevan & Thomas, 2014). These nitroso groups react with sulfhydryl groups, resulting in the formation of S-nitrosothiols (SNO) (Gaston, 1999; Stamler et al., 2001; Aranda et al., 2012). These processes are commonly referred to as nitrosation, nitrosylation, and nitration (Lundberg & Weitzberg, 2022).

In biological systems, NO is formed during the conversion of L-arginine to citrulline by nitric oxide synthase (NOS). There are three isoforms of NOS: neuronal NOS (nNOS or NOS3), Ca^{2+}-endothelial NOS (eNOS or NOS1), and an inducible NOS (iNOS or NOS2). NOS1 and NOS3 are the constitutive isoforms and produce a small amount of NO in the picomolar to nanomolar range, while NOS2 is an inducible isoform that responds to inflammatory stimuli and/or hypoxia and can produce higher amount of NO in the micromolar range (Wang & Hussain, 2017).

High expression of NOS2 is characteristic of many human tumors including breast, prostate, and colon cancers (Erlandsson et al., 2018; Gao et al., 2019; Glynn et al., 2010; Liao et al., 2019; Wang et al., 2003; Zafirellis et al., 2010). A continuous increase of NO production is known to be involved in tumor progression, metastasis, angiogenesis, antiapoptotic signaling, genomic instability, chemoresistance, and metabolic reprogramming (Khan et al., 2020; Reddy et al., 2023) (Fig. 6.2). S-nitrosylation is a key mechanism for NO-mediated signaling and regulation of protein function (Fernando et al., 2019; Salimian Rizi et al., 2017). Recent studies, performed in different kind of tumors, have identified several processes involved in S-nitrosylation and denitrosylation of proteins (Stomberski et al., 2019).

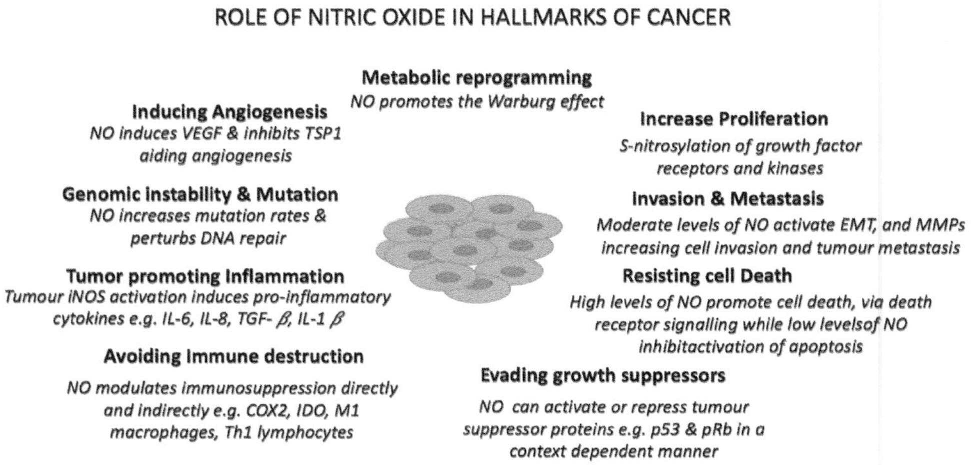

FIGURE 6.2 Nitric oxide and cancer. NO affects several features of cancer. *Figure modified from Reddy, T. P., Glynn, S. A., Billiar, T. R., Wink, D. A., & Chang, J. C. (2023). Targeting nitric oxide: Say NO to metastasis. Clinical Cancer Research, 29(10), 15573265, 1855–1868. American Association for Cancer Research Inc.: United States, https://aacrjournals.org/clincancerres/article/29/10/1855/726245/Targeting-Nitric-Oxide-Say-NO-to, https://doi.org/10.1158/1078-0432.CCR-22-2791.*

In this context, S-nitrosoglutathione reductase (GSNOR), a denitrosylase enzyme encoded by the human gene alcohol dehydrogenase class-3 (ADH5), plays a crucial role in several tumors. GSNOR regulates S-nitrosylation of proteins by affecting the cellular balance between S-nitrosylated proteins and S-nitrosoglutathione (GSNO), and its dysregulation significantly alters cellular homeostasis and leads to progression of different kind of tumors (Barnett & Buxton, 2017).

Several proteins involved in tumorigenesis and other processes are nitrosylated, such as B cell lymphoma-2 (Bcl-2) (Chanvorachote et al., 2006; Leon-Bollotte et al., 2011; Lim et al., 2008), Ras, Fas, mitogen-activated protein kinase phosphatases-1 (MPK-1) (Guan et al., 2012), tumor necrosis factor receptor-associated protein 1 (TRAP1) (Rizza et al., 2016), and phosphatase and tensin homolog (PTEN) (Gupta et al., 2017). In particular, nitrosylation of wild-type Ras protein is required for the development and maintenance of pancreatic ductal adenocarcinoma (PDAC) (Tan et al., 2019).

Nitric oxide and chemoresistance in cancer cells

There are several studies describing the role of NO in chemoresistance, but some of them are contradictory. For example, some studies reported that NO released from NO donors can sensitize cisplatin activity, while other studies demonstrated that NO contributes to the development of cisplatin resistance in "in vitro" tumor models (Chanvorachote et al., 2006; Godoy et al., 2012). One of the mechanisms involved appears to be nitrosylation of the Bcl2 protein, which leads to protein stabilization (Chanvorachote et al., 2006), inhibition of apoptosis, and cisplatin resistance.

It was reported in human melanoma tumor cells that endogenous or exogenous NO nitrosylates caspase-3 and prolylhydroxylase-2 (Godoy et al., 2012). The latter, in turn, degrades hypoxia-inducible factor (HIF-1α) by hydroxylation. The alteration/inhibition of the activities of these enzymes led to cisplatin resistance (Godoy et al., 2012). It has also been reported that nitrosylation can either stabilize or induce degradation of HIF-1α (Li et al., 2007; Mateo et al., 2003). At high NO concentrations (>1 mM), HIF-1α is stabilized, while at lower concentrations (<400 nM), HIF-1α is rapidly degraded (Mateo et al., 2003). In human breast tumor cells MCF-7 and colon tumor cells HT-29 treated with propylamine nonoate (PPNO), an NO donor, chemoresistance correlated with nytrosylation of topoisomerase I and topoisomerase II (Kumar et al., 2016; Sharma et al., 2015) (Scheme 6.1).

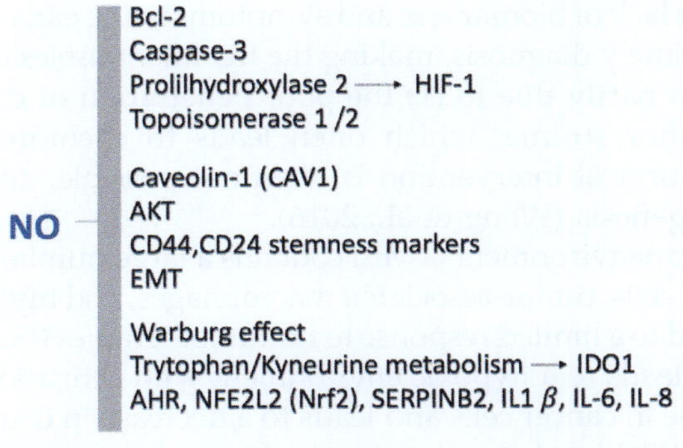

SCHEME 6.1 Scheme of the signaling pathways modulated by NO in chemoresistance.

However, some contradictory effects of NO on the sensitization or the development of drug resistance probably depend on the dose and duration of exposure to NO (Napoli et al., 2013; Ridnour et al., 2006), on the cellular sensitivity to NO, activity, and localization of NOS (Murad, 2006).

After a long exposure of NO over 7–14 days, lung cancer cells became resistant to cisplatin, doxorubicin, and etoposide in a concentration- and time-dependent manner. The resistance was characterized by an increase in caveolin-1 (CAV1), antiapoptotic Bcl2, and activated protein kinase B (AKT) (Wongvaranon et al., 2013), which lead to cancer cell survival.

The metabolic adaptations of cancer cells have been shown to be important events influencing tumor initiation, progression, and spread (You et al., 2023). In addition, they also control cancer stem cells and favor evasion of the immune system through reduced immunogenicity or recognition (Bogdan, 2015). In this context, NO has been shown to play an important role as a mediator of the metabolic phenotype of cancer (Chang et al., 2015; Salimian Rizi et al., 2017), including its involvement in the various strategies employed by tumors to evade detection by the immune system (PeÑarando et al., 2019) and promote stemness features and epithelial–mesenchymal transition (EMT) (Belgorosky et al., 2020; Maiuthed et al., 2018; Monteiro et al., 2019) (Scheme 6.1). NO has been found to increase glycolysis and lactate production in breast cancer (De Vitto et al., 2013), and it has been recognized as a positive regulator of the Warburg effect in ovarian cancer (Caneba et al., 2014). Recently, GSNOR deficiency has been shown to lead to metabolic reprogramming resulting in the acquisition of an aggressive and immune evasive phenotype in colorectal cancer (Mena-Osuna et al., 2023). In pancreatic cancer, levels of kynurenine, a tryptophan metabolite, were associated with high NOS2 expression, and higher levels of kynurenine predicted poorer patient survival. Moreover, gene expression analysis showed a positive correlation between NOS2 expression and tryptophan/kynurenine metabolism genes, including indoleamine 2,3-dioxygenase 1 (IDO1) and several aryl hydrocarbon receptor (AHR) target genes, including nuclear factor erythroid 2-related factor 2 NFE2L2 (NRF2), plasminogen activator inhibitor-2 (SERPINB2), IL-1β, IL-6, and IL-8, which play a role in pancreatic cancer (Wang et al., 2020).

Characteristics of pancreatic cancer

Pancreatic cancer is the fourth leading cause of death in the Western world (Partyka et al., 2023) and is expected to become the second leading cause in the United States by 2030 (Rahib et al., 2021). Unfortunately, due to the lack of biomarkers and symptoms in the early stages of the disease, it is very difficult to make a timely diagnosis, making the treatment useless (Swords et al., 2015).

The poor prognosis is partly due to (1) the poor penetration of drugs into the dense and poorly vascularized tumor stroma, which often leads to chemoresistance (Neesse et al., 2015); (2) the fact that surgical intervention is often not possible; and (3) the crucial role of inflammation in tumorigenesis (Wang et al., 2016).

Indeed, the tumor microenvironment (TME) contains a large number of immunosuppressive cells such as regulatory T cells, tumor-associated macrophages, and myeloid-derived suppressor cells (MDSCs), which lead to a limited response to immunotherapies (Ho et al., 2020). In addition, low perfusion of PDAC leads to a hypoxic environment with activation of HIF-1α, which promotes an EMT phenotype in cancer cells and leads to a decrease in drug transporters.

An increase in NOS2/NO has been associated with poor survival in patients with resected early-stage PDAC. Experiments by the Hussain's group (Wang et al., 2016) showed that primary tumor cells isolated from NOS2-deficient KPC mice exhibited reduced proliferation and invasiveness. The increase in apoptotic cell death was highlighted by cleaved caspase-3, higher expression of E-cadherin, and a reduction in cell migration and invasion. All these events are associated with inhibition of ERK signaling and decreased inactivation of the transcription factor Forkhead Box-O (FOXO), a tumor suppressor (Arden, 2006; Paik et al., 2007; Tothova et al., 2007).

Gemcitabine, the most important treatment for pancreatic cancer

Gemcitabine (2′,2′-difluoro-2′-deocycytidine (dFdC, GEM)) is the first-line treatment for PDAC. It is a deoxycytidine analog with different cellular mechanisms. The main mechanism of action is the inhibition of DNA synthesis (Huang et al., 1991). GEM triphosphate competes with deoxycytidine triphosphate, inhibiting the activity of DNA polymerase and allowing the incorporation of a single downstream nucleotide that prevents chain elongation (Gandhi & Plunkett, 1990). The penultimate position of GEM, known as "masked chain termination," leads to DNA polymerase arrest and also prevents the removal of GEM by DNA repair enzymes (Huang et al., 1991). GEM also acts through self-amplification by inhibiting enzymes involved in deoxynucleotide metabolism. GEM is also able to induce apoptosis in response to cellular stress in tumour cells, but not in normal cells, through caspase signaling (Ferreira et al., 2000; Habiro et al., 2004) or activation of the p38 mitogen-activated protein kinase (MAPK) pathway (Habiro et al., 2004; Kummer et al., 1997) (Fig. 6.3).

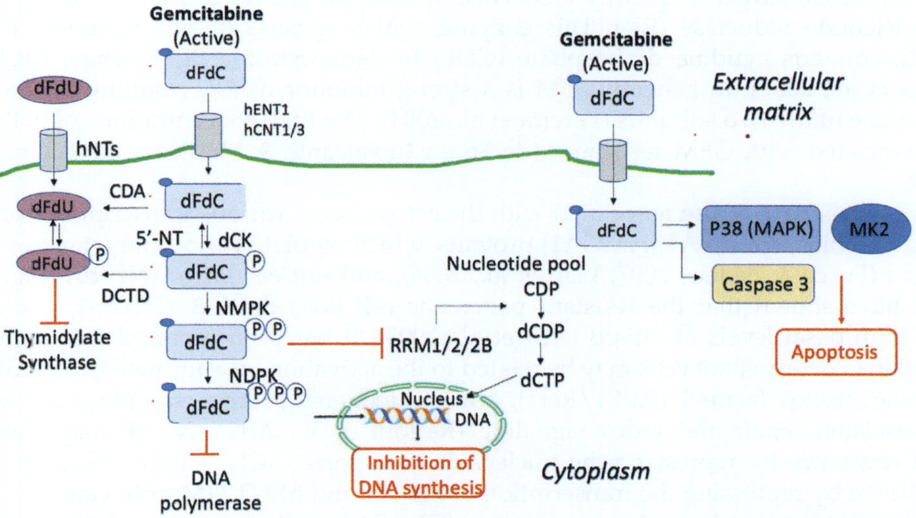

FIGURE 6.3 Mechanisms of action of gemcitabine: inhibition of DNA synthesis and apoptosis GEM (dFdC) enters the cells via nucleoside transporters (hCNT and hENT) at the tumor cell membrane. It is then phosphorylated three times by the enzymes dCK, NMPK, and NDPK to form the active metabolites gem-diphosphate and gem-triphosphate. Gem diphosphate inhibits ribonucleotide reductase (RR). This enzyme produces other deoxynucleotides, including deoxycytidines, which can be incorporated into DNA, and gem triphosphate competes directly with deoxycytidine diphosphate for incorporation into the DNA strand, where it inhibits DNA polymerase and terminates the elongation of the DNA strand. The enzymes cytidine deaminase (CDA) and deoxycytidylate deaminase (DCTD) inhibit GEM metabolism by converting GEM and Gem monophosphate into the inactive metabolites difluorodeoxyuridine (DfDu) (Natu & Nagaraju, 2023). On the other hand, GEM induces apoptosis by activating both p38 (MAPK) with the effector protein MK2 and the caspase-3 pathway.

Mechanisms of action of gemcitabine

As with most chemotherapeutic agents, the efficacy of GEM is limited by both intrinsic and acquired chemoresistance of tumour cells after long-term treatment (de Sousa Cavalcante & Monteiro, 2014). Various cellular signaling pathways, transcription factors, and enzymes of nucleotide metabolism have been associated with GEM resistance and sensitivity although some data are conflicting. For example, expression and phosphorylation of heat shock protein 27 (Hsp27) has been associated with GEM resistance (Kuramitsu et al., 2012) and also with antiproliferative effects (Nakashima et al., 2011). The role of nucleoside transporters also appears to play an important role in both sensitivity and resistance to GEM. Expression of hENT1, the major cellular GEM transporter, is considered a prognostic factor for GEM-based chemotherapies: high levels of hENT1 are generally associated with longer overall survival of pancreatic cancer patients, while low levels are associated with GEM resistance and shorter overall survival (Farrell et al., 2009; Giovannetti et al., 2006). However, Nakano and colleagues found no reduction in hENT1 expression in some GEM-resistant pancreatic cell lines (Nakano et al., 2007).

Once GEM enters into the cell, it must be activated by phosphorylation by deoxycytidine kinase (dCK). For this reason, the efficacy of GEM is often associated with dCK activity. Indeed, GEM-resistant ovarian and pancreatic cancer cell lines have been reported to lack dCK mRNA (Ohhashi et al., 2008).

A molecular target frequently associated with GEM chemoresistance is the holoenzyme ribonucleotide reductase (RR). This enzyme, which consists of two subunits, RRM1 and RRM2, converts cytidine diphosphate (CDP) to deoxycytidine diphosphate (dCDP) (Cerqueira et al., 2007). In general, GEM is a strong inhibitor of RR, resulting in the complete loss of one of the two subunits (Pereira et al., 2004). The high concentrations of both subunits are associated with GEM resistance (de Sousa Cavalcante & Monteiro, 2014; Ohtaka et al., 2008).

GEM resistance is also associated with the activation of various transcription factors, such as high mobility group A1 (HMGA1) proteins, which regulate various target genes in human cancers (Fusco & Fedele, 2007; Liau et al., 2006), and nuclear factor (NF-κB) (Fig. 6.4). Arlt et al. have shown that the resistant pancreatic cell lines BxPC-3, Capan-1, and PancTu-1 have high basal levels of NF-κB (Arlt et al., 2003). It has been reported that high levels of NF-κB in GEM-resistant cells may be related to the activation by apurinic/apyrimidinic endonuclease1/redox factor-1 (APE1/Ref1), a multifunctional protein that plays a role in DNA base excision repair and redox signaling (Raffoul et al., 2012). NF-κB may contribute to GEM resistance by repressing the nucleoside transporter hCNT1 (Skrypek et al., 2015), or via HIF-1α by repressing the transcription of hENT1 and hENT2 (Morote-Garcia et al., 2009).

There is strong evidence that the CXCL12/CXCR4 signaling pathway (C-X-C motif chemokine12, also known as stromal cell-derived factor1(SDF-1)/C-X-C motif chemokine receptor4), which is also associated with NF-κB and HIF expressions (Arora et al., 2013), mediates GEM resistance in pancreatic cancer cells by activating different signaling pathways: the PI3K/Akt, extracellular signal–regulated kinase (ERK), and focal adhesion kinase (FAK) pathways (Singh et al., 2010).

A distinctive feature of GEM-resistant cells is that they adopt an EMT phenotype and the characteristics of cancer stem cells. In these cells, the expressions of Notch2 and Jagged-1 are

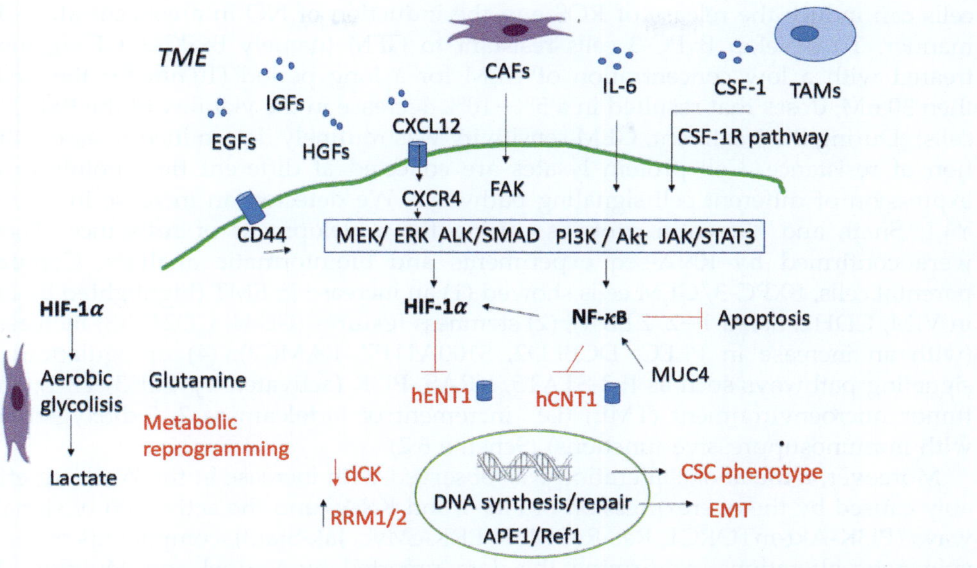

FIGURE 6.4 Mechanism of gemcitabine resistance in pancreatic cancer cells considering TME contribution. Stromal cells (such as CAFs, TAMs) secrete cytokines or chemokines and interact with pancreatic cancer cells (PDAC) by activating prosurvival signaling pathways to promote EMT and CSC, inhibit apoptosis, and induce GEM resistance. NF-κB and HIF- 1α promote chemoresistance by limiting GEM uptake and inhibiting apoptosis. Hypoxic TME induces PDAC cells to compete with stromal cells and immune cells for nutrients, triggering adaptive meabolic reprogramming of energy substances such as glucose, lactate, and glutamine to maintain metabolic homeostasis and promote GEM resistance (Gu et al., 2021). *Black arrows* indicate activation or upregulation; *red arrows* indicate inhibition or downregulation. *APE/Ref1*, human apurinic (apyrimidinic) endonuclease/redox factor 1; *CAFs*, cancer-associated fibroblasts; *CSCs*, cancer stem cells; *CSF-1*, colony-stimulating factor 1; *CXCL12*, C-X-C motif chemokine 12; *CXCR4*, C-X-C motif chemokine receptor 4; *dCK*, deoxycytidine kinase; *EFGs*, epidermal growth factors; *EMT*, epithelial–mesenchymal transition; *FAK*, focal adhesion kinase; *hCNT1*, human concentrative nucleoside transporter 1; *hENT1*, human equilibrative nucleoside transporter 1; *HGFs*, hematopoietic growth factors; *HIF-1α*, hypoxia-inducible factor 1α; *IGFs*, insulin-like growth factors I–II; *IL-6*, interleukin 6; *MUC4*, mucin 4; *NF-κB*, nuclear factor-κB; *PI3K*, phosphatidylinositol 3-kinase; *RRM1/2*, catalytic subunit M1 and M2 of ribonucleoside diphosphate reductase; *TAMS*, tumor-associated macrophages; *TME*, tumor microenvironment.

upregulated (Wang et al., 2009). It has been described that EMT-induced Gem resistance is partly due to hypoxia (Takagi et al., 2023).

Development and characterization of GEM-resistant BXPC-3 pancreatic cell line

It is known that most chemotherapeutic agents such as tamoxifen, 5-FU, doxorubicin, paclitaxel, and GEM (Yokoyama et al., 2017) can lead to an increase in intracellular ROS and RNS by direct or indirect mechanisms (Deavall et al., 2012; Jiang et al., 2023). This leads to an accumulation of oxidized proteins, DNA damage, and cell cycle arrest.

Since our aim is to investigate the role of NO generated by GEM treatment in the development of resistance, we first confirmed that GEM treatment in BxPC-3 pancreatic cancer cells can induce the release of ROS and the induction of NO in a concentration-dependent manner. To develop BxPC-3 cells resistant to GEM (namely BxPC-3/GEM), the cells are treated with a low concentration of GEM for a long period (10 nm for the first months, then 30 nM, doses that resulted in a 5%–10% decrease in the viability of the BxPC-3 parental cells). During the treatment, GEM sensitivity was routinely determined to assess the acquisition of resistance. Cell protein lysates are collected at different time points to check the expression of different cell signaling pathways. We detected an increase in KRAS, NF-κB, YY1, Snail, and Vimentin proteins during the development of resistance. These results were confirmed by RNA-seq experiments and bioinformatic analysis. Compared with parental cells, BXPC-3/GEM cells showed (1) an increase in EMT (highlighted by an increase in VIM, CDH2, SNAl 1–2, ZEB 1); (2) stemness features (CD44, CD24); (3) increase of HIF-1 (with an increase in PLEC, DCBLD2, S100A11P7, LAMC2); (4) upregulation of several signaling pathways such as IL2-STAT5, KRAS, PI3K (activated by CD63/TIM), variation of tumor microenvironment (TME) (i.e., increment of indoleamine 2, 3-dioxygenases (IDO1) with immunosuppressive functions) (Scheme 6.2).

Moreover, a metabolic alteration was observed with increase in the Warburg effect, probably caused by the overexpression of HIF-1 and KRAS and the activation of signaling pathways (PI3K-Akt-mTORC1, Ras-Raf-MEK-ERK-cMyc, Jak-Stat3), components of the TME and epigenetic alterations, confirming the data reported by Vaupel and Multhoff (Vaupel & Multhoff, 2021).

An increase in hexokinase HK2 was also observed. This enzyme increases the glycolytic flux, and the ATP level. HK2 has been reported to interact with HIF-1α and induce chemoresistance associated with survival (Dai et al., 2020; Liu et al., 2023). However, an increase

Gene Set Name	Gene Set (K)	Overlap (k)	k/K	p-value	FDR q-value
EMT	200	17	0.086	1.54 E-12	7.7 E-11
IL2 STAT5 Signaling	199	11	0.0553	1.11 E-6	1.94 E-5
Hypoxia	200	11	0.055	1.16 E-6	1.94 E-5
Apical Junction	200	9	0.045	6.34 E-5	6,34 E-4
KRAS signaling UP	200	9	0.045	6.34 E-5	6,34 E-4
Myogenesis	200	8	0.04	3.08 E-4	0.00257
Estrogen Response Early	200	7	0.036	0.00158	0.00986
Inflammatory Response	200	7	0.036	0.00158	0.00986
Complement	200	6	0.03	0.00706	0.0375
UV Response DN	144	5	0.0347	0.00751	0.0375
Interferon Alpha Response	97	4	0.0412	0.00914	0.0415

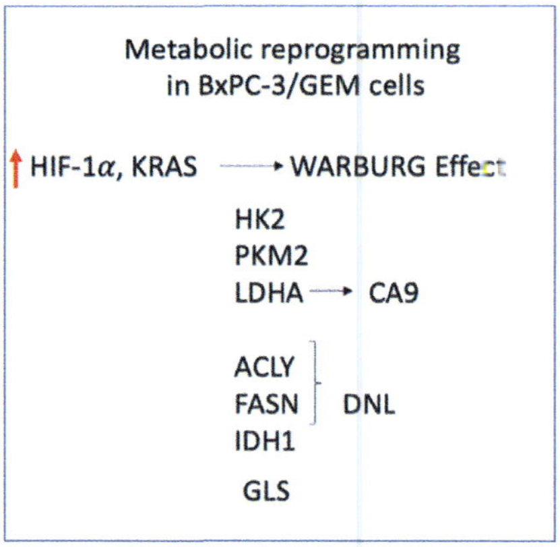

SCHEME 6.2 RNAseq analysis of BxPC-3/GEM resistent cells compared with parental cells. The table shows the RNAseq analysis with the most important upregulated signaling pathways in BxPC-3/GEM cells compared with parental cells. The right panel shows the list of metabolic enzymes that are upregulated in the GEM-resistant cell line, illustrating metabolic reprogramming. *ACLY*, ATP citrate lyase; *CA9*, carbonic anhydrase IX; *DNL*, de novo lipogenesis; *FASN*, fatty acid synthase; *GLS*, glutaminase; *HK*, hesokinase; *IDH1*, isocitrate dehydrogenase; *LDHA*, lactate dehydrogenase A; *PKM2*, pyruvate kinase M2.

in pyruvate kinase PKM2 was also observed in BxPC-3/GEM cells, consistent with a possible switch to aerobic glycolysis during tumorigenesis (Calabretta et al., 2016).

Other data from RNA-Seq analysis concern the increase in lactate dehydrogenase A (LDHA), which has been reported to promote the expression of antiapoptotic proteins and the reduction of ROS production, thereby limiting apoptosis and promoting chemoresistance (Comandatore et al., 2022; Liu et al., 2023). Furthermore, an increase in the expression of carbonic anhydrase 9 (CA9), which is consistently associated with tumor invasion and migration, was observed in association with LDHA (McDonald et al., 2019). In addition to the increase in glycolytic flux, RNA-seq revealed that the "de novo" lipogenesis (DNL) pathway is associated with GEM resistance (Daemen et al., 2015; Swierczynski et al., 2014), with increased expression of some lipogenic enzymes regulated by KRAS, such as ATP citrate lyase (ACLY) and fatty acid synthases (FASN). In turn, overexpression of FASN can upregulate the mRNA and protein levels of PKM2 (Tian et al., 2018). In Bx-PC-3/GEM cells, a higher level of isocitrate dehydrogenase I (IDH1) was present, which catalyzes the transformation of isocitrate into alpha-ketoglutarate, producing NADPH for lipogenesis and detoxification of ROS (Calvert et al., 2017). In BxPC-3/GEM cells, the increase in EMT is also associated with an increase in glutaminase (GLS) and the glycolytic switch (Mukhopadhyay et al., 2020; Park et al., 2022).

Possible interactions between NO and metabolic reprogramming in BxPC-3/GEM pancreatic cancer cells

Compared with the parental cell line, BxPC-3/GEM cells show a lower increase in ROS and RNS after GEM treatment. Even though the resistant cells show an increase in NOS2 protein compared with the parental cells, the increase in GSNOR levels probably keeps the RNS level low.

It has been reported that in the range of 100–500 nM, NO is involved in protumorigenesis and activates various mechanisms (Thomas et al., 2015) such as RAS, stabilization of HIF-1α, epidermal growth factor receptor (EGFR), and protooncogenic tyrosine protein kinase Src (SRC) and interaction between CD63 and TIMP1 leading to activation of PI3K. In addition, all these signals are known to alter metabolic function by stimulating glycolysis and increasing fatty acid synthesis from cytoplasmic citrate (De Berardinis & Chandel, 2016; Somasundaram et al., 2019). For example, stabilization of HIF-1α increases glycolysis and stimulates NF-κB, which increase glucose uptake by inducing glucose transporters (GLUTs) (Liu et al., 2023). In ovarian cancer, NOS2-produced nitric oxide has been observed to promote glycolysis by inducing PKM2 nuclear translocation in an EGFR/ERK2-dependent manner and increasing HK2 expression (Li et al., 2017). PKM2 is known to lead to the Warburg effect by controlling the pyruvate flux (Yang & Lu, 2013), and it also can regulate HIF-1α via NF-κB (Zahra et al., 2020). Phosphorylation of PKM2 by ERK leads to the formation of a complex with β-catenin, which in turn induces LDHA and promotes lactate formation. In addition, PKM2 is a gatekeeper and possesses a redox-active cysteine that, when oxidized or modified, diverts glucose from glycolysis to the pentose phosphate pathway. Thus, an increase in PKM2 not only promotes the Warburg effect but also stimulates numerous anabolic pathways for protein, fatty acid, and ribose synthesis as well as NADPH.

In addition, PI3K activates mTORC1 via Akt, which leads to an increase in fatty acid and protein syntheses (De Bernardinis & Chandal, 2016; Somasundaram et al., 2019). NO can also modulate glutamine metabolism. Activation of ERK metabolism, which increases both c-Myc and PI3K/Akt/mTOR, increases the conversion of glutamine to glutamate, which can be fed into the tricarboxylic acid (TCA) cycle via alpha-ketoglutarate (Guo et al., 2016). However, in the presence of moderate NO, aconitase can be inhibited, preventing the conversion of citrate to isocitrate and α-ketoglutarate. Glucose and glutamine can be used as carbon sources to stimulate proliferation at the level of nitrosative stress. Interestingly, these cells show increased expression of glutaminase (GLS) during EMT and tend to be more glycolytic (Lee et al., 2016; Lucena et al., 2016). As NO triggers EMT, there may be a shift from glutamine utilization and proliferation to glucose-dependent migration and cancer stem cell–like behavior.

Many of these aspects determined by NO are present in BxPC-3/GEM cell line (Fig. 6.5). During the development of GEM resistance, a higher increase in NOS2 and GSNOR protein levels was observed compared with the parental cell line. The increase of GSNOR could cause the activation of NF-κB, YY1 and Snail by denitrosylation (Bonavida & Baritaki, 2012),

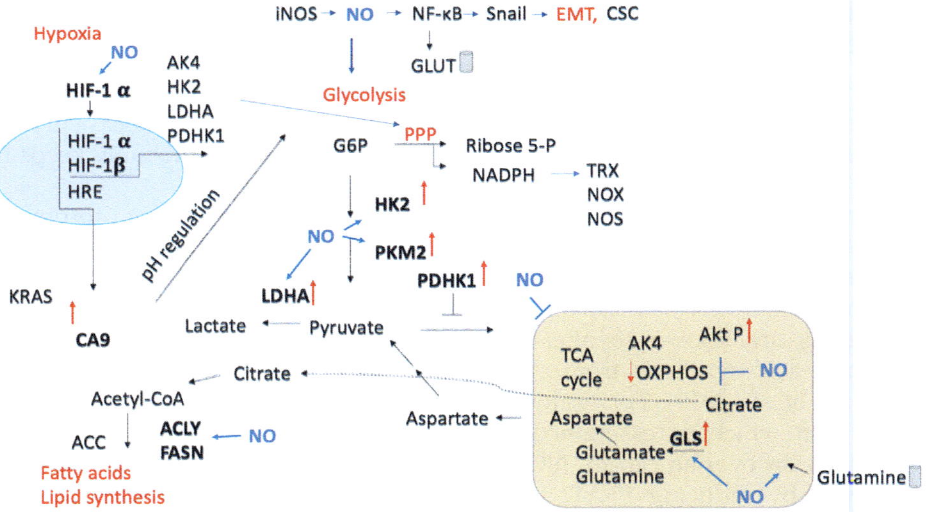

FIGURE 6.5 Simple schematic of metabolic reprogramming in BxPC-3/GEM pancreatic cancer cells with highlighted NO interaction. In BxPC-3/GEM cells, upregulation of the AK4-HIF-1α axis and KRAS leads to stimulation of glycolysis through an increase in HK2, PKM2, LDHA, and CA9 enzymes and inhibition of OXPHOS. The increase in glucose uptake (which is also due to an increase in NF-κB) stimulates PPP and generates NADPH, which in turn is used in de novo lipogenesis along with the conversion of citrate to acetyl-CoA. In addition, increased glutamine uptake leads to an increase in pyruvate. All these metabolic pathways are also favored by a moderate increase in NO. *ACLY*, ATP citrate lyase; *AK4*, adenylate kinase 4; *CA9*, carbonic anhydrase IX; *FASN*, fatty acid synthase; *G6P*, glucose phosphate; *GLS*, glutaminase; *GLUT*, glucose transporter; *HK2*, hesokinase 2; *HRE*, hypoxia response elements; *LDHA*, lactate dehydrogenase A; *NOS*, nitric oxide synthase; *NOX*, NADPH oxidase; *PDHK1*, pyruvate dehydrogenase kinase 1; *PKM2*, pyruvate kinase M2; *PPP*, pentose phosphate pathway; *TRX*, thioredoxin.

leading to tumor progression and EMT. NO-mediated signaling could affect glucose utilization in BxPC-3/GEM cells by directly interacting with central enzymes or by activating oncogenic pathways. Activation of oncogenic signaling pathways such as Ras (Son et al., 2013), mammalian target of rapamycin (mTOR), an important downstream effector of active PI3K/Akt signaling (Csibi et al., 2013), and c-Myc (Gao et al., 2009; López-Sánchez et al., 2020) by NO can also modulate glutamine metabolism, which is used as a carbon source by the TCA cycle at nitrosative NO concentrations to generate essential building blocks for proliferation (Somasundaram et al., 2019). DNL, which is characterized by ACLY, acetyl-CoA carboxylase (ACC), and FASN, can also be regulated by NO. A proteomics-based study has shown that S-nitrosylation is the molecular mechanism linking eNOS activity and regulation of mitochondrial fatty acid metabolism (Doulias et al., 2013). In addition, ACC is endogenously S-nitrosylated (Wang et al., 2015; Buckley et al., 2017). Many enzymes of glycolysis, the TCA cycle, and fatty acid metabolism are involved in S-nitrosylation (Doulias et al., 2013). In addition, an integrated metabolomic approach has shown that endogenous S-nitrosylation inhibits physiologically relevant metabolic enzymes (Bruegger et al., 2018). The link between NO and mitochondrial biogenesis has been demonstrated in leukemia cells (Carew et al., 2004). In particular, mitochondrial fission has been associated with highly activated glycolysis in many cancers (Dai & Jiang, 2019). This process is mainly regulated by optic atrophy 1 (OPA1) and mitofusins 1 and 2 (Mfn1 and Mfn2), while Dynamin-related protein 1 (Drp1) and Dynamin2 (Dnm2) are the main proteins driving mitochondrial fission (Tilokani et al., 2018). Among these proteins, NO nytrosylates OPA1 and Drp1. Notably, OPA1, MFN1/2, and DnM2 are more highly expressed in BxPC-3/Gem cells than in parental cells.

Discussion and future perspectives

Resistance to chemotherapy is one of the major complications in the treatment of cancer. There are numerous studies addressing the different molecular mechanisms that lead to resistance, and many of them are linked with each other. This review focuses on pancreatic cancer, which is one of the leading causes of death in the world for a variety of reasons. Prolonged treatment with gemcitabine, the first drug used to treat pancreatic cancer, inevitably leads to resistance, which can arise in several steps related to the mechanism of action, uptake, and the different molecular pathways involved in the tumor's response to treatment. Since none of the different mechanisms alone lead to resistance, we have focused on NO, a gaseous messenger produced during GEM treatment that is known to be associated with cancer progression and chemoresistance. During the development of GEM resistance in BxPC-3 pancreatic cancer cells, several signaling pathways are overexpressed compared with the parental cell line. Most of these pathways have been reported to be modulated by NO during tumorigenesis. The aim of our investigations was to highlight a possible role of NO in the development of GEM resistance in PDAC.

There are studies that report the ability of exogenously administered NO to reduce resistance in tumors. The research group of Ji (Chen et al., 2020) used a sequential delivery strategy based on the nitric oxide (NO) donor S-nitroso-N-acetylpenicillamine (SNAP) loaded with liposomes (Lip-SNAP). The release of NO was found to inhibit the formation of dense stroma in pancreatic cancer and enhance the intratumoral penetration of liposomes loaded with GEM (Lip-GEM), thereby increasing the efficiency of drug delivery. Using the same strategy, NO nanogenerators and NO-releasing molecules were synthesized (Wang et al., 2023). Liposome-encapsulated NO-releasing gemcitabine prodrugs (NO-GEMs) increased intracellular NO levels and the nitration process with inhibition of multidrug resistance–associated protein 5 (MRP5; ABCC5). This led to intracellular accumulation of GEM and increased apoptotic cell death in GEM-resistant PDAC cells, which express MRP5 to a higher degree than the GEM-sensitive cells (Masetto et al., 2020). A 6-week treatment with low-dose etoposide induced resistance in the pancreatic cancer cell line PT45-P1, increased IL-6B, and NO levels and inhibited caspase-3/-7/-8/-9 activity. Caspase activation was restored by either the iNOS inhibitor 1400 W, the reducing agent dithiothreitol or the IL-1 receptor antagonist, resulting in greater sensitivity to anticancer drug treatment (Müerköster et al., 2006).

Targeting NOS has gained interest as a promising therapeutic strategy to reduce metastatic ability and improve the efficacy of chemo- and radiotherapies in various solid malignant tumors. In 2021, the first phase 1/2 trial with an iNOS inhibitor was conducted in patients with chemorefractory locally advanced breast cancer (LABC) or metastatic triple-negative breast cancer (TNBC). The cotreatment with the pan-NOS inhibitor NG-monomethyl-L-arginine (L-NMMA) and taxane reduced tumor growth and prolonged survival (Chung et al., 2021).

In our BxPC-3 pancreatic cancer cell model, a single cotreatment with GEM and L-NG-nitroarginine methyl ester (L-NAME) or N-{[3-(aminomethyl)phenyl]methyl}ethenimidamide (1400 W) (an aspecific or specific NOS inhibitor, respectively) enhanced the sensitization of BxPC-3 cells to GEM treatment. A comparison of the mechanisms leading to GEM resistance in BxPC-3 pancreatic cancer cells with the signaling pathways modulated by NO allows to find new therapeutic strategies to prevent the development of resistance. Therefore, in our future work, BxPC-3 cells will be treated with GEM and a NOS2 inhibitor to study the major signaling pathways that are altered in the treated cells and identify potential targets for new therapeutic strategies to prevent the development of resistance.

Acknowledgments

AIRC (Associazione Italiana per la Ricerca sul Cancro), project PI 98989.

References

Aranda, E., Lopez-Pedrera, C., R. De La Haba-Rodriguez, J., & Rodriguez-Ariza, A. (2012). Nitric oxide and cancer: The emerging role of S-nitrosylation. *Current Molecular Medicine, 12*(1), 50–67. https://doi.org/10.2174/156652412798376099

Arden, K. C. (2006). Multiple roles of FOXO transcription factors in mammalian cells point to multiple roles in cancer. *Experimental Gerontology, 41*(8), 709–717. https://doi.org/10.1016/j.exger.2006.05.015

Arlt, A., Gehrz, A., Müerköster, S., Vorndamm, J., Kruse, M. L., Fölsch, U. R., & Schäfer, H. (2003). Role of NF-κB and Akt/PI3K in the resistance of pancreatic carcinoma cell lines against gemcitabine-induced cell death. *Oncogene, 22*(21), 3243–3251. https://doi.org/10.1038/sj.onc.1206390

Arora, S., Bhardwaj, A., Singh, S., Srivastava, S. K., McClellan, S., Nirodi, C. S., Piazza, G. A., Grizzle, W. E., Owen, L. B., & Singh, A. P. (2013). An undesired effect of chemotherapy: Gemcitabine promotes pancreatic cancer cell invasiveness through reactive oxygen species-dependent, nuclear factorκb- and hypoxia-inducible factor 1α-mediated up-regulation of CXCR4. *Journal of Biological Chemistry, 288*(29), 21197–21207. https://doi.org/10.1074/jbc.M113.484576

Barnett, S. D., & Buxton, I. L. O. (2017). The role of S-nitrosoglutathione reductase (GSNOR) in human disease and therapy. *Critical Reviews in Biochemistry and Molecular Biology, 52*(3), 340–354. https://doi.org/10.1080/10409238.2017.1304353

Belgorosky, D., Girouard, J., Langle, Y. V., Hamelin-Morrissete, J., Marino, L., Agüero, E. I., Malagrino, H., Reyes-Moreno, C., & Eiján, A. M. (2020). Relevance of iNOS expression in tumor growth and maintenance of cancer stem cells in a bladder cancer model. *Journal of Molecular Medicine, 98*(11), 1615–1627. https://doi.org/10.1007/s00109-020-01973-0

Bogdan, C. (2015). Nitric oxide synthase in innate and adaptive immunity: An update. *Trends in Immunology, 36*(3), 161–178. https://doi.org/10.1016/j.it.2015.01.003

Bonavida, B., & Baritaki, S. (2012). Inhibition of epithelial-to-mesenchymal transition (EMT) in cancer by nitric oxide: Pivotal roles of nitrosylation of NF-κB, YY1 and Snail. *Forum on Immunopathological Diseases and Therapeutics, 3*(2), 125–133. https://doi.org/10.1615/forumimmundisther.2012006065

Bruegger, J. J., Smith, B. C., Wynia-Smith, S. L., & Marletta, M. A. (2018). Comparative and integrative metabolomics reveal that S-nitrosation inhibits physiologically relevant metabolic enzymes. *Journal of Biological Chemistry, 293*, 6282–6296.

Buckley, D., Duke, G., Heuer, T. S., O'Farrell, M., Wagman, A. S., McCulloch, W., & Kemble, G. (2017). Fatty acid synthase – modern tumor cell biology insights into a classical oncology target. *Pharmacology and Therapeutics, 177*, 23–31.

Calabretta, S., Bielli, P., Passacantilli, I., Pilozzi, E., Fendrich, V., Capurso, G., Fave, G. D., & Sette, C. (2016). Modulation of PKM alternative splicing by PTBP1 promotes gemcitabine resistance in pancreatic cancer cells. *Oncogene, 35*(16), 2031–2039. https://doi.org/10.1038/onc.2015.270

Calvert, A. E., Chalastanis, A., Wu, Y., Hurley, L. A., Kouri, F. M., Bi, Y., Kachman, M., May, J. L., Bartom, E., Hua, Y., Mishra, R. K., Schiltz, G. E., Dubrovskyi, O., Mazar, A. P., Peter, M. E., Zheng, H., James, C. D., Burant, C. F., Chandel, N. S., … Stegh, A. H. (2017). Cancer-associated IDH1 promotes growth and resistance to targeted therapies in the absence of mutation. *Cell Reports, 19*(9), 1858–1873. https://doi.org/10.1016/j.celrep.2017.05.014

Caneba, C. A., Yang, L., Baddour, J., Curtis, R., Win, J., Hartig, S., Marini, J., & Nagrath, D. (2014). Nitric oxide is a positive regulator of the Warburg effect in ovarian cancer cells. *Cell Death and Disease, 5*(6). https://doi.org/10.1038/cddis.2014.264

Carew, J. S., Nawrocki, S. T., Xu, R. H., Dunner, K., McConkey, D. J., Wierda, W. G., Keating, M. J., & Huang, P. (2004). Increased mitochondrial biogenesis in primary leukemia cells: The role of endogenous nitric oxide and impact on sensitivity to fludarabine. *Leukemia, 18*, 1934–1940.

Cerqueira, N. M. F. S. A., Fernandes, P. A., & Ramos, M. J. (2007). Understanding ribonucleotide reductase inactivation by gemcitabine. *Chemistry - A European Journal, 13*(30), 8507–8515. https://doi.org/10.1002/chem.200700260

Chang, C. F., Diers, A. R., & Hogg, N. (2015). Cancer cell metabolism and the modulating effects of nitric oxide. *Free Radical Biology and Medicine, 79*, 324–336. https://doi.org/10.1016/j.freeradbiomed.2014.11.012

Chanvorachote, P., Nimmannit, U., Stehlik, C., Wang, L., Jiang, B. H., Ongpipatanakul, B., & Rojanasakul, Y. (2006). Nitric oxide regulates cell sensitivity to cisplatin-induced apoptosis through S-nitrosylation and inhibition of Bcl-2 ubiquitination. *Cancer Research, 66*(12), 6353–6360. https://doi.org/10.1158/0008-5472.CAN-05-4533

Chen, X., Jia, F., Li, Y., Deng, Y., Huang, Y., Liu, W., Jin, Q., & Ji, J. (2020). Nitric oxide-induced stromal depletion for improved nanoparticle penetration in pancreatic cancer treatment. *Biomaterials, 246*, 119999.

Chung, A. W., Anand, K., Anselme, A. C., Chan, A. A., Gupta, N., Venta, L. A., Schwartz, M. R., Qian, W., Xu, Y., Zhang, L., Kuhn, J., Patel, T., Rodriguez, A. A., Belcheva, A., Darcourt, J., Ensor, J., Bernicker, E., Pan, P. Y., Chen, S. H., … Chang, J. C. (2021). A phase 1/2 clinical trial of the nitric oxide synthase inhibitor L-NMMA and taxane for treating chemoresistant triple-negative breast cancer. *Science Translational Medicine, 13*, 624.

Comandatore, A., Franczak, M., Smolenski, R. T., Morelli, L., Peters, G. J., & Giovannetti, E. (2022). Lactate Dehydrogenase and its clinical significance in pancreatic and thoracic cancers. *Seminars in Cancer Biology, 86*, 93–100. https://doi.org/10.1016/j.semcancer.2022.09.001

Csibi, A., Fendt, S. M., Li, C., Poulogiannis, G., Choo, A. Y., Chapski, D. J., Jeong, S. M., Dempsey, J. M., Parkhitko, A., Morrison, T., Henske, E. P., Haigis, M. C., Cantley, L. C., Stephanopoulos, G., Yu, J., & Blenis, J. (2013). The mTORC1 pathway stimulates glutamine metabolism and cell proliferation by repressing SIRT4. *Cell, 153*(4), 840–854. https://doi.org/10.1016/j.cell.2013.04.023

Daemen, A., Peterson, D., Sahu, N., McCord, R., Du, X., Liu, B., Kowanetz, K., Hong, R., Moffat, J., Gao, M., Boudreau, A., Mroue, R., Corson, L., O'Brien, T., Qing, J., Sampath, D., Merchant, M., Yauch, R., Manning, G., … Evangelista, M. (2015). Metabolite profiling stratifies pancreatic ductal adenocarcinomas into subtypes with distinct sensitivities to metabolic inhibitors. *Proceedings of the National Academy of Sciences, 112*(32). https://doi.org/10.1073/pnas.1501605112

Dai, S., Peng Y,Zhu, Y., Xu, D., Zhu, F., Xu, W., Chen, Q., Zhu, X., Liu, T., Hou, C., Wu, J., & Miao, Y. (2020). Glycolysis promotes the progression of pancreatic cancer and reduces cancer cell sensitivity to gemcitabine. *Biomedicine and Pharmacotherapy, 121*, 109521.

Dai, W., & Jiang, L. (2019). Dysregulated mitochondrial dynamics and metabolism in obesity,diabetes, and cancer. *Frontiers in Endocrinology, 10*, 570.

De Berardinis, R. J., & Chandel, N. S. (2016). 2016. Fundamentals of cancer metabolism. *Science Advances, 2*. e1600200.

de Sousa Cavalcante, L., & Monteiro, G. (2014). Gemcitabine: Metabolism and molecular mechanisms of action, sensitivity and chemoresistance in pancreatic cancer. *European Journal of Pharmacology, 741*, 8–16. https://doi.org/10.1016/j.ejphar.2014.07.041

De Vitto, H., Mendonca, B. S., Elseth, K. M., Onul, A., Xue, J., Vesper, B. J., Gallo, C. V. M., Rumjanek, F. D., Paradise, W. A., & Radosevich, J. A. (2013). Molecular chenges induced by high nitric oxide adaptation in human breast cancer cell line BT-20 (BT-20-HNO): A switch from aerobic to anaerobic metabolism. *Tumour Biology, 34*, 403–413.

Deavall, D. G., Martin, E. A., Horner, J. M., & Roberts, R. (2012). Drug-induced oxidative stress and toxicity. *Journal of Toxicology, 2012*. https://doi.org/10.1155/2012/645460

Doulias, P. T., Tenopoulou, M., Greene, J. L., Raju, K., & Ischiropoulos, H. (2013). Nitric oxide regulates mitochondrial fatty acid metabolism through reversible protein S-nitrosylation. *Science Signaling, 6*(256). https://doi.org/10.1126/scisignal.2003252

Erlandsson, A., Carlsson, J., Andersson, S.-O., Vyas, C., Wikström, A. O., Davidsson, S., & Rider, J. R. (2018) High inducible nitric oxide synthase in prostate tumor epithelium is associated with lethal prostate cancer. *Scandian Journal of Urology, 52*, 129–133.

Farrell, J. J., Elsaleh, H., Garcia, M., Lai, R., Ammar, A., Regine, W. F., Abrams, R., Benson, A. B., Macdonald, J., Cass, C. E., Dicker, A. P., & Mackey, J. R. (2009). Human equilibrative nucleoside transporter 1 levels predict response to gemcitabine in patients with pancreatic cancer. *Gastroenterology, 136*(1), 187–195. https://doi.org/10.1053/j.gastro.2008.09.067

Fernando, V., Zheng, X., Walia, Y., Sharma, V., Letson, J., & Furuta, S. (2019). S-Nitrosylation: An emerging paradigm of redox signaling. *Antioxidants, 8*(9). https://doi.org/10.3390/antiox8090404

Ferreira, C. G., Span, S. W., Peters, G. J., Kruyt, F. A. E., & Giaccone, G. (2000). Chemotherapy triggers apoptosis in a caspase-8-dependent and mitochondria-controlled manner in the non-small cell lung cancer cell line NCI-H460. *Cancer Research, 60*(24), 7133–7141.

Fusco, A., & Fedele, M. (2007). Roles of HMGA proteins in cancer. *Nature Reviews Cancer, 7*, 899–910.

Gandhi, V., & Plunkett, W. (1990). Modulatory activity of 2′,2′-difluorodeoxycytidine on the phosphorylation and cytotoxicity of arabinosyl nucleosides. *Cancer Research, 50*(12), 3675–3680.

Gao, P., Tchernyshyov, I., Chang, T. C., Lee, Y. S., Kita, K., Ochi, T., Zeller, K. I., De Marzo, A. M., Van Eyk, J. E., Mendell, J. T., & Dang, C. V. (2009). C-Myc suppression of miR-23a/b enhances mitochondrial glutaminase expression and glutamine metabolism. *Nature, 458*(7239), 762–765. https://doi.org/10.1038/nature07823

Gao, Y., Zhou, S., Xu, Y., Sheng, S., Qian, S. Y., & Huo, X. (2019). Nitric oxide synthase inhibitors 1400W and L-NIO inhibit angiogenesis pathway of colorectal cancer. *Nitric Oxide. Biology and Chemistry, 83*, 33–39.

Gaston, B. (1999). Nitric oxide and thiol groups. *Biochimica et Biophysica Acta (BBA) - Bioenergetics, 1411*(2–3), 323–333. https://doi.org/10.1016/s0005-2728(99)00023-7

Giovannetti, E., Del Tacca, M., Mey, V., Funel, N., Nannizzi, S., Ricci, S., Orlandini, C., Boggi, U., Campani, D., Del Chiaro, M., Iannopollo, M., Bevilacqua, G., Mosca, F., & Danesi, R. (2006). Transcription analysis of human equilibrative nucleoside transporter-1 predicts survival in pancreas cancer patients treated with gemcitabine. *Cancer Research, 66*(7), 3928–3935. https://doi.org/10.1158/0008-5472.can-05-4203

Glynn, S. A., Boersma, B. J., Dorsey, T. H, Yi, M., Yfantis, H. G., Ridnour, L. A., Martin, D. N., Switzer, C. H., Hudson, R. S., Wink, D. A., Lee, D. H., Stephens, R. M., & Ambs, S. (2010). Increased NOS2 predicts poor survival in estrogen receptor-negative breast cancer patients. *Journal of Clinical Investigation, 120*, 3843–3854.

Godoy, L. C., Anderson, C. T. M., Chowdhury, R., Trudel, L. J., & Wogan, G. N. (2012). Endogenously produced nitric oxide mitigates sensitivity of melanoma cells to cisplatin. *Proceedings of the National Academy of Sciences, 109*(50), 20373–20378. https://doi.org/10.1073/pnas.1218938109

Gu, Z., Du, Y., Zhao, X., & Wang, C. (2021). Tumor microenvironment and metabolic remodeling in gemcitabine-based chemoresistance of pancreatic cancer. *Cancer Letters, 521*, 98–108.

Guan, W., Sha, J., Chen, X., Xing, Y., Yan, J., & Wang, Z. (2012). S-Nitrosylation of mitogen activated protein kinase phosphatase-1 suppresses radiation-induced apoptosis. *Cancer Letters, 314*(2), 137–146. https://doi.org/10.1016/j.canlet.2011.09.022

Guo, Y. Y., Deng, Y. J., Li, X. Q., Ning, Y., Lin, X. P., Guo, S. M., Chen, M. X., Han, M., & Feng, Y.-M. (2016). Glutaminolysis was induced by TGF-β1 through PP2Ac regulated Raf-MEK-ERK signaling in endothelial cells. *PLoS One, 11*(9). https://doi.org/10.1371/journal.pone.0162658

Gupta, A., Anjomani-Virmouni, S., Koundouros, N., Dimitriadi, M., Choo-Wing, R., Valle, A., Zheng, Y., Chiu, Y. H., Agnihotri, S., Zadeh, G., Asara, J. M., Anastasiou, D., Arends, M. J., Cantley, L. C., & Poulogiannis, G. (2017). PARK2 depletion connects energy and oxidative stress to PI3K/Akt activation via PTEN S-nitrosylation. *Molecular Cell, 65*(6), 999–1013. https://doi.org/10.1016/j.molcel.2017.02.019

Habiro, A., Tanno, S., Koizumi, K., Izawa, T., Nakano, Y., Osanai, M., Mizukami, Y., Okumura, T., & Kohgo, Y. (2004). Involvement of p38 mitogen-activated protein kinase in gemcitabine-induced apoptosis in human pancreatic cancer cells. *Biochemical and Biophysical Research Communications, 316*(1), 71–77. https://doi.org/10.1016/j.bbrc.2004.02.017

Ho, W. J., Jaffee, E. M., & Zheng, L. (2020). The tumour microenvironment in pancreatic cancer — clinical challenges and opportunities. *Nature Reviews Clinical Oncology, 17*(9), 527–540. https://doi.org/10.1038/s41571-020-0363-5

Holohan, C., Van Schaeybroeck, S., Longley, D. B., & Johnston, P. G. (2013). Cancer drug resistance: An evolving paradigm. *Nature Reviews Cancer, 13*(10), 714–726. https://doi.org/10.1038/nrc3599

Huang, P., Chubb, S., Hertel, L. W., Grindey, G. B., & Plunkett, W. (1991). Action of 20,20-difluorodeoxycytidine on DNA synthesis. *Cancer Research, 51*, 6110–6117.

Jiang, H., Zuo, J., Li, B., Chen, R., Luo, K., Xiang, X., Lu, S., Huang, C., Liu, L., Tang, J., & Gao, F. (2023). Drug-induced oxidative stress in cancer treatments: Angel or devil? *Redox Biology, 63*. https://doi.org/10.1016/j.redox.2023.102754

Khan, F. H., Dervan, E., Bhattacharyya, D. D., McAuli, J. D., Miranda, K. M., Sharon, A., & Glynn, S. A. (2020). The role of nitric oxide in cancer: Master regulator or NOt? *International Journal of Molecular Sciences, 21*, 9393.

Kumar, A., Ehrenshaft, M., Tokar, E. J., Mason, R. P., & Sinha, B. K. (2016). Nitric oxide inhibits topoisomerase II activity and induces resistance to topoisomerase II-poisons in human tumor cells. *Biochimica et Biophysica Acta - General Subjects, 1860*(7), 1519–1527. https://doi.org/10.1016/j.bbagen.2016.04.009

Kummer, J. L., Rao, P. K., & Heidenreich, K. A. (1997). Apoptosis induced by withdrawal of trophic factors is mediated by p38 mitogen-activated protein kinase. *Journal of Biological Chemistry, 272*(33), 20490–20494. https://doi.org/10.1074/jbc.272.33.20490

Kuramitsu, Y., Wang, Y., Taba, K., Suenaga, S., Ryozawa, S., Kaino, S., Sakaida, I., & Nakamura, K. (2012). Heat-shock protein 27 plays the key role in gemcitabine-resistance of pancreatic cancer cells. *Anticancer Research, 32*(6), 2295–2299.

López-Sánchez, L. M., Aranda, E., & Rodríguez-Ariza, A. (2020). Nitric oxide and tumor metabolic reprogramming. *Biochemical Pharmacology, 176*. https://doi.org/10.1016/j.bcp.2019.113769

Lee, S. Y., Jeon, H. M., Ju, M. K., Jeong, E. K., Kim, C. H., Park, H. G., Han, S. I., & Kang, H. S. (2016). Dlx-2 and glutaminase upregulate epithelial-mesenchymal transition and glycolytic switch. *Oncotarget, 7*(7), 7925–7939. https://doi.org/10.18632/oncotarget.6879

Leon-Bollotte, L., Subramaniam, S., Cauvard, O., Plenchettecolas, S., Paul, C., Godard, C., Martinezruiz, A., Legembre, P., Jeannin, J., & Bettaieb, A. (2011). S-Nitrosylation of the death receptor Fas promotes Fas ligandmediated apoptosis in cancer cells. *Gastroenterology, 140*(7), 2009–2018. https://doi.org/10.1053/j.gastro.2011.02.053

Li, F., Sonveaux, P., Rabbani, Z. N., Liu, S., Yan, B., Huang, Q., Vujaskovic, Z., Dewhirst, M. W., & Li, C. Y. (2007). Regulation of HIF-1alpha stability through S-nitrosylation. *Molecular Cell, 26*, 63–74.

Li, L., Zhu, L., Hao, B., Gao, W., Wang, Q., Li, K., Wang, M., Huang, M., Liu, Z., Yang, Q., Li, X., Zhong, Z., Huang, W., Xiao, G., Xu, Y., Yao, K., & Liu, Q. (2017). iNOS-derived nitric oxide promotes glycolysis by inducing pyruvate kinase M2 nuclear translocation in ovarian cancer. *Oncotarget, 8*(20), 33047–33063. https://doi.org/10.18632/oncotarget.16523

Liao, W., Ye, T., & Liu, H. (2019). Prognostic value of inducible nitric oxide synthase (iNOS) in human cancer: A systematic review and meta-analysis. *BioMed Research International, 2019*, 9.

Liau, S., Ashley, S., & Whang, E. (2006). Lentivirus-mediated RNA interference of HMGA1 promotes chemosensitivity to gemcitabine in pancreatic adenocarcinoma. *Journal of Gastrointestinal Surgery, 10*, 1254–1263.

Lim, K. H., Ancrile, B. B., Kashatus, D. F., & Counter, C. M. (2008). Tumour maintenance is mediated by eNOS. *Nature, 452*(7187), 646–649. https://doi.org/10.1038/nature06778

Liu, C., Li, C., & Liu, Y. (2023). The role of metabolic reprogramming in pancreatic cancer chemoresistance. *Frontiers in Pharmacology, 13*. https://doi.org/10.3389/fphar.2022.1108776

Lucena, M. C., Carvalho-Cruz, P., Donadio, J. L., Oliveira, I. A., De Queiroz, R. M., Marinho-Carvalho, M. M., Sola-Penna, M., De Paula, I. F., Gondim, K. C., McComb, M. E., Costello, C. E., Whelan, S. A., Todeschini, A. R., & Dias, W. B. (2016). Epithelial mesenchymal transition induces aberrant glycosylation through hexosamine biosynthetic pathway activation. *Journal of Biological Chemistry, 291*(25), 12917–12929. https://doi.org/10.1074/jbc.M116.729236

Lundberg, J. O., & Weitzberg, E. (2022). Nitric oxide signaling in health and disease. *Cell, 185*(16), 2853–2878. https://doi.org/10.1016/j.cell.2022.06.010

Maiuthed, A., Bhummaphan, N., Luanpitpong, S., Mutirangura, A., Aporntewan, C., Meepraser, A., Rungrotmongkol, T., Rojanasakul, Y., & Chanvorachote, P. (2018). Nitric oxide promotes cancer cell dedifferentiation by disrupting an Oct4:caveolin-1 complex: A new regulatory mechanism for cancer stem cell formation. *Journal of Biological Chemistry, 293*(35), 13534–13552. https://doi.org/10.1074/jbc.ra117.000287

Masetto, F., Chegaev, K., Gazzano, E., Mullappilly, N., Rolando, B., Arpicco, S., Fruttero, R., Riganti, C., & Donadelli, M. (2020). MRP5 nitration by NO-releasing gemcitabine encapsulated in liposomes confers sensitivity in chemoresistant pancreatic adenocarcinoma cells. *Biochimica et Biophysica Acta, Molecular Cell Research, 1867*, 118824.

Mateo, J., García-Lecea, M., Cadenas, S., Hernández, C., & Moncada, S. (2003). Regulation of hypoxia-inducible factor-1α by nitric oxide through mitochondria-dependent and -independent pathways. *Biochemical Journal, 376*(2), 537–544. https://doi.org/10.1042/bj20031155

McDonald, P. C., Chafe, S. C., Brown, W. S., Saberi, S., Swayampakula, M., Venkateswaran, G., Nemirovsky, O., Gillespie, J. A., Karasinska, J. M., Kalloger, S. E., Supuran, C. T., Schaeffer, D. F., Bashashati, A., Shah, S. P., Topham, J. T., Yapp, D. T., Li, J., Renouf, D. J., Stanger, B. Z., & Dedhar, S. (2019). Regulation of pH by carbonic anhydrase 9 mediates survival of pancreatic cancer cells with activated KRAS in response to hypoxia. *Gastroenterology, 157*(3), 823–837. https://doi.org/10.1053/j.gastro.2019.05.004

Mena-Osuna, R., Mantrana, A., Guil-Luna, S., Sánchez-Montero, M. T., Navarrete-Sirvent, C., Morales-Ruiz, T., Rivas-Crespo, A., Toledano-Fonseca, M., García-Ortíz, M. V., García-Jurado, G., Gómez-España, M. A., González-Fernández, R., Villar, C., Medina-Fernández, F. J., Villalba, J. M., Aranda, E., & Rodríguez-Ariza, A. (2023). Metabolic shift underlies tumor progression and immune evasion in S-nitrosoglutathione reductase-deficient cancer. *The Journal of Pathology, 260*(3), 261–275. https://doi.org/10.1002/path.6080

Monteiro, H. P., Rodrigues, E. G., Amorim Reis, A. K. C., Longo, L. S., Ogata, F. T., Moretti, A. I. S., da Costa, P. E., Teodoro, A. C. S., Toledo, M. S., & Stern, A. (2019). Nitric oxide and interactions with reactive oxygen species in the development of melanoma, breast, and colon cancer: A redox signaling perspective. *Nitric Oxide - Biology and Chemistry, 89*, 1–13. https://doi.org/10.1016/j.niox.2019.04.009

Morote-Garcia, J. C., Rosenberger, P., Nivillac, N. M. I., Coe, I. R., & Eltzschig, H. K. (2009). Hypoxia-inducible factor-dependent repression of equilibrative nucleoside transporter 2 attenuates mucosal inflammation during intestinal hypoxia. *Gastroenterology, 136*(2), 607–618. https://doi.org/10.1053/j.gastro.2008.10.037

Müerköster, S. S., Lust J,Arlt, A., Häsler, R., Witt, M., Sebens, T., Schreiber, S., Fölsch, U. R., & Schäfer, H. (2006). Acquired chemoresistance in pancreatic carcinoma cells: Induced secretion of IL-1beta and NO lead to inactivation of caspases. *Oncogene, 25*, 3973–3981.

Mukhopadhyay, S., Goswami, D., Adiseshaiah, P. P., Burgan, W., Yi, M., Guerin, T. M., Kozlov, S. V., Nissley, D. V., & McCormick, F. (2020). Undermining glutaminolysis bolsters chemotherapy while NRF2 promotes chemoresistance in KRAS-driven pancreatic cancers. *Cancer Research, 80*(8), 1630–1643. https://doi.org/10.1158/0008-5472.CAN-19-1363

Murad, F. (1998). Nitric oxide signaling: Would you believe that a simple free radical could be a second messenger, autacoid, paracrine substance, neurotransmitter, and hormone? *Recent Progress in Hormone Research, 53*, 43–59.

Murad, F. (2006). Nitric oxide and cyclic GMP in cell signaling and drug development. *New England Journal of Medicine, 355*, 2003–2011.

Nakano, Y., Tanno, S., Koizumi, K., Nishikawa, T., Nakamura, K., Minoguchi, M., Izawa, T., Mizukami, Y., Okumura, T., & Kohgo, Y. (2007). Gemcitabine chemoresistance and molecular markers associated with gemcitabine transport and metabolism in human pancreatic cancer cells. *British Journal of Cancer, 96*(3), 457–463. https://doi.org/10.1038/sj.bjc.6603559

Nakashima, M., Adachi, S., Yasuda, I., Yamauchi, T., Kawaguchi, J., Itani, M., Yoshioka, T., Matsushima-Nishiwaki, R., Hirose, Y., Kozawa, O., & Moriwaki, H. (2011). Phosphorylation status of heat shock protein 27 plays a key role in gemcitabine-induced apoptosis of pancreatic cancer cells. *Cancer Letters, 313*(2), 218–225. https://doi.org/10.1016/j.canlet.2011.09.008

Napoli, C., Paolisso, G., Casamassimi, A., Al-Omran, M., Barbieri, M., Sommese, L., Infante, T., & Ignarro, L. J. (2013). Effects of nitric oxide on cell proliferation: Novel insights. *Journal of the American College of Cardiology, 62*(2), 89–95. https://doi.org/10.1016/j.jacc.2013.03.070

Natu, J., & Nagaraju, G. P. (2023). Gemcitabine effects on tumor microenvironment of pancreatic ductal adenocarcinoma: Special focus on resistanec mechanisms and metronomic therapies. *Cancer Letters, 573*, 216382.

Neesse, A., Algül, H., Tuveson, D. A., & Gress, T. M. (2015). Stromal biology and therapy in pancreatic cancer: A changing paradigm. *Gut, 64*(9), 1476–1484. https://doi.org/10.1136/gutjnl-2015-309304

Ohhashi, S., Ohuchida, K., Mizumoto, K., Fujita, H., Egami, T., Yu, J., Toma, H., Sadatomi, S., Nagai, E., & Tanaka, M. (2008). Down-regulation of deoxycytidine kinase enhances acquired resistance to gemcitabine in pancreatic cancer. *Anticancer Research, 28*(4 B), 2205–2212.

Ohtaka, K., Kohya, N., Sato, K., Kitajima, Y., Ide, T., Mitsuno, M., & Miyazaki, K. (2008). Ribonucleotide reductase subunit M1 is a possible chemoresistance marker to gemcitabine in biliary tract carcinoma. *Oncology Reports, 20*(2), 279–286. https://doi.org/10.3892/or_00000004

Paik, J. H., Kollipara, R., Chu, G., Ji, H., Xiao, Y., Ding, Z., Miao, L., Tothova, Z., Horner, J. W., Carrasco, D. R., Jiang, S., Gilliland, D. G., Chin, L., Wong, W. H., Castrillon, D. H., & DePinho, R. A. (2007). FoxOs are lineage-restricted redundant tumor suppressors and regulate endothelial cell homeostasis. *Cell, 128*(2), 309–323. https://doi.org/10.1016/j.cell.2006.12.029

Park, J. M., Peng, J. M., Shen, Y. S., Lin, C. Y., Hsu, T. W., Su, Y. H., Chen, H. A., Saengboonmee, C., Chang, J. S., Chiu, C. F., & Shan, Y. S. (2022). Phosphomimetic Dicer S1016E triggers a switch to glutamine metabolism in gemcitabine-resistant pancreatic cancer. *Molecular Metabolism, 65*. https://doi.org/10.1016/j.molmet.2022.101576

Partyka, O., Pajewska, M., Kwaśniewska, D., Czerw, A., Deptała, A., Budzik, M., Cipora, E., Gaska, I., Gazdowicz, L., Mielnik, A., Sygit, K., Sygit, M., Krzych-Fałta, E., Schneider-Matyka, D., Grochans, S., Cybulska, A. M., Drobnik, J., Bandurska, E., Ciećko, W., … Kozłowski, R. (2023). Overview of pancreatic cancer epidemiology in Europe and recommendations for screening in high-risk populations. *Cancers, 15*(14). https://doi.org/10.3390/cancers15143634

PeÑarando, J., Aranda, E., & RodrÍguez-Ariza, A. (2019). Immunomodulatory roles of nitric oxide in cancer: Tumor microenvironment says "NO" to antitumor immune response. *Translational Research, 210*, 99–108. https://doi.org/10.1016/j.trsl.2019.03.003

Pereira, S., Fernandes, P. A., & Ramos, M. J. (2004). Mechanism for ribonucleotide reductase inactivation by the anticancer drug gemcitabine. *Journal of Computational Chemistry, 25*(10), 1286–1294. https://doi.org/10.1002/jcc.20054

Raffoul, J.J., Heydari, A.R., & Hillman, G.G. (2012). DNA repair and cancert herapy: Targeting APE1/Ref-1 using dietary agents. *Journal of Oncology, 2012*, 370481.

Rahib, L., Wehner, M. R., Matrisian, L. M., & Nead, K. T. (2021). Estimated projection of US cancer incidence and death to 2040. *JAMA Network Open, 4*, Article e214708.

Reddy, T. P., Glynn, S. A., Billiar, T. R., Wink, D. A., & Chang, J. C. (2023). Targeting nitric oxide: Say NO to metastasis. *Clinical Cancer Research, 29*(10), 1855–1868. https://doi.org/10.1158/1078-0432.CCR-22-2791

Ridnour, L. A., Thomas, D. D., Donzelli, S., Espey, M. G., Roberts, D. D., Wink, D. A., & Isenberg, J. S. (2006). The biphasic nature of nitric oxide responses in tumor biology. *Antioxidants and Redox Signaling, 8*(7–8), 1329–1337. https://doi.org/10.1089/ars.2006.8.1329

Rizza, S., Montagna, C., Cardaci, S., Maiani, E., Di Giacomo, G., Sanchez-Quiles, V., Blagoev, B., Rasola, A., De Zio, D., Stamler, J. S., Cecconi, F., & Filomeni, G. (2016). S -nitrosylation of the mitochondrial chaperone TRAP1 sensitizes hepatocellular carcinoma cells to inhibitors of succinate dehydrogenase. *Cancer Research, 76*(14), 4170–4182. https://doi.org/10.1158/0008-5472.can-15-2637

Salimian Rizi, B., Achreja, A., & Nagrath, D. (2017). Nitric oxide: The forgotten child of tumor metabolism. *Trends in Cancer, 3*(9), 659–672. https://doi.org/10.1016/j.trecan.2017.07.005

Sharma, N. K., Kumar, A., Kumari, A., Tokar, E. J., Waalkes, M. P., Bortner, C. D., Williams, J., Ehrenshaft, M., Mason, R. P., & Sinha, B. K. (2015). Nitric oxide down-regulates topoisomerase i and induces camptothecin resistance in human breast MCF-7 tumor cells. *PLoS One, 10*(11). https://doi.org/10.1371/journal.pone.0141897

Singh, S., Srivastava, S. K., Bhardwaj, A., Owen, L. B., & Singh, A. P. (2010). CXCL12–CXCR4 signalling axis confers gemcitabine resistance to pancreatic cancer cells: A novel target for therapy. *British Journal of Cancer, 103*(11), 1671–1679. https://doi.org/10.1038/sj.bjc.6605968

Skrypek, N., Duchêne, B., Hebbar, M., Leteurtre, E., van Seuningen, I., & Jonckheere, N. (2013). The MUC4 mucin mediates gemcitabine resistance of human pancreatic cancer cells via the Concentrative Nucleoside Transporter family. *Oncogene, 32*(13), 1714–1723. https://doi.org/10.1038/onc.2012.179

Somasundaram, V., Basudhar, D., Bharadwaj, G., No, J. H., Ridnour, L. A., Cheng, R. Y. S., Fujita, M., Thomas, D. D., Anderson, S. K., McVicar, D. W., & Wink, D. A. (2019). Molecular mechanisms of nitric oxide in cancer progression, signal transduction, and metabolism. *Antioxidants and Redox Signaling, 30*(8), 1124–1143. https://doi.org/10.1089/ars.2018.7527

Son, J., Lyssiotis, C. A., Ying, H., Wang, X., Hua, S., Ligorio, M., Perera, R. M., Ferrone, C. R., Mullarky, E., Shyh-Chang, N., Kang, Y., Fleming, J. B., Bardeesy, N., Asara, J. M., Haigis, M. C., Depinho, R. A., Cantley, L. C., & Kimmelman, A. C. (2013). Glutamine supports pancreatic cancer growth through a KRAS-regulated metabolic pathway. *Nature, 496*(7443), 101–105. https://doi.org/10.1038/nature12040

Stamler, J. S., Lamas, S., & Fang, F. C. (2001). Nitrosylation: The prototypic redox-based signaling mechanism. *Cell, 106*(6), 675–683. https://doi.org/10.1016/S0092-8674(01)00495-0

Stomberski, C. T., Hess, D. T., & Stamler, J. S. (2019). Protein S-nitrosylation: Determinants of specificity and enzymatic regulation of S-nitrosothiol-based signaling. *Antioxidants and Redox Signaling, 30*(10), 1331–1351. https://doi.org/10.1089/ars.2017.7403

Swierczynski, J., Hebanowska, A., & Sledzinski, T. (2014). Role of abnormal lipid metabolism in development, progression, diagnosis and therapy of pancreatic cancer. *World Journal of Gastroenterology, 20*(9), 2279–2303. https://doi.org/10.3748/wjg.v20.i9.2279

Swords, D. S., Mone, M. C., Zhang, C., Presson, A. P., Mulvihill, S. J., & Scaife, C. L. (2015). Initial misdiagnosis of proximal pancreatic adenocarcinoma is associated with delay in diagnosis and advanced stage at presentation. *Journal of Gastrointestinal Surgery, 19*(10), 1813–1821. https://doi.org/10.1007/s11605-015-2923-z

Takagi, T., Fujiwara-Tani, R., Mori, S., Kishi, S., Nishiguchi, Y., Sasaki, T., Ogata, R., Ikemoto, A., Sasaki, R., Ohmori, H., Luo, Y., Bhawal, U. K., Sho, M., & Kuniyasu, H. (2023). Lauric acid overcomes hypoxia-induced gemcitabine chemoresistance in pancreatic ductal adenocarcinoma. *International Journal of Molecular Sciences, 24*(8). https://doi.org/10.3390/ijms24087506

Tan, C., Li, Y., Huang, X., Wei, M., Huang, Y., Tang, Z., Huang, H., Zhou, W., Wang, Y., & Hu, J. (2019). Extensive protein S-nitrosylation associated with human pancreatic ductal adenocarcinoma pathogenesis. *Cell Death and Disease, 10*(12). https://doi.org/10.1038/s41419-019-2144-6

Thomas, D. D., Heinecke, J. L., Ridnour, L. A., Cheng, R. Y., Kesarwala, A. H., Switzer, C. H., McVicar, D. W., Roberts, D. D., Glynn, S., Fukuto, J. M., Wink, D. A., & Miranda, K. M. (2015). Signaling and stress: The redox landscape in NOS2 biology. *Free Radical Biology and Medicine, 87*, 204–225. https://doi.org/10.1016/j.freeradbiomed.2015.06.002

Tian, S., Li, P., Sheng, S., & Jin, X. (2018). Upregulation of pyruvate kinase m2 expression by fatty acid synthase contributes to gemcitabine resistance in pancreatic cancer. *Oncology Letters, 15*(2), 2211–2217. https://doi.org/10.3892/ol.2017.7598

Tilokani, L., Nagashima, S., Paupe, V., & Prudent, J. (2018). Mitochondrial dynamics: Overview of molecular mechanisms. *Essays in Biochemistry, 62*, 341–360.

Tothova, Z., Kollipara, R., Huntly, B. J., Lee, B. H., Castrillon, D. H., Cullen, D. E., McDowell, E. P., Lazo-Kallanian, S., Williams, I. R., Sears, C., Armstrong, S. A., Passegué, E., DePinho, R. A., & Gilliland, D. G. (2007). FoxOs are critical mediators of hematopoietic stem cell resistance to physiologic oxidative stress. *Cell, 128*(2), 325–339. https://doi.org/10.1016/j.cell.2007.01.003

Vasudevan, D., & Thomas, D. D. (2014). Insights into the diverse effects of nitric oxide on tumor biology. *Vitamins and Hormones, 96*, 265–298. https://doi.org/10.1016/B978-0-12-800254-4.00011-8

Vaupel, P., & Multhoff, G. (2021). The Warburg effect: Historical dogma versus current rationale. *Advances in Experimental Medicine and Biology, 1269*, 169–177. https://doi.org/10.1007/978-3-030-48238-1_27

Wang, L., Tang, W., Yang, S., He, P., Wang, J., Gaedcke, J., Ströbel, P., Azizian, A., Ried, T., Gaida, M. M., Yfantis, H. G., Lee, D. H., Lal, A., Van den Eynde, B. J., Alexander, H. R., Ghadimi, B. M., Hanna, N., & Hussain, S. P. (2020). NO/RUNX3/kynurenine metabolic signaling enhances disease aggressiveness in pancreatic cancer. *International Journal of Cancer, 146*, 3160–3169.

Wang, J., Torbenson, M., Wang, Q., Ro, J. Y., & Becich, M. (2003). Expression of inducible nitric oxide synthase in paired neoplastic and non-neoplastic primary prostate cell cultures and prostatectomy specimen. *Urologic Oncology: Seminars and Original Investigations, 21*(2), 117–122.

Wang, X., Zhang, H., & Chen, X. (2019). Drug resistance and combating drug resistance in cancer. *Cancer Drug Resistance, 2*(2), 141–160. https://doi.org/10.20517/cdr.2019.10

Wang, J., He, P., Gaida, M., Yang, S., Schetter, A. J., Gaedcke, J., Ghadimi, B. M., Ried, T., Yfantis, H., Lee, D., Weiss, J. M., Stauffer, J., Hanna, N., Alexander, H. R., & Hussain, S. P. (2016). Inducible nitric oxide synthase enhanecs disease aggressiveness in pancreatic cancer. *Oncotarget, 7*, 52993–53004.

Wang, J., & Hussain, S. P. (2017). NO and pancreatic cancer: A complex interaction with therapeutic potential. *Antioxidants & Redox Signaling, 26*, 1000–1008.

Wang, Z., Li, Y., Kong, D., Banerjee, S., Ahmad, A., Azmi, A., Ali, S., Abbruzzese, J. L., Gallick, G. E., & Sarkar, F. H. (2009). Acquisition of epithelial-mesenchymal transition phenotype of gemcitabine-resistant pancreatic cancer cells is linked with activation of the notch signaling pathway. *Cancer Research, 69*, 2400–2407.

Wang, C., Ma, J., Zhang, N., Yang, Q., Jin, Y., & Wang, Y. (2015). The acetyl-CoA carboxylase enzyme: a target for cancer therapy? *Expert Review of Anticancer Therapy, 15*, 667–676.

Wang, Y., Tang, Q., Wu, R., Sun, S., Zhang, J., Chen, J., Gong, M., Chen, C., & Liang, X. (2023). Ultrasound-triggered piezocatalyst for selectively controlled NO gas and chemodrug release to enhance drug penetration in pancreatic cancer. *ACS Nano, 17*, 3557–3573.

Wongvaranon, P., Pongrakhananon, V., Chunhacha, P., & Chanvorachote, P. (2013). Acquired resistance to chemotherapy in lung cancer cells mediated by prolonged nitric oxide exposure. *Anticancer Research, 33*(12), 5433–5444.

Yang, W., & Lu, Z. (2013). Nuclear PKM2 regulates the Warburg effect. *Cell Cycle, 12*, 3154–3158.

Yokoyama, C., Sueyoshi, Y., Ema, M., Mori, Y., Takaishi, K., & Hisatomi, H. (2017). Induction of oxidative stress by anticancer drugs in the presence and absence of cells. *Oncology Letters, 14*(5), 6066–6070. https://doi.org/10.3892/ol.2017.6931

You, M., Xie, Z., Zhang, N., Zhang, Y., Xiao, D., Liu, S., Zhuang, W., Li, L., & Tao, Y. (2023). Signaling pathways in cancer metabolism: Mechanisms and therapeutic targets. *Signal Transduction and Targeted Therapy, 8*(1). https://doi.org/10.1038/s41392-023-01442-3

Zafirellis, K., Zachaki, A., Agrogiannis, G., & Gravani, K. (2010). Inducible nitric oxide synthase expression and its prognostic significance in colorectal cancer. *Acta Pathologica, Microbiologica et Immunologica Scandinavica, 118*(2), 115–124. https://doi.org/10.1111/j.1600-0463.2009.02569.x

Zahra, K., Dey, T., Ashish, Mishra, S. P., & Pandey, U. (2020). Pyruvate kinase M2 and cancer: The role of PKM2 in promoting tumorigenesis. *Frontiers in Oncology, 10*. https://doi.org/10.3389/fonc.2020.00159

Zheng, H.-C. (2017). The molecular mechanisms of chemoresistance in cancers. *Oncotarget, 8*, 59950–59964.

CHAPTER

7

Exploring the dual role of nitric oxide in glioblastoma

Inesa Navasardyan[1,2] *and Benjamin Bonavida*[1]

[1]Department of Microbiology, Immunology and Molecular Genetics, David Geffen School of Medicine, Jonsson Comprehensive Cancer Center, University of California, Los Angeles, CA, United States; [2]College of Osteopathic Medicine of the Pacific, Western University of Health Sciences, Pomona, CA, United States

Abbreviations

BBB blood–brain barrier
CAMs cell adhesion molecules
CAR-T chimeric antigen receptor T cell
ECM extracellular matrix
EGFR epidermal growth factor receptor
EMT epithelial–mesenchymal transition
eNOS endothelial nitric oxide synthase
GBM glioblastoma multiforme
GSCs glioblastoma stem cells
HIF-1α hypoxia-inducible factor 1-alpha
IFN-γ interferon-gamma
IL-1β interleukin-1 beta
iNOS inducible nitric oxide synthase
L-NAME Nω-nitro-L-arginine methyl ester
MGMT O6-methylguanine-DNA methyltransferase
MMPs matrix metalloproteinases
mTOR mammalian target of rapamycin
NIC nanoscale immunoconjugates
nNOS neuronal nitric oxide synthase
NO nitric oxide
NOS nitric oxide synthase
PHDs prolyl hydroxylases
PI3K phosphoinositide 3-kinase
RKIP Raf-1 kinase inhibitor protein
RNS reactive nitrogen species

Therapeutic Applications of Nitric Oxide in Cancer and Inflammatory Disorders
https://doi.org/10.1016/B978-0-443-22354-9.00007-3

ROS reactive oxygen species
sEVs serum-derived small extracellular vesicles
STAT3 signal transducer and activator of transcription 3
TGF-β transforming growth factor-beta
TIMP-1 tissue inhibitor of metalloproteinase-1
TME tumor microenvironment
TMZ temozolomide
TNF-α tumor necrosis factor-alpha
TSP-1 thrombospondin-1
VEGF vascular endothelial growth factor
VHL von Hippel-Lindau

Introduction

Glioblastoma multiforme (GBM) represents a highly aggressive and rapidly growing form of brain cancer comprising approximately 50% of all gliomas (Hanif et al., 2017). Gliomas are primary brain tumors originating from glial cells, whereas GBMs specifically arise from astrocytes, a specialized type of glial cell that support nerve cells; therefore, GBM is also referred to as a grade IV astrocytoma (Walker et al., 2011). Initially manifesting as a growth of cells in the brain or spinal cord, GBM rapidly grows to infiltrate and destroy the nearby tissue (Seker-Polat et al., 2022). GBM constitutes the most common form of high-grade primary malignant brain tumor with approximately 12,000 new cases diagnosed annually in the United States (Mesfin & Al-Dhahir, 2023).

GBM is believed to arise partly due to genetic and epigenetic alterations in normal astroglial cells, leading to their transformation into a malignant state (Clarke et al., 2013). Such alterations are largely due to molecular signaling pathways such as the activation of the epidermal growth factor receptor (EGFR) or the downregulation of the tumor suppressor genes *p53* and *PTEN* (Hatanpaa et al., 2010; Zheng et al., 2008). Amplification and overexpression of EGFR is a significant feature of GBM and that occurs through both ligand-dependent as well as ligand-independent mechanisms (Xu et al., 2017). EGFRvIII is the most common EGFR mutation in GBM and is the result of the deletion of exons 2 through 7, resulting in a constitutively active EGFR protein (Pan & Magge, 2020). Moreover, concomitant deletion of the *p53* and *PTEN* genes in the CNS of mice resulted in a presentation resembling primary GBM in humans (Zheng et al., 2008). Further mutational analysis in primary human GBM revealed frequent inactivating mutations of the tumor suppressors p53 and PTEN, leading to increased tumor growth (Zheng et al., 2008). The intricate nature of GBM may be attributed to the interplay and cross-talk among various signaling pathways, rendering it difficult to understand the tumor behavior comprehensively and difficulty to effectively treat GBM.

Treatment of GBM

Current treatment modalities in patients with GBM include surgery, radiotherapy, and combination therapy involving chemotherapy with concomitant temozolomide (TMZ) (Fernandes et al., 2017). Despite the availability of such treatments, individuals diagnosed with GBM often have a poor prognosis and median survival of approximately 15 months despite aggressive therapeutic approaches (Tamimi et al., 2017). Immunotherapy has emerged as a promising avenue for GBM research, encompassing strategies such as immune checkpoint inhibitors and chimeric antigen receptor (CAR) T-cell therapy (Rocha Pinheiro et al., 2023). A phase I clinic trial involving intracerebral administration of PD-1 and CTLA-4 immune checkpoint blocking monoclonal antibodies following maximal safe resection in patients with recurrent GBM revealed increased overall survival rates (Duerinck et al., 2021). Moreover, CAR T cell therapy involves the extraction and modification of autologous T cells to express a specific receptor against a GBM antigen that are then infused back into the patient. Preclinical studies using CAR T cell therapies have yielded promising results and are now being tested in clinical trials of GBM (Luksik et al., 2023; Wang et al., 2023). Despite recent advances in immunotherapeutics, the efficacy of targeted therapies against GBM is limited by the ability of certain drugs to effectively penetrate the blood–brain barrier (BBB) (Rui & Green, 2021). The limited success of checkpoint inhibitor antibodies to CTLA-4 and PD-1 is primarily attributed to their inability to cross the BBB, and thus, the use of targeted nanoscale immunoconjugates (NIC) was proposed to increase systemic delivery of immune checkpoint inhibitors across the BBB to activate the antitumor immune response in the brain (Galstyan et al., 2019). The development of novel strategies to overcome the BBB, such as the use of nanotherapeutics, is currently being studied to allow for increased drug delivery across the BBB to effectively target brain cells.

NO production and its biological process

Nitric oxide (NO) is a highly reactive free radical that acts as a signaling molecule in both physiological as well as pathological processes. NO is produced from a group of enzymes termed nitric oxide synthases (NOS) via the conversion of arginine into citrulline. The three NOS isoforms include endothelial NOS (eNOS), inducible NOS (iNOS), and neuronal NOS (nNOS) (Förstermann & Sessa, 2012). iNOS, also known as NOS2, is often expressed by T cells, macrophages, and mature dendritic cells and is thought to be transcriptionally activated by inflammatory cytokines (Förstermann & Sessa, 2012). iNOS/NOS2 is a well-studied component of the immune response, particularly in cancer, where it may exert either an anti- or protumorigenic response (Vannini et al., 2015). Increased iNOS activity is associated with malignancy in gastric cancer, squamous cell carcinoma, hepatocellular carcinoma, and melanoma, among others (Vannini et al., 2015). Conversely, in other forms of cancer, such as ovarian and lung cancer, increased iNOS activity suggests a favorable prognosis (Vannini et al., 2015). Given its dual role among various cancers, the direct role of iNOS activity in glioma, and particularly in GBM, is an area of ongoing research and interest.

The current literature suggests that low and moderate concentrations of NO (50−100 nM) exhibit a protumorigenic effect, whereas higher concentrations (>300 nM) of NO are antitumorigenic and exert cytotoxic effects on neoplastic cells (Mocellin et al., 2007). Herein, we discuss the role of NOS and NO production as a double-edged sword in the pathogenesis and treatment of cancer, namely in GBM.

Role of the INOS/NO pathway in GBM

NO is a crucial signaling molecule implicated in many biological processes that have dual effects in cancer, exerting either tumor-suppressing or tumor-promoting effects, depending on its time, location, and respective concentration (Khan et al., 2020). iNOS expression may be induced by various cells in the tumor microenvironment (TME) including proinflammatory cytokines such as interleukin-1 beta (IL-1β), tumor necrosis factor-alpha (TNF-α), and interferon-gamma (IFN-γ) (Tran et al., 2017). Increased expression of NOS2 was observed in human A-172 and T98G GBM cell lines following treatment with IL-1β, TNF-α, and IFN-γ (Fujisawa et al., 1995; Hokari et al., 1994). The direct effect of NO is largely dependent on its concentration in the TME, and therefore, it is important to consider the relative concentration of iNOS/NO in GBM to understand its role as either a facilitator or suppressor of tumor progression.

Role of NO as tumor progressor

Reduced levels of NO are often associated with tumor progression (Khan et al., 2020). Notably, low levels of NO (<100 nM) promote increased proliferation and angiogenesis, whereas moderate levels of NO (100−500 nM) are associated with reduced apoptosis and increased invasion and metastasis of tumor cells (Burke et al., 2013). The complex role of NO in the progression of cancer, and more specifically in GBM, involves an intricate interplay of signaling pathways and molecules discussed in the following.

Mutation resulting in the overexpression of the epidermal growth factor receptor (EGFR) is a common finding in individuals with GBM, and consequently, this increased expression of EGFR leads to the overactivation of downstream pathways involved in the pathogenesis of GBM (Xu et al., 2017). EGFRvIII is a constitutively active receptor in GBM that operates independently from the EGF ligand and is frequently encountered in GBM (Xu et al., 2017). Activation of the signal transducer and activator of transcription 3 (STAT3), a downstream target of EGFRvIII, is implicated in the survival of GBM through increased differentiation, invasion, angiogenesis, and immunosuppression; thus, EGFRvIII-mediated overactivation of STAT3 is a crucial regulator of GBM pathogenesis (Kim et al., 2014). iNOS has been identified as a target of STAT3 in EGFRvIII-expressing mouse astrocytes (Puram et al., 2012). STAT3 directly binds the iNOS promoter and upregulates its transcription, leading to glial proliferation and transformation, suggesting a protumorigenic role of NO in GBM (Puram et al., 2012). Moreover, EGFRvIII mutations in GBM result in the loss of PTEN, a negative regulator of the phosphatidylinositol 3-kinase (PI3K)/Akt/rapamycin-sensitive mTOR-complex (mTOR) pathway (Smith et al., 2001). Stimulation of the PI3K/Akt pathway in GBM is associated with tumor cell survival, growth, motility, angiogenesis, and metabolism (Zhao et al., 2017). The loss of PTEN has prompted the use of PI3K inhibitors in both preclinical and clinical studies of GBM, which have yielded promising results (Zhao et al., 2017). Buparlisib, a PI3K inhibitor, is able to effectively penetrate the BBB and exert its antitumor effects regardless of the PTEN or EGFR status in the GBM cell lines and xenograft models (Fan et al., 2017). Buparlisib also demonstrated synergistic effects when used in combination with TMZ in GBM mouse xenografts deficient in PTEN (Wen et al., 2020). Interestingly, NO has been found to suppress the function of PTEN and, in turn, stimulate the PI3K/Akt pathway involved in tumor survival and progression (Ding et al., 2021). Conversely, NO inhibition of YY1, a negative regulator of PTEN expression, may lead to increased expression of PTEN (Hays & Bonavida, 2019). The role of NO on PTEN expression remains conflicting and may vary across types of cancers; thus, the role of NO must be studied in GBM cancer cell lines to determine its specific role in GBM pathogenesis.

Matrix metalloproteinases (MMPs) are a family of enzymes that catalyze the degradation of extracellular matrix (ECM) components such as fibronectin, laminin, collagen, and gelatin. The degradation of the ECM by MMPs allows tumor cells to migrate and invade surrounding tissues allowing for metastasis (Elgundi et al., 2020). MMP-9 and its overexpression in GBM are well known for promoting tumor cell growth and increased clonogenic potential in GBM cell lines (Xue et al., 2017). ELISA measurements of MMP-9 levels of serum-derived small extracellular vesicles (sEVs) revealed a positive correlation between MMP-9 levels and aggressiveness of GBM, suggesting MMP-9 level of serum-sEVs as a negative prognostic biomarker for GBM (Dobra et al., 2023). Low levels of exogenous NO are found to suppress the tissue inhibitor of metalloproteinase (TIMP)-1 and, in turn, enhance MMP activity in a cGMP-dependent manner (Ridnour et al., 2007). Interestingly, the concentration-dependent activation and inactivation of MMP-9 occurred at low and high levels of exogenous NO, respectively (Ridnour et al., 2007).

The Wnt/β-catenin signaling pathway is largely implicated in the cellular differentiation and proliferation of cancer cells. The role of the Wnt signaling pathway in GBM is in the maintenance of glioblastoma stem cells (GSCs), promotion of invasion and metastasis, as well as the induction of multidrug resistance (Lee et al., 2016). The neural stem cells require Wnt signaling; however, it is the overamplification of this pathway that results in the proliferation and invasiveness observed in aggressive GBM tumors (Latour et al., 2021). iNOS expression was found to be inversely correlated with expression of the Wnt/β-catenin regulator DKK1; thus, induced NO synthesis increases the Wnt/β-catenin signaling via downregulation of the *DKK1* gene to promote tumor progression (Du et al., 2013). Inhibition of iNOS resulted in restored levels of DKK1 expression and, thus, inhibition of Wnt signaling to regulate tumor growth (Du et al., 2013).

GBM is a highly vascularized form of brain tumor with glioma growth dependent on the formation of new blood vessels; therefore, angiogenesis plays a crucial role in the pathogenesis of GBM (Ahir et al., 2020). NO is known to stabilize the hypoxia-inducible factor 1-alpha (HIF-1α) and stimulate the secretion of vascular endothelial growth factor (VEGF), both of which are implicated in the regulation of angiogenesis and, subsequently, tumor progression (Yamamoto et al., 2020). Under conditions of normoxia, prolyl hydroxylases (PHDs) hydroxylate specific proline residues on HIF-1α, leading to recognition and degradation by von Hippel-Lindau (VHL) tumor-suppressor protein. NO inhibits the action of PHDs and, thus,

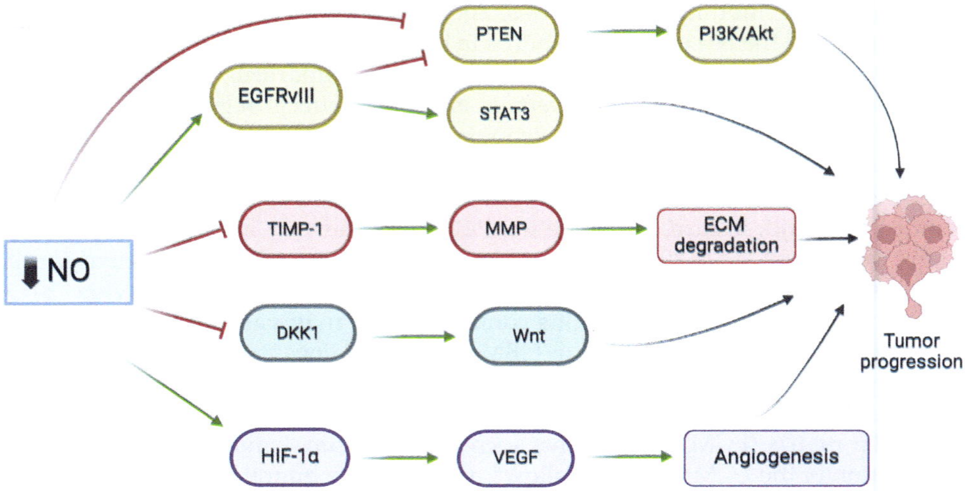

FIGURE 7.1 Role of nitric oxide (NO) in glioblastoma multiforme (GBM) tumor progression. Reduced levels of NO are associated with GBM progression through various processes. EGFRvIII overexpression in GBM activates STAT3, promoting survival, invasion, and angiogenesis. NO also directly suppresses PTEN expression, activating the PI3K/Akt pathway. NO affects MMP-9 and ECM degradation, influencing tumor cell migration and invasion. NO-driven inhibition of DKK1 enhances Wnt signaling, promoting tumor growth. Lastly, NO stabilizes HIF-1α, leading to VEGF secretion and angiogenesis, crucial for GBM growth.

results in the stabilization of HIF-1α (Olson & van der Vliet, 2011). Moreover, direct modification of HIF-1α via NO-mediated S-nitrosylation of a single cysteine residue, preventing its interaction with VHL and allowing HIF-1α to escape degradation (Li et al., 2007). Knockdown of HIF-1α was found to inhibit invasion and migration of human and murine glioma cells in vivo (Méndez et al., 2010). Following stabilization and activation of HIF-1α, it translocates to the nucleus where it targets and upregulates VEGF expression, a crucial proangiogenic factor in GBM (Kaur et al., 2005). Fig. 7.1 schematically summarizes the role of NO in GBM as a tumor progressor.

Reduced nitric oxide (NO) levels in glioblastoma multiforme (GBM) are associated with activation of various protumorigenic pathways, including PI3K/Akt, Wnt, HIF-1α/VEGF, and STAT3 signaling, which promote GBM cell survival, invasion, and angiogenesis. Additionally, overexpression of the EGFRvIII mutant in GBM further activates STAT3, further driving GBM progression.

Role of NO as tumor suppressor

Whereas lower concentrations of iNOS/NO are associated with cytoprotective effects through the activation of various molecules and pathways, increased levels of iNOS and subsequent NO production exhibit cytotoxic effects on tumor cells associated with tumor suppression (Xu et al., 2002). High levels of NO (>500 nM) exert its anticancer effects through DNA damage, free radical formation, cytotoxicity, and apoptosis (Burke et al., 2013). Herein, we describe several mechanisms in which NO exerts its antitumorigenic effects in GBM

The imbalance between the production and clearance of free radicals results in a state of oxidative and nitrosative stress, ultimately resulting in harmful effects in cancer cells through protein modification, lipid peroxidation, and DNA damage (Wink & Mitchell, 1998). High concentrations of NO have been found to induce cell cycle arrest, apoptosis, and senescence in cancer cells through the production of reactive oxygen species (ROS) and reactive nitrogen species (RNS), resulting in reduced tumor growth (Vahora et al., 2016). Nitrosative DNA damage may cause cells to undergo cell cycle arrest in an effort to undergo DNA repair; however, irreparable damage may cause these cells to undergo senescence or apoptosis if the damage is too severe (Nakanishi et al., 2006). Thus, impairments in tumor cell growth and/or its destruction is one mechanism in which NO may act as a tumor suppressor.

NO has also been shown to downregulate various signaling pathways involved in tumor growth and survival, such as transforming growth factor-beta (TGF-β). TGF-β is intimately involved in the regulation of proliferation, differentiation, and survival of various cancer cells. In the context of GBM, TGF-β was identified to act through various intracellularly pathways resulting in the phosphorylation of Smad2/3 proteins and formation of a complex with Smad4 that is capable of translocating to the nucleus to bind DNA and promote transcription of genes involved in glioma pathogenesis (Kaminska et al., 2013). NO has been shown to suppress tumor growth by inhibiting TGF-β activation via interference of Smad signaling, leading to degradation of Smad2 via the ubiquitin proteasome pathway (Saura et al., 2005).

Moreover, NO acts as an inhibitor of angiogenesis when found in high concentrations (Isenberg et al., 2009). Thrombospondin-1 (TSP-1) is a well-known endogenous inhibitor of angiogenesis mediated by NO (Isenberg et al., 2009). High concentrations of TSP-1 are required to modulate endothelial and vascular smooth muscle behaviors; however, only low levels are found functional in vivo (Isenberg et al., 2005). In the presence of NO, however, the potency of TSP-1 increased more than 100-fold (Isenberg et al., 2005). TSP-1 binding to CD47, a signaling receptor, results in reduced cGMP synthesis and activation of cGMP-dependent protein kinase, thus inhibiting downstream NO signaling (Isenberg et al., 2008). Sustained inhibition of NO signaling through this mechanism allows TSP-1-mediated angiogenic inhibition and, thus, inhibits tumor growth (Isenberg et al., 2008).

Cell adhesion molecules (CAMs) and the extracellular matrix (ECM) as a whole are largely implicated in cancer cell invasion and metastasis (Okegawa et al., 2004). In breast cancer, NO has been shown to inhibit cell motility and strengthen cell adhesion, thereby preventing metastasis (Lahiri & Martin, 2009). Another study in metastatic prostate cells revealed that high levels of NO inhibited epithelial–mesenchymal transition (EMT) and simultaneously reversed the mesenchymal cell phenotype, inhibiting invasion and metastasis (Baritaki et al., 2010). The authors proposed that increased NO expression via the NO donor DETA-NONOate inhibits Snail, an inducer of metastasis, while simultaneously inducing Raf-1 kinase inhibitor protein (RKIP), a suppressor of metastasis (Baritaki et al., 2010). EMT has been suggested to be a crucial regulator of GBM pathogenesis, contributing to the ability of cancer cells to invade and metastasize elsewhere; thus, NO-mediated inhibition of EMT may prove to be of therapeutic significance in such cells. Fig. 7.2 schematically summarizes the role of NO in GBM as a tumor suppressor.

Increased levels of nitric oxide (NO) are associated with tumor suppression through multiple mechanisms, including the induction of DNA damage and cell cycle arrest leading to apoptosis, interference with the protumorigenic TGF-β signaling pathway, inhibition of

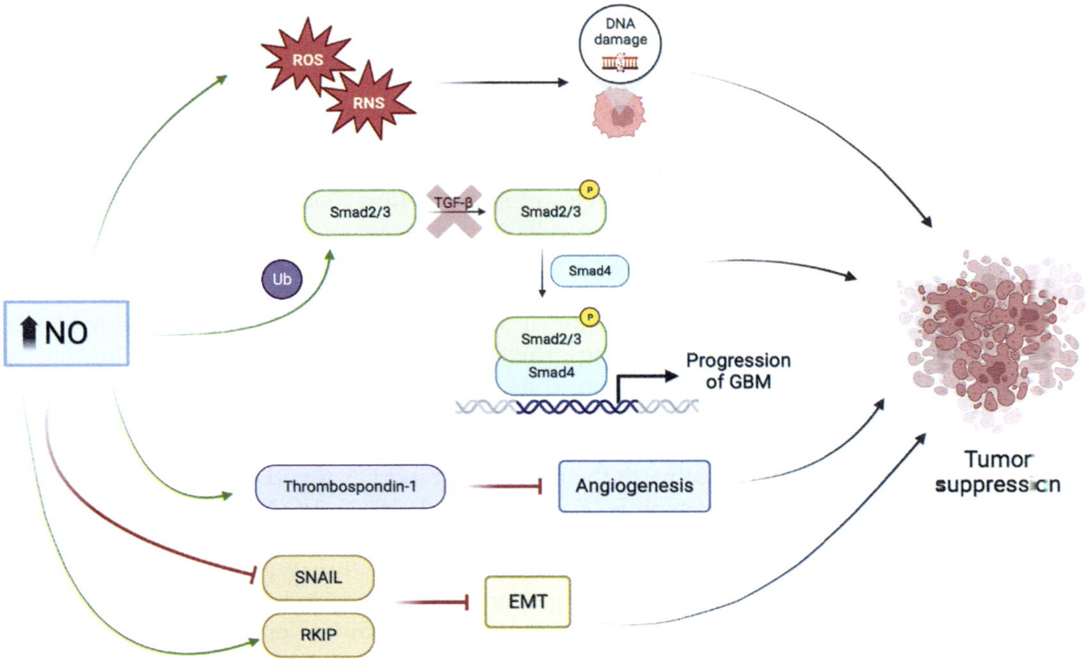

FIGURE 7.2 Role of nitric oxide (NO) in glioblastoma multiforme (GBM) tumor suppression. Increased levels of NO are associated with tumor suppression. High concentrations of NO lead to the generation of reactive oxygen species (ROS) and reactive nitrogen species (RNS), inducing DNA damage and triggering cell cycle arrest and apoptosis, contributing to the suppression of tumor growth. High levels of NO also interfere with the Smad signaling pathway, resulting in the degradation of Smad2. This inhibition of TGF-β activation suppresses the protumorigenic effects of this pathway, contributing to the reduction of tumor growth. NO enhances the potency of thrombospondin-1 (TSP-1), an endogenous inhibitor of angiogenesis. This leads to the inhibition of angiogenesis, ultimately preventing the formation of new blood vessels that are crucial for tumor growth and metastasis. NO mediates the inhibition of epithelial–mesenchymal transition (EMT) through the simultaneous inhibition of Snail protein and stimulation Raf-1 kinase inhibitor protein (RKIP), hindering cancer cell invasion and metastasis. The combined effects of NO on DNA damage, TGF-β downregulation, angiogenesis inhibition, and metastasis suppression contribute to impaired tumor growth and reduced metastatic potential, highlighting the multifaceted role of NO as a tumor suppressor.

angiogenesis, and suppression of EMT and metastasis. The combined effects of NO on these various pathways contribute to impaired tumor growth and reduced metastatic potential, highlighting the multifaceted role of NO as a tumor suppressor.

NO-targeted therapies in GBM

Oral administration of TMZ in combination with radiation therapy is the current standard treatment therapy for GBM (Fernandes et al., 2017). Based on the dual effects of NO in glioblastoma, researchers have explored various targeted therapies using NO donors or NOS inhibitors to either induce or suppress NO expression, respectively. Moreover, NO-targeted therapies have also been used in combination with chemotherapy, radiation therapy, or immunotherapy to sensitize tumor cells and enhance the efficacy of standard therapies, such as TMZ (Salvatori et al., 2023).

Inhibition of NO

NOS inhibitors are compounds that reduce NOS-mediated NO production to limit its protumorigenic effects. Nω-Nitro-L-arginine methyl ester (L-NAME), a nonselective inhibitor of NOS, was used in the treatment of rat C6 gliomas and resulted in reduced tumor blood flow and, thus, reduced tumor volumes via increased tumor necrosis (Tran et al., 2017). Similarly, another study demonstrated the effects of different concentrations of a specific NOS2 inhibitor (1400W) on the U87 MG human GBM cell line (Palumbo et al., 2018). Treatment with 1400W significantly reduced the proliferation, migration, colony formation, and generation of neurospheres in glioma cells and, thus, highlights the role of NO production in the pathogenesis of GBM (Palumbo et al., 2018). Further studies suggested that the mechanism of action by which 1400 W is able to impede tumor growth and progression is via the induction of autophagy and increased extracellular vesicle release in GBM cell lines (Palumbo et al., 2019). Treatment with a NOS2 inhibitor (1400W or S-MIU) in EGFR-vIII-expressing astrocytes led to reduced proliferation, whereas the use of an NO donor (SNAP) increased cell proliferation (Puram et al., 2012). Interestingly, NOS2 did not significantly affect normal astrocyte cell growth, rendering it as a safe therapeutic target with limited toxicity in normal brain cells (Puram et al., 2012). Therefore, patients with identified EGFRvIII mutations may benefit from therapies involving iNOS inhibitors. Moreover, another inhibitor of iNOS (OKN007) has been shown to increase apoptosis and decrease tumor hypoxia, angiogenesis, and cell proliferation in human GBM cells (Towner et al., 2013). One mechanism in which OKN007 mediates suppression of iNOS is via downregulation of cytokines, such as TNF-α and IFN-γ (Pogrebniak et al., 1992). Overall, the wide success in the use of NOS inhibitors in the treatment of GBM further demonstrates its protumorigenic role in GBM.

Induction of NO

Conversely, NO donors are compounds that release NO under specific conditions to enhance its antitumor effects via induction of cell death, radiosensitization, and chemosensitization. JS-K, a well-studied NO donor, was found to induce cell death in GBM cells at high doses via mitotic catastrophe associated with necrosis rather than apoptosis (Günzle et al., 2016). Furthermore, NO is thought to sensitize tumor cells to chemotherapeutics through alterations in the TME via NO-mediated vasodilation, which improves blood flow and vascular permeability, leading to increased drug penetration through the BBB, and enhances drug delivery to the tumor site (Hersh et al., 2016). The effects of diethylenetriamine/NO adduct and TMZ were investigated both alone and in combination in patient-derived GSCs and it was found that the use of an NO donor reduced proliferation and increased susceptibility of GSCs to standard TMZ treatment (Salvatori et al., 2023). Moreover, another mechanism by which NO is thought to sensitize tumor cells to chemotherapeutics is through the inhibition of DNA repair mechanisms used by the tumor cells to defend against the toxic effects of the drug. Resistance to chemotherapeutic agents, such as TMZ, is largely attributed to O6-methylguanine-DNA methyltransferase (MGMT), a DNA repair protein that repairs alkylation at the O6 position of guanine (Fan et al., 2013). MGMT-mediated repair of TMZ-induced DNA damage inhibits apoptosis and, thus, plays a critical role in resistance to chemotherapeutics (Fan et al., 2013). CTX-NO, an NO-releasing compound, is able to specifically target glioma cells to release NO and inhibit the action of DNA repair action of MGMT and, in turn, induce apoptosis of tumor cells (Safdar et al., 2013).

Discussion

The dichotomous role of NO in cancer is largely dependent on its concentration, with its specific role in either promoting tumor progression or suppression contingent upon the levels of NO specific to the cancer type under consideration (Choudhari et al., 2013). Namely, reduced NO levels are linked to protumorigenic effects, whereas increased NO concentrations are associated with antitumorigenic effects (Mintz et al., 2021). The present report primarily strengthens the evidence for NO's involvement in driving tumor progression in the case of GBM compared with its potential role in tumor suppression. Intriguingly, GBM cells exhibit low expression of iNOS-derived NO, thus further corroborating its propensity to promote tumorigenesis in the context of GBM (Tran et al., 2017). Furthermore, studies of glioma cells have been shown to exhibit low levels of NO as a means of evading apoptosis, increasing proliferation and migration, and promoting resistance against radiotherapy or chemotherapy (Tran et al., 2017). Therefore, conducting quantitative investigations to assess the levels of nitric oxide within GBM cells would serve as a valuable tool in validating the specific contribution of NO to GBM pathogenesis. This, in turn, would aid in determining the appropriateness of utilizing either an NO donor or an iNOS inhibitor as a potential therapeutic approach for individuals with GBM. Collectively, these findings underscore the dual nature of NO in the context of cancer progression, highlighting its potential both as a facilitator and as an inhibitor. Consequently, these results emphasize the need for additional research in the regulation of NO production and its associated signaling pathways in GBM and in guiding efficacious strategies for cancer treatment.

Perspectives and future directions

The dual role of NO within the context of GBM presents intriguing yet challenging inquiries warranting exploration through a diverse set of methodologies, including the following:

1. To investigate the levels of NOS2/NO within the context of GBM pathogenesis using animal models through the manipulation of NOS2/NO levels to study the resulting effects on tumor progression. NOS2/NO levels can be correlated with histological and pathological features of the tumors, such as tumor size, invasiveness, angiogenesis, and immune cell infiltration to provide insights into the potential roles of NOS2/NO in GBM pathogenesis.

2. To examine the dynamics of NOS2/NO concentrations within human GBM across various stages of tumor development. If differences in NOS2/NO concentrations vary across tumor stages, further studies may be employed to investigate the role of various signaling pathways, immune responses, or angiogenic processes responsible for such differences.
3. To explore NOS2/NO levels at the single-cell level, thereby allowing the determination of their overarching impact. Furthermore, the division of GBM cells into distinct subsets based on varying NOS2 expression levels should be undertaken to determine the functional implications associated with the respective use of NO donors and/or inhibitors.
4. To employ animal models afflicted with GBM, wherein NOS2 levels are either modulated or knocked-down, to determine the overall effect and responsiveness to treatment interventions involving NO donors and/or inhibitors.

Collectively, the insights obtained from these investigations hold promising potential for translational applications within the context of GBM treatment.

Acknowledgments

We acknowledge the Department of Microbiology, Immunology & Molecular Genetics and the David Geffen School of Medicine for their support.

References

Ahir, B. K., Engelhard, H. H., & Lakka, S. S. (May, 2020). Tumor development and angiogenesis in adult brain tumor: Glioblastoma. *Molecular Neurobiology, 57*(5), 2461–2478.

Baritaki, S., Huerta-Yepez, S., Sahakyan, A., Karagiannides, I., Bakirtzi, K., Jazirehi, A., & Bonavida, B. (December 15, 2010). Mechanisms of nitric oxide-mediated inhibition of EMT in cancer: Inhibition of the metastasis-inducer Snail and induction of the metastasis-suppressor RKIP. *Cell Cycle, 9*(24), 4931–4940.

Burke, A. J., Sullivan, F. J., Giles, F. J., & Glynn, S. A. (March, 2013). The yin and yang of nitric oxide in cancer progression. *Carcinogenesis, 34*(3), 503–512.

Choudhari, S. K., Chaudhary, M., Bagde, S., Gadbail, A. R., & Joshi, V. (May 30, 2013). Nitric oxide and cancer: A review. *World Journal of Surgical Oncology, 11*, 118.

Clarke, J., Penas, C., Pastori, C., Komotar, R. J., Bregy, A., Shah, A. H., Wahlestedt, C., & Ayad, N. G. (August, 2013). Epigenetic pathways and glioblastoma treatment. *Epigenetics, 8*(8), 785–795.

Ding, Z., Ogata, D., Roszik, J., Qin, Y., Kim, S. H., Tetzlaff, M. T., Lazar, A. J., Davies, M. A., Ekmekcioglu, S., & Grimm, E. A. (February 12, 2021). iNOS associates with poor survival in melanoma: A role for nitric oxide in the PI3K-AKT pathway stimulation and PTEN S-nitrosylation. *Frontiers in Oncology, 11*, 631766.

Dobra, G., Gyukity-Sebestyén, E., Bukva, M., Harmati, M., Nagy, V., Szabó, Z., Pankotai, T., Klekner, Á., & Buzás, K. (January 24, 2023). MMP-9 as prognostic marker for brain tumours: A comparative study on serum-derived small extracellular vesicles. *Cancers, 15*(3), 712.

Du, Q., Zhang, X., Liu, Q., Zhang, X., Bartels, C. E., & Geller, D. A. (November 1, 2013). Nitric oxide production upregulates Wnt/β-catenin signaling by inhibiting Dickkopf-1. *Cancer Research, 73*(21), 6526–6537.

Duerinck, J., Schwarze, J. K., Awada, G., Tijtgat, J., Vaeyens, F., Bertels, C., Geens, W., Klein, S., Seynaeve, L., Cras, L., D'Haene, N., Michotte, A., Caljon, B., Salmon, I., Bruneau, M., Kockx, M., Van Dooren, S., Vanbinst, A. M., Everaert, H., Forsyth, R., & Neyns, B. (June, 2021). Intracerebral administration of CTLA-4 and PD-1 immune checkpoint blocking monoclonal antibodies in patients with recurrent glioblastoma: a phase I clinical trial. *Journal of Immunotherapy Cancer, 9*(6). e002296.

Elgundi, Z., Papanicolaou, M., Major, G., Cox, T. R., Melrose, J., Whitelock, J. M., & Farrugia, B. L. (January 17, 2020). Cancer metastasis: The role of the extracellular matrix and the heparan sulfate proteoglycan perlecan. *Frontiers in Oncology, 9*, 1482.

Förstermann, U., & Sessa, W. C. (April, 2012). Nitric oxide synthases: Regulation and function. *European Heart Journal, 33*(7), 829–837, 837a-837d.

Fan, C. H., Liu, W. L., Cao, H., Wen, C., Chen, L., & Jiang, G. (October 24, 2013). O6-methylguanine DNA methyltransferase as a promising target for the treatment of temozolomide-resistant gliomas. *Cell Death Disease, 4*(10), e876.

Fan, Q., Aksoy, O., Wong, R. A., Ilkhanizadeh, S., Novotny, C. J., Gustafson, W. C., Truong, A. Y., Cayanan, G., Simonds, E. F., Haas-Kogan, D., Phillips, J. J., Nicolaides, T., Okaniwa, M., Shokat, K. M., & Weiss, W. A. (March 13, 2017). A kinase inhibitor targeted to mTORC1 drives regression in glioblastoma. *Cancer Cell, 31*(3), 424–435.

Fernandes, C., Costa, A., Osório, L., Lago, R. C., Linhares, P., Carvalho, B., & Caeiro, C. (September 27, 2017). Current standards of care in glioblastoma therapy. In S. De Vleeschouwer (Ed.), *Glioblastoma*. Brisbane (AU): Codon Publications.

Fujisawa, H., Ogura, T., Hokari, A., Weisz, A., Yamashita, J., & Esumi, H. (January 1995). Inducible nitric oxide synthase in a human glioblastoma cell line. *Journal of Neurochemistry, 64*(1), 85–91.

Günzle, J., Osterberg, N., Saavedra, J. E., & Weyerbrock, A. (September 1, 2016). Nitric oxide released from JS-K induces cell death by mitotic catastrophe as part of necrosis in glioblastoma multiforme. *Cell Death Disease, 7*(9), e2349.

Galstyan, A., Markman, J. L., Shatalova, E. S., Chiechi, A., Korman, A. J., Patil, R., Klymyshyn, D., Tourtellotte, W. G., Israel, L. L., Braubach, O., Ljubimov, V. A., Mashouf, L. A., Ramesh, A., Grodzinski, Z. B., Penichet, M. L., Black, K. L., Holler, E., Sun, T., Ding, H., Ljubimov, A. V., & Ljubimova, J. Y. (August 28, 2019). Blood-brain barrier permeable nano immunoconjugates induce local immune responses for glioma therapy. *Nature Communications, 10*(1), 3850.

Hanif, F., Muzaffar, K., Perveen, K., Malhi, S. M., & Simjee, S. U. (January 1, 2017). Glioblastoma multiforme: A review of its epidemiology and pathogenesis through clinical presentation and treatment. *Asian Pacific Journal of Cancer Prevention, 18*(1), 3–9.

Hatanpaa, K. J., Burma, S., Zhao, D., & Habib, A. A. (September, 2010). Epidermal growth factor receptor in glioma: Signal transduction, neuropathology, imaging, and radioresistance. *Neoplasia, 12*(9), 675–684.

Hays, E., & Bonavida, B. (September 17, 2019). Nitric oxide-mediated enhancement and reversal of resistance of anticancer therapies. *Antioxidants, 8*(9), 407.

Hersh, D. S., Wadajkar, A. S., Roberts, N., Perez, J. G., Connolly, N. P., Frenkel, V., Winkles, J. A., Woodworth, G. F., & Kim, A. J. (2016). Evolving drug delivery strategies to overcome the blood brain barrier. *Current Pharmaceutical Design, 22*(9), 1177–1193.

Hokari, A., Zeniya, M., & Esumi, H. (September, 1994). Cloning and functional expression of human inducible nitric oxide synthase (NOS) cDNA from a glioblastoma cell line A-172. *Journal of Biochemistry, 116*(3), 575–581.

Isenberg, J. S., Frazier, W. A., & Roberts, D. D. (March, 2008). Thrombospondin-1: A physiological regulator of nitric oxide signaling. *Cellular and Molecular Life Sciences, 65*(5), 728–742.

Isenberg, J. S., Martin-Manso, G., Maxhimer, J. B., & Roberts, D. D. (March, 2009). Regulation of nitric oxide signalling by thrombospondin 1: Implications for anti-angiogenic therapies. *Nature Reviews Cancer, 9*(3), 182–194.

Isenberg, J. S., Ridnour, L. A., Perruccio, E. M., Espey, M. G., Wink, D. A., & Roberts, D. D. (September 13, 2005). Thrombospondin-1 inhibits endothelial cell responses to nitric oxide in a cGMP-dependent manner. *Proceedings of the National Academy of Sciences of the United States of America, 102*(37), 13141–13146.

Kaminska, B., Kocyk, M., & Kijewska, M. (2013). TGF beta signaling and its role in glioma pathogenesis. *Advances in Experimental Medicine and Biology, 986*, 171–187.

Kaur, B., Khwaja, F. W., Severson, E. A., Matheny, S. L., Brat, D. J., & Van Meir, E. G. (April, 2005). Hypoxia and the hypoxia-inducible-factor pathway in glioma growth and angiogenesis. *Neuro-Oncology, 7*(2), 134–153.

Khan, F. H., Dervan, E., Bhattacharyya, D. D., McAuliffe, J. D., Miranda, K. M., & Glynn, S. A. (December 10, 2020). The role of nitric oxide in cancer: Master regulator or NOt? *International Journal of Molecular Sciences, 21*(24), 9393.

Kim, J. E., Patel, M., Ruzevick, J., Jackson, C. M., & Lim, M. (February 10, 2014). STAT3 activation in glioblastoma: Biochemical and therapeutic implications. *Cancers, 6*(1), 376–395.

Lahiri, M., & Martin, J. H. (February, 2009). Nitric oxide decreases motility and increases adhesion in human breast cancer cells. *Oncology Reports, 21*(2), 275–281.

Latour, M., Her, N. G., Kesari, S., & Nurmemmedov, E. (August 5, 2021). WNT signaling as a therapeutic target for glioblastoma. *International Journal of Molecular Sciences, 22*(16), 8428.

Lee, Y., Lee, J. K., Ahn, S. H., Lee, J., & Nam, D. H. (February, 2016). WNT signaling in glioblastoma and therapeutic opportunities. *Laboratory Investigation, 96*(2), 137–150.

Li, F., Sonveaux, P., Rabbani, Z. N., Liu, S., Yan, B., Huang, Q., Vujaskovic, Z., Dewhirst, M. W., & Li, C. Y. (April 13, 2007). Regulation of HIF-1alpha stability through S-nitrosylation. *Molecules and Cells, 26*(1), 63–74.

Luksik, A. S., Yazigi, E., Shah, P., & Jackson, C. M. (February 23, 2023). CAR T cell therapy in glioblastoma: Overcoming challenges related to antigen expression. *Cancers, 15*(5), 1414.

Méndez, O., Zavadil, J., Esencay, M., Lukyanov, Y., Santovasi, D., Wang, S. C., Newcomb, E. W., & Zagzag, D. (June 1, 2010). Knock down of HIF-1alpha in glioma cells reduces migration in vitro and invasion in vivo and impairs their ability to form tumor spheres. *Molecular Cancer, 9*, 133.

Mesfin, F. B., & Al-Dhahir, M. A. (May 20, 2023). Gliomas. In *StatPearls. Treasure island (FL)*. StatPearls Publishing, 2023 Jan–.

Mintz, J., Vedenko, A., Rosete, O., Shah, K., Goldstein, G., Hare, J. M., Ramasamy, R., & Arora, H. (January 27, 2021). Current advances of nitric oxide in cancer and anticancer therapeutics. *Vaccines, 9*(2), 94.

Mocellin, S., Bronte, V., & Nitti, D. (May, 2007). Nitric oxide, a double edged sword in cancer biology: Searching for therapeutic opportunities. *Medicinal Research Reviews, 27*(3), 317–352.

Nakanishi, M., Shimada, M., & Niida, H. (October, 2006). Genetic instability in cancer cells by impaired cell cycle checkpoints. *Cancer Science, 97*(10), 984–989.

Okegawa, T., Pong, R. C., Li, Y., & Hsieh, J. T. (2004). The role of cell adhesion molecule in cancer progression and its application in cancer therapy. *Acta Biochimica Polonica, 51*(2), 445–457.

Olson, N., & van der Vliet, A. (August 1, 2011). Interactions between nitric oxide and hypoxia-inducible factor signaling pathways in inflammatory disease. *Nitric Oxide, 25*(2), 125–137.

Palumbo, P., Lombardi, F., Augello, F. R., Giusti, I., Luzzi, S., Dolo, V., Cifone, M. G., & Cinque, B. (June 20, 2019). NOS2 inhibitor 1400W induces autophagic flux and influences extracellular vesicle profile in human glioblastoma U87MG cell line. *International Journal of Molecular Sciences, 20*(12), 3010.

Palumbo, P., Lombardi, F., Siragusa, G., Dehcordi, S. R., Luzzi, S., Cimini, A., Cifone, M. G., & Cinque, B. (September 17, 2018). Involvement of NOS2 activity on human glioma cell growth, clonogenic potential, and neurosphere generation. *International Journal of Molecular Sciences, 19*(9), 2801.

Pan, P. C., & Magge, R. S. (November 11, 2020). Mechanisms of EGFR resistance in glioblastoma. *International Journal of Molecular Sciences, 21*(22), 8471.

Pogrebniak, H. W., Merino, M. J., Hahn, S. M., Mitchell, J. B., & Pass, H. I. (August 1992). Spin trap salvage from endotoxemia: The role of cytokine down-regulation. *Surgery, 112*(2), 130–139. discussion 138-9.

Puram, S. V., Yeung, C. M., Jahani-Asl, A., Lin, C., de la Iglesia, N., Konopka, G., Jackson-Grusby, L., & Bonni, A. (June 6, 2012). STAT3-iNOS signaling mediates EGFRvIII-induced glial proliferation and transformation. *Journal of Neuroscience, 32*(23), 7806–7818.

Ridnour, L. A., Windhausen, A. N., Isenberg, J. S., Yeung, N., Thomas, D. D., Vitek, M. P., Roberts, D. D., & Wink, D. A. (October 23, 2007). Nitric oxide regulates matrix metalloproteinase-9 activity by guanylyl-cyclase-dependent and -independent pathways. *Proceedings of the National Academy of Sciences of the United States of America, 104*(43), 16898–16903.

Rocha Pinheiro, S. L., Lemos, F. F. B., Marques, H. S., Silva, L. M., de Oliveira Silva, L. G., Faria Souza Mendes Dos Santos, C., da Costa Evangelista, K., Calmon, M. S., Sande Loureiro, M., & Freire de Melo, F. (April 24, 2023). Immunotherapy in glioblastoma treatment: Current state and future prospects. *World Journal of Clinical Oncology, 14*(4), 138–159.

Rui, Y., & Green, J. J. (December, 2021). Overcoming delivery barriers in immunotherapy for glioblastoma. *Drug Delivery and Translative Research, 11*(6), 2302–2316.

Safdar, S., Payne, C. A., Tu, N. H., & Taite, L. J. (April, 2013). Targeted nitric oxide delivery preferentially induces glioma cell chemosensitivity via altered p53 and O(6) -methylguanine-DNA methyltransferase activity. *Biotechnology and Bioengineering, 110*(4), 1211–1220.

Salvatori, L., Malatesta, S., Illi, B., Somma, M. P., Fionda, C., Stabile, H., Fontanella, R. A., & Gaetano, C. (July 10, 2023). Nitric oxide prevents glioblastoma stem cells' expansion and induces temozolomide sensitization. *International Journal of Molecular Sciences, 24*(14), 11286.

Saura, M., Zaragoza, C., Herranz, B., Griera, M., Diez-Marqués, L., Rodriguez-Puyol, D., & Rodriguez-Puyol, M. (November 25, 2005). Nitric oxide regulates transforming growth factor-beta signaling in endothelial cells. *Circulation Research, 97*(11), 1115–1123.

Seker-Polat, F., Pinarbasi Degirmenci, N., Solaroglu, I., & Bagci-Onder, T. (January 17, 2022). Tumor cell infiltration into the brain in glioblastoma: From mechanisms to clinical perspectives. *Cancers, 14*(2), 443.

Smith, J. S., Tachibana, I., Passe, S. M., Huntley, B. K., Borell, T. J., Iturria, N., O'Fallon, J. R., Schaefer, P. L., Scheithauer, B. W., James, C. D., Buckner, J. C., & Jenkins, R. B. (August 15, 2001). PTEN mutation, EGFR amplification, and outcome in patients with anaplastic astrocytoma and glioblastoma multiforme. *Journal of the National Cancer Institute, 93*(16), 1246–1256.

Tamimi, A. F., & Juweid, M. (September 27, 2017). Epidemiology and outcome of glioblastoma. In S. De Vleeschouwer (Ed.), *Glioblastoma*. Brisbane (AU): Codon Publications.

Towner, R. A., Gillespie, D. L., Schwager, A., Saunders, D. G., Smith, N., Njoku, C. E., Krysiak, R. S., 3rd, Larabee, C., Iqbal, H., Floyd, R. A., Bourne, D. W., Abdullah, O., Hsu, E. W., & Jensen, R. L. (March, 2013). Regression of glioma tumor growth in F98 and U87 rat glioma models by the Nitrone OKN-007. *Neuro-Oncology, 15*(3), 330–340.

Tran, A. N., Boyd, N. H., Walker, K., & Hjelmeland, A. B. (June 10, 2017). NOS expression and NO function in glioma and implications for patient therapies. *Antioxidants and Redox Signaling, 26*(17), 986–999.

Vahora, H., Khan, M. A., Alalami, U., & Hussain, A. (March, 2016). The potential role of nitric oxide in halting cancer progression through chemoprevention. *Journal of Cancer Preventive, 21*(1), 1–12.

Vannini, F., Kashfi, K., & Nath, N. (December, 2015). The dual role of iNOS in cancer. *Redox Biology, 6*, 334–343.

Walker, C., Baborie, A., Crooks, D., Wilkins, S., & Jenkinson, M. D. (December, 2011). Biology, genetics and imaging of glial cell tumours. *British Journal of Radiology, 84*(2), S90–S106. Spec No 2.

Wang, C., Li, Y., Gu, L., Chen, R., Zhu, H., Zhang, X., Zhang, Y., Feng, S., Qiu, S., Jian, Z., & Xiong, X. (April 18, 2023). Gene targets of CAR-T cell therapy for glioblastoma. *Cancers, 15*(8), 2351.

Wen, P. Y., Rodon, J. A., Mason, W., Beck, J. T., DeGroot, J., Donnet, V., Mills, D., El-Hashimy, M., & Rosenthal, M. (July, 2020). Phase I, open-label, multicentre study of buparlisib in combination with temozolomide or with concomitant radiation therapy and temozolomide in patients with newly diagnosed glioblastoma. *ESMO Open, 5*(4). e000673.

Wink, D. A., & Mitchell, J. B. (September, 1998). Chemical biology of nitric oxide: Insights into regulatory, cytotoxic, and cytoprotective mechanisms of nitric oxide. *Free Radical Biology and Medicine, 25*(4–5), 434–456.

Xu, H., Zong, H., Ma, C., Ming, X., Shang, M., Li, K., He, X., Du, H., & Cao, L. (July, 2017). Epidermal growth factor receptor in glioblastoma. *Oncology Letters, 14*(1), 512–516.

Xu, W., Liu, L. Z., Loizidou, M., Ahmed, M., & Charles, I. G. (December, 2002). The role of nitric oxide in cancer. *Cell Research, 12*(5–6), 311–320.

Xue, Q., Cao, L., Chen, X. Y., Zhao, J., Gao, L., Li, S. Z., & Fei, Z. (March, 2017). High expression of MMP9 in glioma affects cell proliferation and is associated with patient survival rates. *Oncology Letters, 13*(3), 1325–1330.

Yamamoto, N., Oyaizu, T., Enomoto, M., Horie, M., Yuasa, M., Okawa, A., & Yagishita, K. (February 17, 2020). VEGF and bFGF induction by nitric oxide is associated with hyperbaric oxygen-induced angiogenesis and muscle regeneration. *Scientific Reports, 10*(1), 2744.

Zhao, H. F., Wang, J., Shao, W., Wu, C. P., Chen, Z. P., To, S. T., & Li, W. P. (June 7, 2017). Recent advances in the use of PI3K inhibitors for glioblastoma multiforme: Current preclinical and clinical development. *Molecular Cancer, 16*(1), 100.

Zheng, H., Ying, H., Yan, H., Kimmelman, A. C., Hiller, D. J., Chen, A. J., Perry, S. R., Tonon, G., Chu, G. C., Ding, Z., Stommel, J. M., Dunn, K. L., Wiedemeyer, R., You, M. J., Brennan, C., Wang, Y. A., Ligon, K. L., Wong, W. H., Chin, L., & DePinho, R. A. (October 23, 2008). p53 and Pten control neural and glioma stem/progenitor cell renewal and differentiation. *Nature, 455*(7216), 1129–1133.

CHAPTER

8

Nitric oxide donating systems and their potential in shaping tumor immunoregulation

Greta Varchi

Italian National Research Council, Institute of Organic Synthesis and Photoreactivity, Bologna, Italy

Introduction

Cancer continues to be a significant health burden worldwide, requiring continuous research and the development of novel therapeutic strategies. Nitric oxide (NO), a small gaseous and highly reactive endogenous molecule, has emerged as a versatile signaling molecule with diverse physiological and pathological roles, including cancer (Khan et al., 2020; Somasundaram et al., 2019). NO is produced by a family of enzymes called nitric oxide synthases (NOS), which catalyze the conversion of the amino acid L-arginine into NO and L-citrulline (Fig. 8.1). There are three isoforms of NOS: neuronal NOS (nNOS or NOS1), inducible NOS (iNOS or NOS2), and endothelial NOS (eNOS or NOS3). Each isoform has distinct tissue distribution, regulatory mechanisms, and physiological functions. nNOS is constitutively found in neurons and is involved in neuronal communication, neurotransmission, and synaptic plasticity. It produces NO in response to calcium influx, which acts as a retrograde messenger, regulating blood flow, synaptic transmission, and neuronal development (Forstermann & Sessa, 2012).

iNOS is typically not expressed under normal physiological conditions but can be induced by inflammatory mediators, cytokines, and immune responses (Mishra et al., 2020). iNOS produces large amounts of NO for an extended period of time, contributing to immune responses, host defense against pathogens, and tissue damage in chronic inflammation (Korde Choudhari et al., 2013). eNOS is primarily expressed in endothelial cells lining blood vessels. It is responsible for the production of NO in response to various stimuli such as shear stress, hormones, and neurotransmitters. eNOS-derived NO plays a critical role in regulating vascular tone, blood pressure, and angiogenesis (Di Lorenzo et al., 2013). NO regulates

Therapeutic Applications of Nitric Oxide in Cancer and Inflammatory Disorders
https://doi.org/10.1016/B978-0-443-22354-9.00008-5

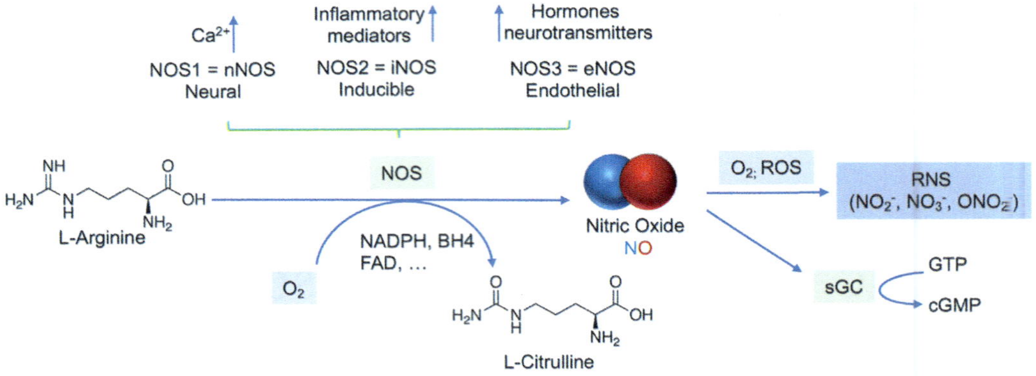

FIGURE 8.1 Schematic representation of nitric oxide biology, including its synthesis, functions, and signaling pathways.

cellular functions such as gene expression, protein modification, and cell survival. It diffuses across cell membranes and interacts with various molecules, including metal ions, proteins, and reactive oxygen species. The main signaling pathway involves NO binding to and activating the enzyme soluble guanylate cyclase (sGC), which leads to the synthesis of cyclic guanosine monophosphate (cGMP). cGMP then activates protein kinase G (PKG), which phosphorylates downstream target proteins, modulating their activity and cellular responses (Fig. 8.1).

Additionally, NO can interact with other molecules, such as oxygen and reactive oxygen species (ROS), leading to the formation of reactive nitrogen species (RNS) such as peroxynitrite, which can modify proteins, lipids, and DNA, influencing cellular processes and contributing to oxidative stress. In recent years, researchers have explored the potential of NO-donating agents as a promising avenue for cancer treatment. This chapter seeks to offer an up-to-date overview (2023–20) of the characteristics and significance of exogenous NO in cancer therapy, particularly its interactions with immune regulation and modulation of the tumor microenvironment (TME), especially when used in conjunction with other therapeutic approaches. It is important to note that a comprehensive exploration of the intricate biological interactions between NO and the immune system falls outside the purview of this contribution. For in-depth insights into this subject, readers are encouraged to refer to the excellent and up-to-date papers and book chapters authored by experts in the field (Kim & Thomas, 2022; McGinity et al., 2021; Mintz et al., 2021; Navasardyan & Bonavida, 2021; Sahebnasagh et al., 2022; Vedenko et al., 2020).

The tumor microenvironment

The TME represents a complex, dynamic milieu where the tumor cells reside, characterized by a diverse set of cell populations and noncellular components. It encompasses an intricate network of tumor cells, immune cells, stromal cells, blood vessels, and extracellular matrix. The TME plays an active role in governing tumor initiation, progression, metastasis, and responsiveness to therapeutic interventions. Its components encompass a range of stromal and immune cells, including fibroblasts, endothelial cells, and pericytes, as well as various immune cells like macrophages, lymphocytes, and dendritic cells (Anderson & Simon, 2020). These components collaborate in a bidirectional communication, influencing the tumor's behavior and response to treatments. The TME plays a critical role in shaping the tumor's features, growth, and survival, fostering a microenvironment conducive to tumor progression through the release of growth factors, inflammatory mediators, and extracellular matrix alterations (Baghban et al., 2020). The interactions between the TME and tumor cells are intricate, forming a complex ecosystem that supports tumor survival and growth while also serving as a significant determinant of treatment outcomes and resistance. Several key processes underlie tumor growth and expansion within the TME, including hypoxia, endothelial-to-mesenchymal transition (EMT), macrophage infiltration, angiogenesis, and the regulatory effects of secreted factors such as reactive oxygen species (ROS) or nitric oxide (Vedenko et al., 2020). Among these components, **tumor-associated macrophages** (TAMs) are a key component of the immune infiltrate within the TME. TAMs are recruited to the tumor site and exhibit a high degree of plasticity, adopting various activation states based on the local cues. Within the tumor, TAMs can polarize into different phenotypes, primarily described as M1-like or M2-like. The M1 phenotype is characterized by proinflammatory and antitumor properties, whereas the M2 phenotype is associated with immune suppression and supports tumor growth and progression (He & Zhang, 2021). TAMs, often skewed toward the M2 phenotype, can release various cytokines, growth factors, and matrix-remodeling enzymes that favor tumor growth, angiogenesis, invasion, and immune evasion. Their functional diversity and the complexity of their interactions within the TME present challenges in targeting these cells for cancer therapy, while also offering potential opportunities for modulating their behavior to hinder tumor progression. Therapeutic strategies aiming to polarize TAMs to a more antitumoral M1-like phenotype or to eliminate them have shown promise, although a deeper understanding of the nuanced roles TAMs play in different cancer types and stages is required to develop effective treatments (Lin et al., 2021).

Hypoxia also plays a fundamental role in cancer development and progression; the oxygenation profiles in solid tumors display heterogeneity, featuring mosaics of hypoxic and normoxic regions, and this diversity results in varying metabolic stress and immune functions across different tumor regions. Hypoxia stems from an imbalance attributed to abnormal vasculature, causing insufficient oxygen supply and heightened oxygen demand due to the rapid proliferation of tumor cells (Emami Nejad et al., 2021). Under hypoxic conditions, various immunosuppressive mechanisms are activated, including the reduction of effector T cell proliferation and survival, as well as the induction of dysfunction in tumor antigen-specific effector T cells and natural killer (NK) cells. Additionally, hypoxia contributes to the recruitment and polarization of immunosuppressive cell types such as myeloid-derived suppressor cells (MDSCs), TAMs, and regulatory T (Treg) cells, ultimately leading to failures in immunotherapy (Kopecka et al., 2021). Overall, hypoxia drives various molecular and cellular changes in the TME, promoting aggressive tumor behavior, metastasis, and resistance to therapy. Therefore, strategies that target hypoxia are actively being explored to improve cancer treatment by mitigating its adverse effects on tumor progression and resistance.

The enigmatic dual nature of nitric oxide in cancer

In recent years, nitric oxide has regained attention, sparking extensive debates about its dual role, particularly its protumorigenic features and its potential to induce apoptosis, suggesting a therapeutic role in cancer beyond its pathological implications (Khan et al., 2020; Vedenko et al., 2020). Abnormal NO release from various enzymatic sources is commonly observed in various tumors. For instance, iNOS, regulated by the tumor suppressor gene p53, is associated with malignancy, impacting hematological and solid tumors. Elevated iNOS activity correlates with decreased survival in glioma patients (Bui et al., 2023; Ekmekcioglu et al., 2017). In central nervous system malignancies, abnormally high levels of NO, particularly attributed to nNOS isoform, indicate tumor malignancy and differentiation in brain tumors (Szeliga & Albrecht, 2021; Tran et al., 2017). nNOS expression has also been identified as an important target in cancer-associated fibroblast therapy and has been negatively associated with patient response to adoptive T cell therapy in melanoma (Yarlagadda et al., 2017). The endothelial isoform of NOS (eNOS) plays a role in tumor processes, including therapy resistance, angiogenesis, and metastasis. Overexpression of eNOS is observed in various cancers, with tumor-expressed eNOS crucial in initiating and sustaining pancreatic tumorigenesis (Lampson et al., 2012). Additionally, upregulation of eNOS is identified in tumor epithelial cells from mesenchymal poor-prognosis colorectal tumors, making it a critical source of NO in this colorectal cancer subtype (Altun et al., 2013).

Beyond tumor cells, diverse cell types in the TME, such as keratinocytes, chondrocytes, fibroblasts, and immune cells such as antigen-presenting cells (APC), macrophages, and natural killer cells, contribute to NO production, exerting autocrine and paracrine effects (Kim & Thomas, 2022; Sahebnasagh et al., 2022). Macrophages, which exert crucial roles in host defense through the activation of proinflammatory responses, are significant sources of NO. Recent studies have shown that macrophage-derived NO has a dual role, suppressing immune responses at high concentrations while promoting endothelial activation and T cell infiltration at low concentrations (Navasardyan & Bonavida, 2021; Sahebnasagh et al., 2022).

NO's role in cancer is intricate and context-dependent, exhibiting both protumorigenic and antitumorigenic effects at different stages of cancer development. Its influence on cancer progression and metastasis involves interactions with various signaling pathways and cell types within the TME (Xu et al., 2002). Understanding these complexities is crucial for developing targeted therapeutic strategies leveraging NO's effects for more effective cancer management (Kim et al., 2022; Li, Yoon, et al., 2022; Ren et al., 2022). For instance, in addressing the challenge of chemotherapy and radiotherapy resistance in metastatic cancer, NO sensitizes cells to subsequent treatments, offering a combinatory approach to cancer therapy (Bonavida, 2020). Beyond its role in promoting essential processes within the TME, such as angiogenesis, metabolism, and apoptosis, NO may play a crucial part in reshaping the immune component of the TME. In particular, high levels of NO can contribute to immunosuppression by inhibiting T cell responses and promoting the generation of MDSCs and Tregs. Conversely, at lower concentrations, NO has positive effects, facilitating the activity of NK cells and aiding in the activation and function of immune cells like dendritic cells (DCs) and macrophages.

The intricate relationship between NO and the immune response in the TME is a delicate interplay between its suppressive and activating effects, modulated by its concentration and the surrounding conditions. Consequently, precise regulation of NO levels holds promise for effectively manipulating the immune response against cancer. Guided by this understanding, the subsequent sections will primarily explore the efficacious utilization of NO donors to favorably influence the TME toward an immune-responsive state.

Nitric oxide donors

Nitric oxide–donating agents have gained significant attention in cancer treatment due to their potential to modulate various aspects of tumor biology and improve therapeutic outcomes. These agents are designed to release controlled amounts of NO in a targeted manner, exploiting the dual nature of NO as both a cytotoxic molecule and a signaling molecule with diverse effects on tumor cells and the TME (Bonavida et al., 2008).

NO-donating agents can directly induce cytotoxicity in cancer cells through multiple mechanisms, including DNA damage and inhibition of DNA repair mechanisms, leading to cell cycle arrest and apoptosis; disruption of mitochondrial function, causing oxidative stress and cell death; inhibition of angiogenesis, impeding the formation of new blood vessels that supply tumors with nutrients and oxygen; inhibition of tumor cell migration and invasion (Mintz et al., 2021). Additionally, NO-donating agents can enhance the efficacy of existing cancer therapies by sensitizing cancer cells to radiation therapy, photodynamic or photothermal therapies by increasing oxidative stress and DNA damage (Li, Yoon, et al., 2022) and enhancing the effects of chemotherapy by targeting drug-resistant cancer cells and improving drug delivery to tumors (Bonavida, 2020). Indeed, NO can induce vasodilation, which can be harnessed to improve drug delivery to tumors. In this scenario, precise dosing and controlled release of NO are critical to achieve therapeutic benefits while avoiding excessive toxicity. The timing of NO release is crucial, as too much NO can lead to harmful effects, whereas the heterogeneity of tumors requires tailored approaches to optimize the use of NO-donating agents. NO donors are compounds designed to release controlled amounts of NO in a targeted and controlled manner. There are several classes of NO-donating agents, each with unique characteristics and mechanisms of NO release (Fig. 8.2).

A brief and nonexhaustive description of the major NO donor classes is reported in the following paragraph, along with references to more focused papers on the specific subject. Among NO-donating compounds, S-nitrosothiols (RSNO) are characterized by the presence

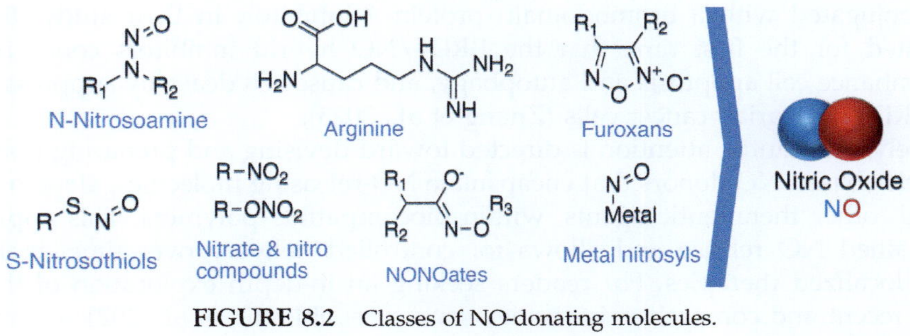

FIGURE 8.2 Classes of NO-donating molecules.

of a sulfur−nitrogen−oxygen (S−N−O) group that can release NO through various pathways, including enzymatic and nonenzymatic processes, such as light, ultrasound, heat, X-ray, and glutathione. Some RSNO, such as S-nitrosoglutathione (GSNO), occur naturally, while others are synthetically produced. However, most of these synthetic derivatives are unstable and difficult to purify, thus requiring the development of alternative strategies for their application in cancer therapy (Reis et al., 2019).

Nitrate esters are also an important class of NO donors: these organic compounds contain a nitrate group (ONO_2) that can undergo enzymatic or chemical reactions to release NO. Nitroglycerin, used to treat angina, is a classic example of a nitrate ester that releases NO to dilate blood vessels and increase blood flow. An important class of hybrid nitrates applied in cancer therapy is represented by NO-donating nonsteroidal antiinflammatory drugs (NO-NSAIDs), conjugating the antiinflammatory potential of NSAIDs, mitigating inflammation-associated pathways that contribute to cancer development and progression with the specific antitumor action of released NO (Dunlap et al., 2008; Kashfi et al., 2002).

Diazeniumdiolates, also known as NONOates, are synthetic compounds designed to release NO upon reaction with water. They represent the most widely studied class of NO donors capable of releasing NO with tunable release kinetics based on their chemical structure (Alimoradi et al., 2019).

NO photodonors, particularly those based on nitroanilines, represent a promising class of compounds designed to release NO upon exposure to light, triggering the release of controlled amounts of NO. This photoactivation provides spatial and temporal control over NO delivery, a crucial aspect of therapeutic applications (Fallica et al., 2021). Sortino's group pioneered this research field, further developing more intriguing molecular systems integrating the light-triggered NO release with fluorescent reporting capabilities (Fraix et al., 2021; Laneri et al., 2022; Malanga et al., 2019). Indeed, the incorporation of fluorescent monitoring enables noninvasive tracking and quantification of NO dosage, which is crucial for understanding its biological effects. In addition, there are several examples of hybrid compounds combining NO-donating moieties with other pharmacophores, allowing for targeted and multifunctional therapeutic effects (Omidkhah & Ghodsi, 2022). An example has been reported by our research group dealing with the synthesis of a multimodal conjugate comprising an NO photodonor moiety and the photosensitizer pheophorbide A for the combined release of NO and ROS. The novel conjugate proved to be effective in vitro against a panel of prostate cancer cell lines, including those resistant to antiandrogen treatment i.e., overexpressing the androgen receptor (Rapozzi et al., 2015). Another example of such hybrid molecules was recently reported by He's group, describing the synthesis of a dual-targeted furoxan conjugated with a bromodomain protein 4 inhibitor. In their study, the authors demonstrated for the first time that the BRD4/NO hybrid inhibitors could reduce cell viability, enhance cell apoptosis and autophagy, and cause cell death by suppressing the activity of BRD4 in ovarian cancer cells (Zhang et al., 2023).

Ultimately, substantial attention is directed toward devising and preparing polymer- and nanoparticles-based NO donors that encapsulate NO-releasing molecules, alone or in combination with other therapeutic agents, within biocompatible polymers. This approach provides sustained NO release and allows for controlled delivery over time, making them useful for localized therapies. For readers seeking an in-depth exploration of this subject, numerous recent and comprehensive reviews are accessible (Jin et al., 2021; Li, Yoon, et al.,

2022; Vong & Nagasaki, 2020; Zhao et al., 2023). Each class of NO-donating agents has its advantages and limitations in terms of NO release kinetics, stability, and specificity. The selection of a specific class depends on the intended application and therapeutic goals. Researchers continue to explore and develop various NO-donating agents to harness the therapeutic potential of NO for a range of medical conditions, including cancer treatment. Research in this field is ongoing, with the goal of developing innovative NO-based therapies that can be used as stand-alone treatments or in combination with other therapies.

Nitric oxide donors and the immune system

Overcoming immunosuppression is critical for effective anticancer treatments, especially in triple-negative breast cancer (TNBC), a challenging breast cancer subtype. The study by Paul and coworkers explores whether combining the standard TNBC chemotherapy, doxorubicin, with the NO donor glyceryltrinitrate (GTN, Fig. 8.3) can improve the anticancer treatment efficacy and conquer chemotherapy resistance in in vivo model of TNBC (Mabrouk et al., 2023). While prior research demonstrated the potentiating effect of NO donors on chemotherapy's efficacy in solid cancers both in vitro and preclinical models, the influence of NO donors on immunomodulation remained ambiguous since tumor cells were xenografted in immune-compromised mice (Huerta et al., 2009). To address this issue, the research by Paul and coworkers focused on immunocompetent Balb/C mice inoculated with murine mammary 4T1 and EMT6 cells. The study established that the enhanced antitumor effect of the doxorubicin/GTN combination in TNBC relied significantly on the immune system, specifically on the recruitment and infiltration of CD8 tumor-infiltrating lymphocytes (TILs). In addition, the expression of PD-1 in CD8 TILs, often seen as a marker of T cell exhaustion (Li et al., 2023), was found to correlate with the enhanced antitumor effect in

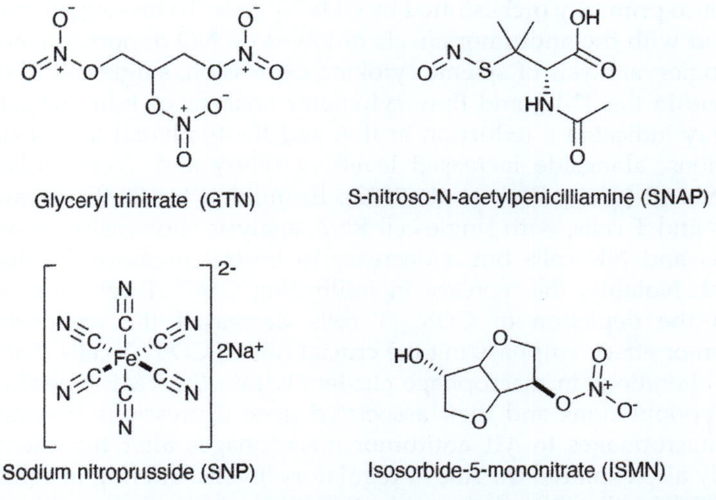

FIGURE 8.3 Chemical structures of glyceryl trinitrate (GTN), S-nitroso-N-acetylpenicilliamine (SNAP), sodium nitroprusside (SNP), and isosorbide-5-mononitrate (ISMN).

the study. Furthermore, while the addition of an anti-PD-1 to doxorubicin had limited effects on chemotherapy, combining doxorubicin with GTN significantly amplified the treatment's efficacy. This combination correlated with decreased levels of PD-L1 on both tumor cells and MDSCs. MDSCs, particularly the polymorphonuclear-MDSCs (PMN-MDSCs), play a pivotal role in immune suppression, but their activity was notably diminished in response to the doxorubicin/GTN treatment.

The study also elucidated the association between CXCR2 expression, PMN-MDSCs, and their pathologically activated state, which usually correlates with poor prognostic outcomes and metastasis. The decreased expression of the S100A9 protein and iNOS in tumors and MDSCs, respectively, in response to the doxorubicin/GTN treatment contributed to a reduction in PMN-MDSCs' immunosuppressive function. Moreover, the research highlighted the impact of GTN on the reprogramming of PMN-MDSCs, significantly reducing the expression of FATP2 and inhibiting STAT5 activation through ROS-dependent mechanisms, diminishing the immunosuppressive activity of these cells. Overall, this research shows that the combination of GTN with doxorubicin holds promise in enhancing the recruitment and activation of CD8 TILs while reprogramming the immunosuppressive activity of PMN-MDSCs in TNBC, thus deserving further investigation for its potential translation into clinical settings (Mabrouk et al., 2023).

A recent study by Lai and Wang delved into the impact of various nitric oxide–releasing compounds, e.g., S-nitroso-N-acetylpenicilliamine (SNAP), sodium nitroprusside (SNP), and isosorbide-5-mononitrate (ISMN) (Fig. 8.3), on tumor progression in immune-competent mouse models, e.g., a subcutaneous lung tumor (LL2 cells in C57BL/6 mice), an orthotopic melanoma model (B16F10 cells in C57BL/6 mice) and a colon cancer model (CT-26 cells in BALB/c mice) (Li et al., 2023). Their investigation encompassed detailed cytokine analysis and single-cell RNA sequencing to scrutinize the immune cell populations and transcriptome modifications following treatment with selected nitric oxide–releasing agents. The authors found that the administration of low-dose exogenous nitric oxide prompted a remarkable antitumor response primarily orchestrated by $CD8^+$ T cells. To investigate the immune mechanisms associated with the antitumor effects of low-dose NO donors, the authors employed three methodologies: analysis of splenic cytokine expression, single-cell RNA sequencing of immune cells within the TME, and flow cytometry analysis of immune cells. Results from the cytokine array indicated a reduction in IL-6 and IL-10, pivotal for T helper cells (Th2)–mediated functions, alongside increased levels of IFN-γ and TNF-α following treatment with low-dose SNAP (Chen, Wei, et al., 2022). Examining the TME revealed a dominance of macrophages and T cells, with single-cell RNA analysis showcasing an increase in $CD8^+$ cytotoxic T cells and NK cells but a decrease in central memory T cells after low-dose SNAP treatment. Notably, the increase in infiltrating $CD8^+$ T cells was verified by flow cytometry, and the depletion of $CD8^+$ T cells decreased the response to the SNAP-mediated antitumor effect, emphasizing the crucial role of $CD8^+$ T cells. The study additionally uncovered alterations in macrophage clusters within the TME. Specifically, changes in macrophage subpopulations and their associated gene expressions suggested a shift from M2 protumor macrophages to M1 antitumor macrophages after treatment with low-dose SNAP. The study also explored the role of regulatory T cells, offering insights into their functions and potential implications for combination therapies with low-dose SNAP and anti-CTLA4 antibodies for inducing a synergetic effect on tumor regression. Finally, the combination treatment of low-dose SNAP or SNP with cisplatin showed improved therapeutic effects compared with cisplatin treatment alone, suggesting potential synergistic benefits in combination therapy for cancer treatment (Li et al., 2023).

Several studies have demonstrated that certain chemotherapeutic agents, along with other treatment modalities such as radiotherapy, photodynamic therapy, and sonodynamic therapy, have the potential to trigger immunogenic cell death (ICD), a unique form of programmed cell death that stimulates an immune response against tumor cells (Li, Lai, et al., 2022; Xie et al., 2022). In ICD, dying cancer cells release damage-associated molecular patterns (DAMPs), such as calreticulin (CRT), a calcium sensor protein typically located in the endoplasmic reticulum (ER), ATP, and high-mobility group box 1 (HMGB1). CRT exposure on the cell surface serves as an "eat me" signal, facilitating the recognition and phagocytosis of the cancer cells by DCs. The subsequent presentation of tumor antigens by DCs to T cells triggers a robust immune response, particularly involving cytotoxic $CD8^+$ T lymphocytes. This orchestrated process not only eliminates the dying cancer cells but also establishes immunological memory, providing long-term protection against recurrent tumors (Ahmed & Tait, 2020).

As mentioned in the previous sections, tumor hypoxia triggers a series of immunosuppressive processes that lead to immunotherapy failure. To address this challenge, Wang and coworkers developed oxygen-switchable ROS/RNS nanogenerators (NGs) and demonstrated their capacity to generate reactive oxygen species (ROS) and reactive nitrogen species (RNS) in response to differing oxygen concentrations (Jiang et al., 2023). The authors describe the preparation of a copolymer formed by tetraphenylporphyrin methyl (TPPM) and a coumarin-based NO donor (Fig. 8.4) used to produce the ROS/RNS NGs. Interestingly, upon light irradiation, these NGs act as photosensitizer, generating ROS under normoxic conditions, and as photocatalysts, activating the coumarin-based NO donor to produce RNS under hypoxic conditions. Both ROS and NO efficiently induced ICD and reshaped the immunosuppressive TME, promoting an anticancer immune response.

In particular, the authors tested their NGs systems in a subcutaneous 4T1 tumor model (Fig. 8.5A), showing an almost complete tumor regression in the RO/NS NGs plus irradiation-treated group (Fig. 8.5B). The authors also observed markedly elevated levels of granzyme B+ $CD8^+$ T cells, indicating that RO/NS NGs plus irradiation could induce adaptive immune activation and a significant increase of activated T cells ($CD8^+$ and $CD4^+$ T) in

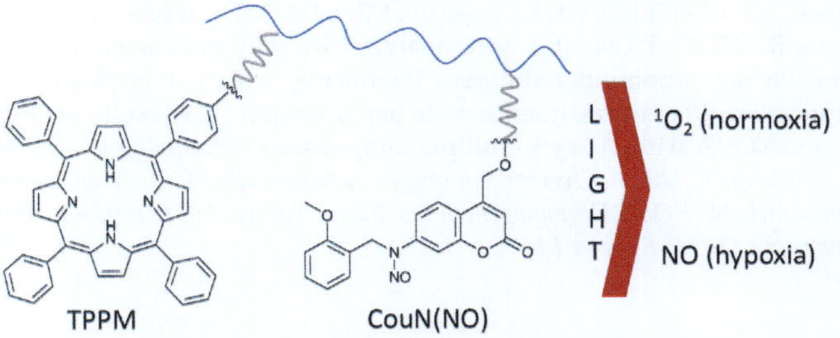

FIGURE 8.4 Schematic representation of the copolymer formed by tetraphenylporphyrin methyl (TPPM) and a coumarin-based NO donor.

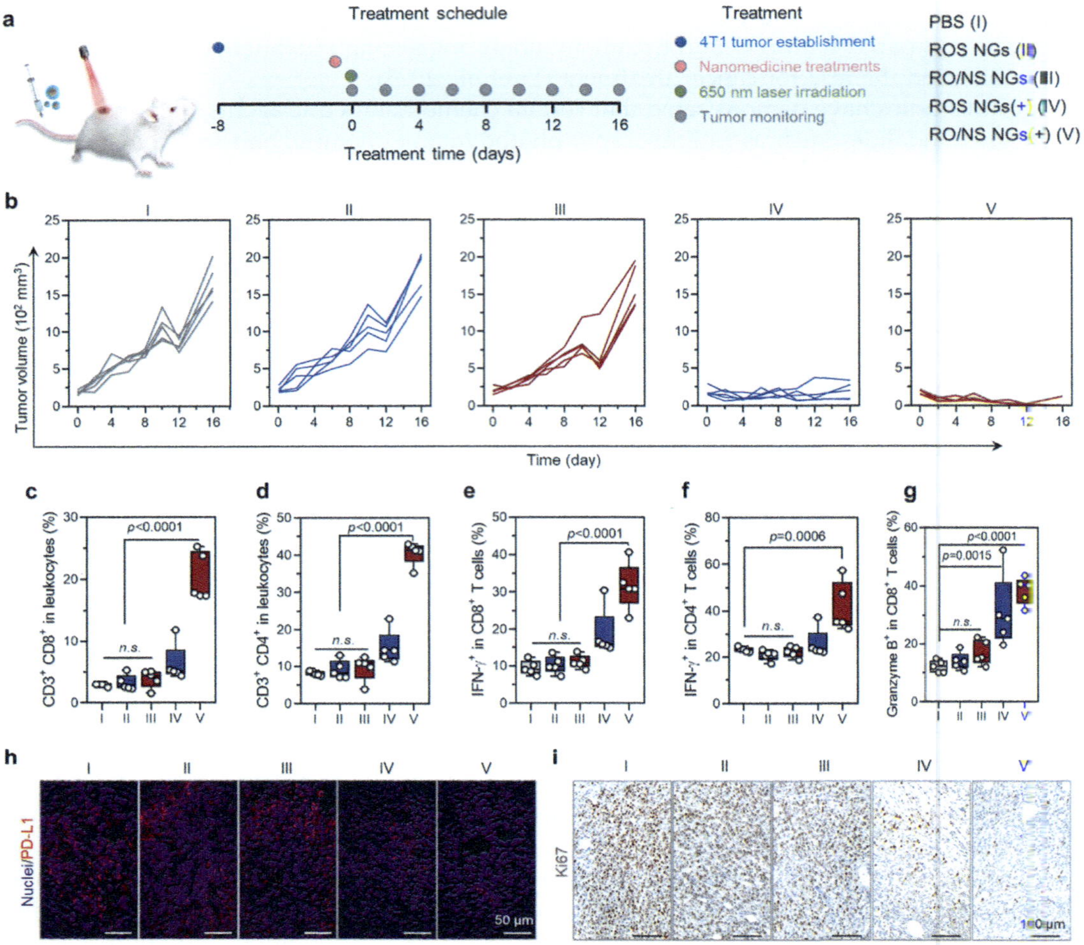

FIGURE 8.5 Oxygen-switchable RO/NS NGs exhibit a synergistic antitumor response in 4T1 tumor models. (A) Schematic showing the process of antitumor therapy. Mice established 4T1 tumors on the right flank were *i.v.* injected with different formulations, following exposure to laser irradiation on tumors. (B) Tumor growth curves of mice during treatments (n = 5). (C–G) After 16 days, the mice were sacrificed and spleens were harvested, and the percentages of $CD8^+$ T cells ($CD3^+CD8^+$) (C), $CD4^+$ T cells ($CD3^+CD4^+$) (D), IFN-γ^+ $CD8^+$ T cells (E), IFN-γ^+ $CD4^+$ T cells (F), and granzyme B^+ $CD8^+$ T cells (G) were analyzed by flow cytometry (n = 5). (H) Immuofluorescent images of PD-L1 expression on tumors after different treatments, scale bar is 50 μm. (I) Immunohistochemical staining of Ki67 in tumors after different treatments, scale bar is 100 μm. The results are expressed as mean ± SEM and analyzed by one-way ANOVA with Tukey's multiple comparisons test. *Modified from Jiang, W., Shen, Z., Guo, Z., Wang, Q., Li, Q., Hu, J., & Wang, Y. (2023). Overcoming oxygen heterogeneity of tumor microenvironments to boost cancer immunotherapy by oxygen-switchable ROS/RNS nanogenerators.* Nano Today, 48, *101696–101696. https://doi.org/10.1016/j.nantod.2022.101696. Copyright (2023) Elsevier Ltd.*

tumor-draining lymph nodes, an increased expression of CD69 on T cells, and a higher percentage of memory CD44+CD62L+ T cells compared with control groups (Fig. 8.5C–G). Additionally, in accordance with other studies (Salimian Rizi et al., 2017), it was found that PD-L1 expression decreased in RO/NS NGs plus irradiation-treated tumors (Fig. 8.5H). Ki67 staining indicated that tumors treated with RO/NS NGs in combination with laser irradiation exhibited substantial apoptosis and decreased proliferation (Fig. 8.5I). This innovative strategy, utilizing a single component (TPPM photosensitizer/photocatalyst) for generating both ROS and RNS, leverages the local oxygen levels within the TME, offering a promising approach to overcome the challenges posed by oxygen heterogeneity in tumor immunotherapy.

Among the chemotherapeutic drugs capable of inducing ICD, doxorubicin has been identified not only for its ability to induce apoptotic cell death in cancer cells but also for its additional capacity to stimulate an immunogenic response within tumor cells. Before inducing tumor cell apoptosis, doxorubicin initiates CRT translocation to the cell's plasma membrane. CRT translocation, following doxorubicin exposure, is thought to trigger the identification and engulfment of dying cancer cells by DCs (Huang et al., 2023). Beyond facilitating CRT translocation, doxorubicin was identified to induce nitric oxide (NO) production, which can instigate a proapoptotic effect on tumor cells (Riganti et al., 2005). The study from Riganti and coworkers focused on the impact of doxorubicin on colon cancer cells, aiming to clarify the relationship between NO synthesis, doxo's cytotoxic and immunogenic effects, and the role of CRT (De Boo et al., 2009). The researchers silenced iNOS in HT29 colon cancer cells to explore whether the effects of reduced NO levels in these cells are similar to those in doxorubicin-resistant HT29-dx cells. Their findings revealed that decreased NO levels in HT29 iNOS cells led to similarities with HT29-dx cells in terms of doxorubicin accumulation and chemoresistance. Further investigations pointed out that the absence of NO synthesis potentially influences drug efflux and the transporters responsible for doxorubicin expulsion. Moreover, NO seemed to affect the exposure of CRT on the cell membrane, highlighting that NO plays a pivotal role in modulating immune- and chemoresistance in colon cancer cells. Indeed, the absence of NO synthesis led to reduced CRT exposure, affecting the efficacy of doxorubicin in triggering immune responses and tumor cell phagocytosis by dendritic cells. Reintroducing NO by administering sodium nitroprusside in iNOS-silenced cells increased CRT exposure and the cell's immunogenicity, thus potentially reversing the immunoresistant phenotype. The study implies that NO levels critically affect the immune and chemoresistance mechanisms in colon cancer cells, suggesting that NO synthesis is closely related to the efficacy of doxorubicin, particularly in eliciting CRT exposure, impacting the immunogenic potential of the tumor cells (De Boo et al., 2009).

On the same topic, Wang and coworkers highlighted the ability of NO to induce a powerful ICD effect in in vitro and in vivo models of breast cancer (Jiang et al., 2022). The authors developed polynitrosated polyesters through a straightforward polycondensation reaction followed by a postsynthetic modification that allows for high NO functionalization. Utilizing this polymer, a polynitrosated NO nanogenerator (NanoNO) was produced by self-assembly (Fig. 8.6), capable of efficiently delivering NO within tumors and resulting in a notable ICD effect. The study's results substantiate NO as a potent inducer of ICD, provoking the release of immunogenic DAMPs from tumor cells and triggering a robust and targeted antitumor response. The immunogenic effects induced by NO involve intertwined pathways of ER

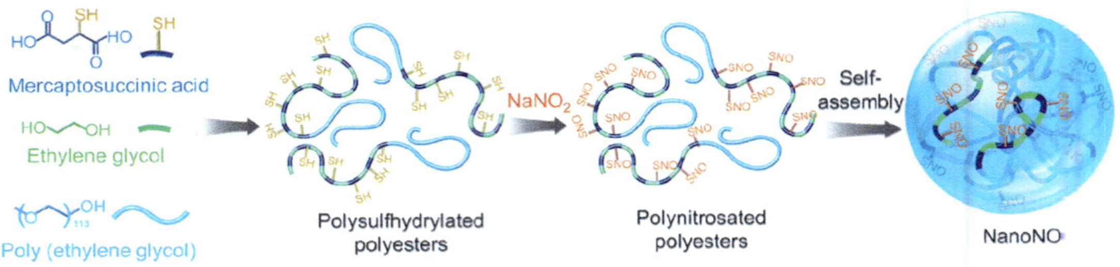

FIGURE 8.6 Synthesis route of *S*-nitrosylation polymers and their self-assembly of the NanoNO. *Modified from Jiang, W., Dong, W., Li, M., Guo, Z., Wang, Q., Liu, Y., Bi, Y., Zhou, H., & Wang, Y. (2022). Nitric oxide induces immunogenic cell death and potentiates cancer immunotherapy.* ACS Nano, 16*(3), 3881–3894. https://doi.org/10.1021/acsnano.1c09048. Copyright (2023) American Chemical Society.*

stress and mitochondria-mediated intrinsic apoptosis (MOMP), as demonstrated in previous works (Theivendran et al., 2022). Notably, dying cells treated with NanoNO exhibit vaccine-like effects, enhancing the proliferation and activation of T cells upon rechallenge with syngeneic living tumor cells in mice. Additionally, the application of NanoNO demonstrates its ability to initiate anticancer immunity and synergize effectively with αPD-1, leading to superior immune infiltration against metastatic tumors.

A recent study from Arora's group investigated the relationship between the colony-stimulating factor 1 receptor (CSF1R) and the NO content regulated by NOS3 (Firdaus et al., 2022). The study findings indicate that the uncoupling of NOS3 in high-grade human prostate cancer leads to a decrease in NO levels and an increase in oxidative stress, suggesting a nitroso-redox imbalance. This imbalance negatively impacts the effectiveness of CSF1R blockade therapy, as key sites on CSF1R cannot undergo effective S-nitrosylation. The research suggests that exogenous NO administration, specifically via S-nitrosoglutathione (GSNO) treatment, effectively S-nitrosylates CSF1R at key sites, enhancing its ability to counter prostate cancer. The combination of NO and CSF1R blockade significantly reduces tumor burden compared with monotherapies. Additionally, the study shows that the immune landscape within tumors is favorably modulated by NO plus CSF1R blockade, leading to a decrease in protumorigenic cell populations and an increase in antitumor immune cells. However, the authors acknowledge the need for further exploration of in vivo tumor burden reduction correlations and the identification of other S-nitrosylated molecules influencing NO-CSF1R blockade therapy. The study's findings align with existing research on nitroso-redox balance in various biological systems, including the cardiovascular system and inflammation control in macrophages (Brüne et al., 2013).

Among the possible strategies in tumor immunotherapy, a current focus is on the inhibition of the indoleamine-2,3-dioxygenase (IDO) enzyme due to its pivotal role in orchestrating tumor immune evasion (Liu et al., 2017). IDO1 acts as the first and rate-limiting step in the kynurenine (Kyn) pathway, where it is responsible for tryptophan (Trp) catabolism, leading to Kyn overproduction (Labadie et al., 2019). Acting as an immune checkpoint molecule, IDO1 leads to a robust immune suppression through negative regulation of the mTOR/P–S6 kinase pathway; the stimulation of the aryl hydrocarbon receptor (AhR) pathway, which promotes further IDO1 expression as well as Treg generation, and the activation of a serine/threonine-protein kinase, GCN2 (general control nonderepressible 2) kinase, which increases IL-6 production and development of Tregs and MDSCs (Labadie et al., 2019). IDO1 overexpression by dendritic cells, myeloid-derived suppressor cells, tumor-associated, macrophages, and cancer cells plays a role in locoregional interference in T cell activation in a variety of human cancers such as TNBC, pancreas, lung, prostate, colorectal, cervical, gastric, ovarian, and head and neck cancers (Kraehenbuehl et al., 2022). Finally, IDO1 expression reprograms APCs towards an immune-suppressive phenotype modifying the local environment from immunogenic to tolerogenic (Ravishankar et al., 2015).

To exploit the potential beneficial role of IDO1 inhibition, Shuai and collaborators developed a GSH-sensitive nanoformulation, e.g., BMS-SNAP-MOF, to enhance immunotherapy in TNBC models (Du et al., 2022). The nanocarrier, a copper-phloroglucinol metal–organic framework (MOF) loaded with the IDO inhibitor BMS-986205 and the NO donor SNAP, was coated with Pluronic F127 to boost the system's biocompatibility in vivo. Upon accumulation in the tumor site, the nanodrug responded specifically to the high GSH concentration of the TME, releasing BMS-986205 and SNAP. Concomitantly, the transformation of Cu^{2+} to Cu^{+} in this microenvironment triggers the release of NO, alleviating immunosuppression in antitumor T lymphocytes and remodeling the immunosuppressive TME (Fig. 8.7).

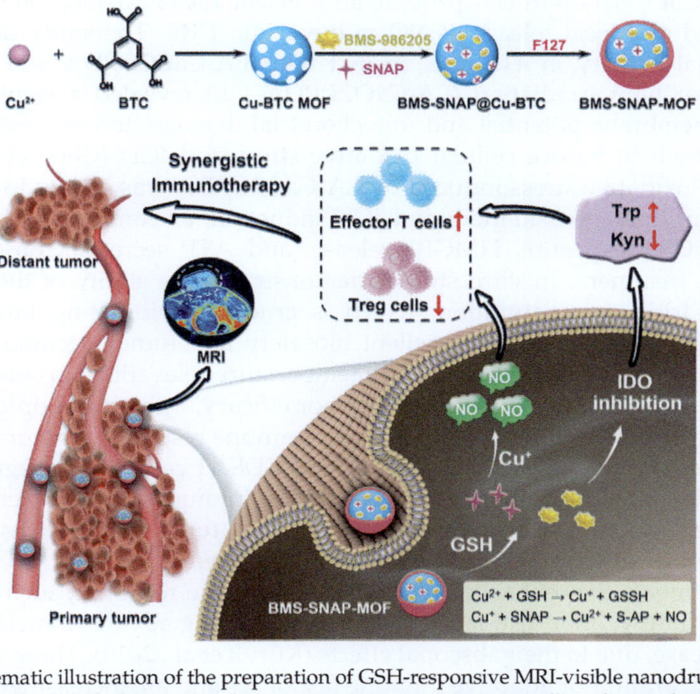

FIGURE 8.7 Schematic illustration of the preparation of GSH-responsive MRI-visible nanodrug BMS-SNAP-MOF and the IDO inhibition and NO therapy combination treatment to synergistically enhance the antitumor immunotherapy. *Reproduced with permission from. Copyright (2023) John Wiley and Sons. From Kraehenbuehl, L., Weng, C.-H., Eghbali, S., Wolchok, J. D., & Merghoub, T. (2022). Enhancing immunotherapy in cancer by targeting emerging immunomodulatory pathways.* Nature Reviews Clinical Oncology, *19(1), 37–50. https://doi.org/10.1038/s41571-021-00552-7*

The assessment of the tumor immune microenvironment modulation by BMS-SNAP-MOF was explored in mice-bearing 4T1 tumors. Treatment with single-component MOF, i.e., BMS-MOF and SNAP-MOF, increased cytotoxic $CD8^+$ T cells in tumor tissues compared with PBS treatment, indicating potential antitumor immune responses through IDO inhibition and NO generation, respectively. Notably, the combination treatment with BMS-SNAP-MOF exhibited the most significant tumor-infiltrating $CD8^+$ T cells, suggesting a synergistic effect of NO therapy and IDO inhibition. Furthermore, the BMS-SNAP-MOF treatment resulted in an immune-responsive tumor phenotype, as indicated by the increased $CD8^+$ T cells/CD4+ T cells ratio. Immunohistochemical analysis revealed the most efficient tumor infiltration of $CD8^+$ T cells in the BMS-SNAP-MOF-treated group. Additionally, BMS-SNAP-MOF induced the highest upregulation of IFN-γ expression in tumor sections, indicating a robust antitumor-immune effect. The treatment also led to a significant reduction in Treg cells and exhibited a potent systemic antitumor immune response against both primary and distant/metastatic tumors, confirming its therapeutic potential in metastatic settings (Du et al., 2022).

Another example of combination therapy, including exogenous NO administration, was recently reported by Hao and coworkers (Hao et al., 2022). The authors report the preparation of polydopamine (PDA) modified and L-arginine (LA)-loaded mesoporous CaO_2 nanoparticles (LA-CaO_2@PDA) for NO-enhanced photothermal immunotherapy of breast cancer. The urchin-like CaO_2 nanoparticles provide an efficient LA encapsulation, in turn allowing for the controlled and local release of NO in the acidic TME. The study demonstrates that the mitochondrial activity in 4T1 cells treated with LA-CaO2@PDA and LA-CaO2@PDA plus 808 nm laser light irradiation (LA-CaO2@PDA + L) revealed a significant decline in mitochondrial membrane potential and mitochondrial damage. Indeed, excessive Ca^{2+} concentration is known to induce cellular oxidative stress and ICD (Chen et al., 2019; Zheng et al., 2021). The oxidative stress induced by LA-CaO2@PDA was shown to cause mitochondrial dysfunction, lipid peroxidation, and ICD induction, as confirmed by the typical ICD hallmarks, e.g., CRT exposure, HMGB1 release, and ATP secretion, observed upon LA-CaO2@PDA + L treatment. In vitro studies demonstrated the ability of these ICD inducers to enhance dendritic cell maturation, which is crucial for initiating immune responses. In vivo, LA-CaO2@PDA exhibited excellent biosafety and tumor accumulation, while the photothermal effects were confirmed by temperature elevation in tumor tissue. LA-CaO2@PDA + L demonstrated superior antitumor efficacy, inducing apoptosis and necrosis, inhibiting lung metastases, and evoking a robust immune response. Immunological analyses revealed increased mature DCs, elevated $CD4^+$ and $CD8^+$ T cells, and upregulated immunostimulatory cytokines, indicating effective antitumor immunity. The potent systemic antitumor response against both primary and distant tumors highlights the promising therapeutic potential of LA-CaO2@PDA in cancer treatment.

Regional treatments that enhance the antitumor immune response, such as radiation and oncolytic virus therapy, offer distinct advantages for treating advanced melanoma, including unresectable disease, due to their abscopal effects (Kuryk et al., 2019). These therapies are successful in eliciting antitumor adaptive immune responses and are increasingly combined with immunotherapies to address various immune regulatory pathways (Havunen et al., 2018). The article from Thomas' group discusses the drug delivery behavior of a thermosensitive hydrogel composed of FDA-approved polymers for the local delivery of the NO donor

GSNO and a monoclonal antibody (mAb) antagonizing CTLA-4 (aCTLA-4) (Kim et al., 2022). The study utilized FDA-approved and highly biocompatible materials, Pluronic F127 and gelatin, to create a copolymer through simple bioconjugation chemistry. This copolymer was then used to produce drug-loaded hydrogels (GSNO + aCTLA-4/HG) of optimal size for lymphatic uptake, resulting in more potent and durable therapeutic efficacy compared to oral delivery. The combination of GSNO and aCTLA-4 coformulated within F127-g-gelatin hydrogel demonstrated decreased growth of the treated primary tumor, with efficacies slightly superior to that of the free drugs in combination. Additionally, GSNO + aCTLA-4/HG substantially diminished the contralateral untreated tumor. The study found that one-time intratumoral (i.t.) administration of GSNO + aCTLA-4/HG showed significantly higher antitumor effects than combined one-time i.t. administration of GSNO and three times intraperitoneal administration of aCTLA-4. These results suggest that prolonging the action of GSNO and aCTLA-4 mAb through a single administration i.t. administration of GSNO + aCTLA-4/HG improves the combination therapy's therapeutic index. The study concludes that this platform represents a technology highly amenable to clinical translation, leveraging NO's immune modulatory functions to enhance the therapeutic outcomes of immune checkpoint blockade therapy.

Combination therapy addressing multiple pathways has become a mainstream in cancer therapy. In this framework, Chen and collaborators devised a near-infrared (NIR)-triggered nitric oxide (NO)-releasing nanoplatforms, e.g., UNTPs, comprising an upconversion nanoparticles (UCNPs) core stabilized by a polymeric NO donor, i.e., mPEG-PNTC-PEI, for remodeling the immunosuppressive TME, enhance tumor immunogenicity, and synergize with immune checkpoint blockade to augment tumor immunotherapy (Fig. 8.8) (Li, Qian, et al., 2022). Upon NIR irradiation, the UCNPs convert NIR into ultraviolet (UV) light, triggering a controlled NO release from UNTPs, which, by inducing ICD, leads to DCs maturation and increased T cell infiltration. Moreover, NO is expected to inhibit the programmed death ligand 1 (PD-L1) expression and promote the polarization of tumor-associated macrophages (M2). When combined with an immune checkpoint inhibitor αPD-1, the nanosystem orchestrates robust immune responses (Fig. 8.8).

The study investigated the in vivo antitumor therapeutic effect of UNTPs plus light (UNTPs + L) in combination with αPD-1 therapy in a melanoma mouse tumor model, with tumors established on both sides (right-side: primary tumor; left-side: distant tumor). Monitoring tumor growth through bioluminescence imaging and volume measurement revealed that UNTPs + L+αPD-1 treatment significantly outperformed other treatments, resulting in notable primary and distant tumor suppression. Remarkably, UNTPs + L+αPD-1 demonstrated a median survival time of 26.5 days, showcasing significant benefits over other groups. Fluorescence staining confirmed substantial NO release in tumors treated with UNTPs + L and UNTPs + L+αPD-1, indicating effective NO delivery under NIR irradiation. Histological analyses further confirmed extensive necrosis and apoptosis in tumors treated with UNTPs + L+αPD-1, emphasizing its outstanding efficacy. Immune-related examinations revealed efficient immunogenic cell death, increased mature dendritic cells, and enhanced infiltration of cytotoxic and helper T lymphocytes in primary and distant tumors treated with UNTPs + L+αPD-1. Cytokine analysis indicated elevated IFN-γ secretion, highlighting the successful activation of systemic immune responses. Additionally, UNTPs + L+αPD-1 downregulated PD-L1 expression and polarized

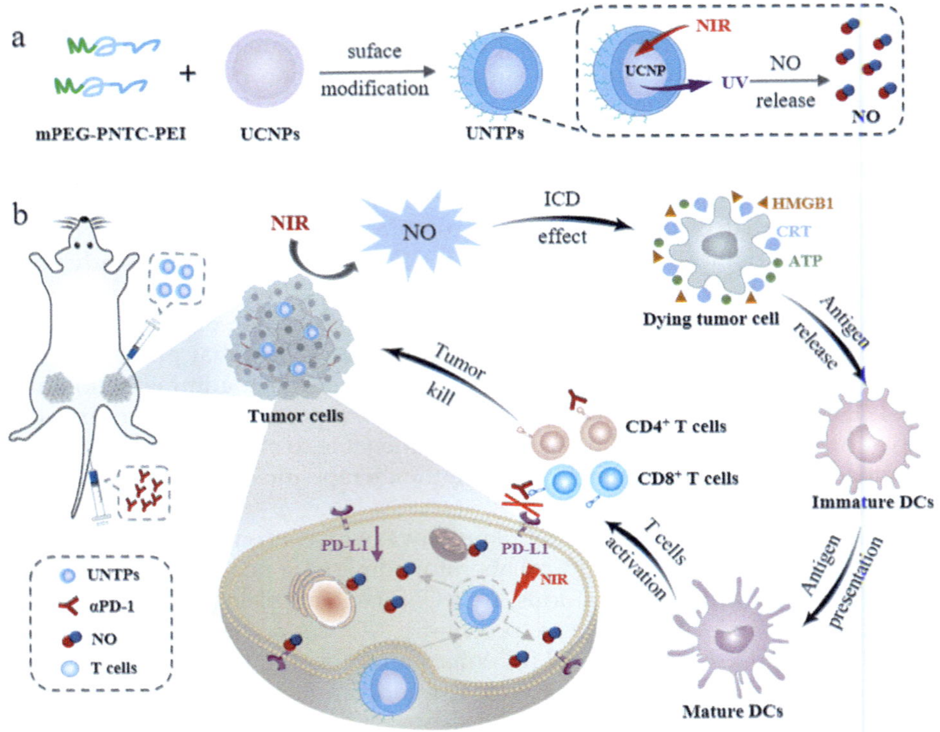

FIGURE 8.8 Schematic illustration of the construction and NIR-regulated NO-releasing procedure for upconverted polymeric NO-stabilized nanogasholder (UNTPs), and the proposed workflow of UNTPs upon NIR-irradiation to inhibit the primary and distant tumors by leveraging the benefits of NO as an immunomodulatory molecule and combining with immune checkpoint blockade. *Reproduced with permission from. From Li, Y., Qian, H., Huang, X., Zhou, X., Zhao, B., Yang, L., Chen, J., Qiao, H., Huang, D., Zhong, Y., & Chen, W. (2022). Up-converted nanogasholder with precise nitric oxide release remodels immunosuppressive microenvironment and potentiates tumor immunotherapy.* Nano Today, *42, 101381–101381. https://doi.org/10.1016/j.nantod.2022.101381*

immunosuppressive M2 macrophages into the immune-stimulatory M1 phenotype in primary tumors. Overall, the UNTPs + L+αPD-1 combination therapy demonstrated impressive antitumor efficacy by leveraging direct tumoricidal effects, augmented systemic immune responses, and TME reprogramming (Li, Qian, et al., 2022).

The irregular tumor blood vessels and the dense tissue surrounding the tumor create a significant physical barrier that severely restricts the infiltration of cytotoxic T lymphocytes into the tumor and diminishes the effectiveness of immune checkpoint inhibitors (ICIs) (Datta et al., 2019). To overcome these challenges, Zhang and coworkers synthesized a reduction-sensitive NO donor conjugate incorporating furoxans and oxaliplatin (FO). Subsequently, they devised a bioinspired lipoprotein system for FO, termed S-LFO, with enhanced accessibility to stromal cells within the TME (Wang et al., 2022). The primary objective of the study was to reshape the intratumor physical barriers, facilitating CTL infiltration, thereby enhancing the synergistic effects of ICI-mediated immunotherapy (Fig. 8.9). The inclusion

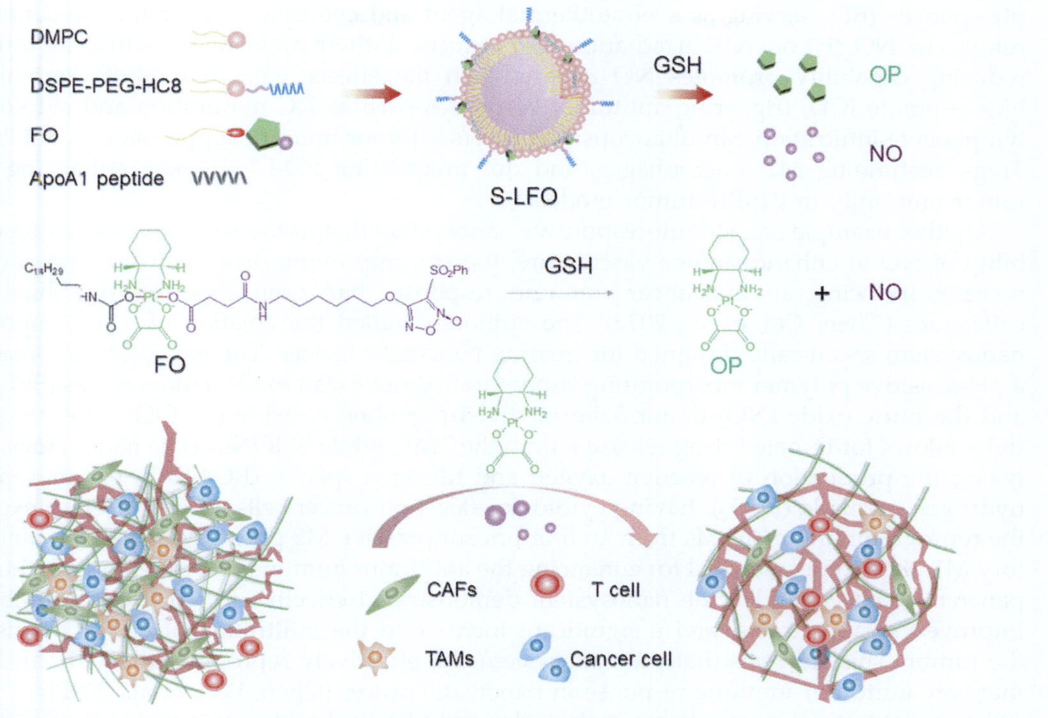

FIGURE 8.9 Schematic illustration of the S-LFO-mediated remodeling of intratumor physical barriers to potentiate CTL infiltration in tumors. In reductive intracellular environments, NO could be responsively released from S-LFO, promoting tumor vessel normalization and disrupting the stromal barriers. This could pave the way for CTL infiltration in tumors to improve the efficacy of anticancer immunotherapy. *Reproduced with permission from. From Wang, Y., Xie, H., Wu, Y., Xu, S., Li, Y., Li, J., Xu, X., Wang, S., Li, Y., & Zhang, Z. (2022). Bioinspired lipoproteins of furoxans–oxaliplatin remodel physical barriers in tumor to potentiate T-cell infiltration.* Advanced Materials, 34(14). *https://doi.org/10.1002/adma.202110614*

of the HTTIPKVC (HC8) peptide in S-LFO aimed to enhance its interaction with endothelial cells (ECs), cancer-associated fibroblasts (CAFs), and TAMs within the tumor milieu.

Results from this research demonstrate that S-LFO treatment effectively reduces the frequency of TAMs, CAFs, and ECs in tumors. Additionally, it shows a decrease in the expression of major extracellular matrix components such as collagen I, fibronectin, and chondroitin sulfate. Furthermore, the treatment leads to an increase in the frequency of $CD3^{+}CD8^{+}$ T cells in the tumor, indicating enhanced CTLs infiltration. The results also indicate the potential of S-LFO treatment in relieving hypoxia in tumors, overall suggesting that S-LFO treatment has the potential to remodel physical barriers in tumors, thereby facilitating T cell infiltration and potentially enhancing antitumor immune responses.

In the context of combination therapy and exploiting the potential of NO to induce ICD, Chen's group presented an approach involving a drug-free and NIR-enabled nanogasholder named NOPS@BP (Huang et al., 2022). The nanogasholder is constituted by an external layer made of a nitrate-containing polymeric NO donor and an internal core containing black phosphorus (BP), serving as a photothermal agent and energy converter for the controlled release of NO. Upon NIR irradiation, BP induces a thermal increase, while its intrinsic reducing capability promotes NO release. Both the effects, e.g., mild photothermia and NO, promote ICD, triggering antitumor responses such as DC maturation and cytotoxic T lymphocyte infiltration. Simultaneously, it reverses tumor immunosuppression by inhibiting Tregs, restraining M2 macrophages, and downregulating PD-L1, thereby enhancing antitumor immunity in B16F10 tumor models.

Another example of a stimuli-responsive nanosystem that harnesses the vasodilating capability of NO to enhance tumor vasculature, thereby improving drug delivery, and simultaneously inducing an anticancer immune response, has been reported by Shuai and colleagues (Chen, Cai, et al., 2022). The authors detailed the creation of a pH-responsive nanosystem specifically designed for treating pancreatic cancer. The nanosystem consists of a pH-sensitive polymer incorporating superparamagnetic iron oxide nanoparticles (SPIONs), and the nitric oxide (NO) donor 5,5-dimethyl-1-pyrroline N-oxide (DMPO). The pH sensitivity allows for targeted drug release within the TME, while SPIONs act as nanozymes, catalyzing the production of reactive oxygen and nitrogen species (RONS) in the presence of hydrogen peroxide (H_2O_2), having cytotoxic effects on cancer cells. The study also explores the reprogramming of TAMs from an immunosuppressive M2 phenotype to a proinflammatory M1 phenotype, essential for enhancing the antitumor immune response. In an orthotopic pancreatic cancer model, the nanosystem demonstrated effective tumor growth inhibition, improved survival rates, and a significant increase in the infiltration of $CD8^+$ T cells into the tumor. This suggests that the nanosystem can effectively reprogram the TME and promote an antitumor immune response in pancreatic cancer (Chen, Wei, et al., 2022).

In an alternative approach, researchers developed mitochondria-targeted and ultrasound-responsive nanoparticles (PIH−NO) to codeliver oxygen and NO, enhancing sonodynamic therapy (SDT) and immune response (Ji et al., 2021). The PIH-NO nanoparticles accumulate in mitochondria, relieving hypoxia, and releasing O_2 and NO upon ultrasound irradiation. This enhances SDT, inhibiting tumor growth and inducing ICD. Furthermore, PIH-NO promotes dendritic cell maturation, increases infiltrating immune cells, polarizes M2 macrophages into the M1 phenotype, and depletes myeloid-derived suppressor cells, reversing immunosuppression for enhanced cancer immunotherapy.

Another system based on a thermal-sensitive NO donor copolymer (PAAV-SNO) was developed for constructing an erythrocyte membrane-camouflaged nanobullet codelivering the NIR II photothermal agent IR1061 and the IDO1 inhibitor 1-methyl-tryptophan (RBCm/PAAV-SNO/IR1061 + 1-MT NPs) (Yang et al., 2020). In vitro experiments showed that the nanobullet had minimal cytotoxicity to breast cancer cells in the absence of laser irradiation but became highly effective when exposed to NIR II laser, primarily due to photothermal therapy. Furthermore, the treatment inhibited cancer cell migration and demonstrated the ability to suppress IDO1 activity and to reduce the expression of PD-L1, potentially enhancing the immune response. In vivo studies revealed that the nanobullet, when used as a vaccination method, inhibited tumor growth and increased the presence of $CD8^+$ cytotoxic T cells while reducing regulatory T cells. The combination therapy, involving PTT, IDO-1 inhibition, and NO gas therapy, was found to be more effective than individual treatments or treatments without laser irradiation. Additionally, the treatment demonstrated efficacy in suppressing breast cancer lung metastasis in a mouse model and appeared to normalize tumor blood vessels, alleviate tumor hypoxia, and stimulate positive immune responses (Yang et al., 2020).

Lastly, the study by Thomas' group aimed to explore the immunological effects of paclitaxel (PTX) and NO, both individually and in combination, in different in vitro tumor models (colorectal cancer, murine lymphoma, murine melanoma, and murine breast cancer), and in an in vivo model of subcutaneous melanoma (Kim et al., 2020). They developed a DDS composed of polymerized PTX (pPTX) and polymerized β-cyclodextrin (pCD) to enable controlled and targeted release of both PTX and NO within cancer cells. The rationale behind this approach was to harness the synergistic cytotoxicity of PTX and the potential immune-modulatory effects of NO for enhanced cancer treatment. Their results showed that the DDS, indeed, induced ICD-activated DCs and facilitated the expansion of T cells when tested in vitro. However, when the DDS was tested in vivo animal models, the findings became more complex. In particular, the DDS was found to expand populations of MDSCs and Tregs in the TME, therefore hindering the antitumor immune response. On the other hand, combining the DDS with anti-CTLA-4 therapy led to significantly improved tumor suppression in both primary and secondary tumors, as well as prolonged survival in experimental animals. This combination strategy appeared to overcome some of the limitations observed with the DDS alone, such as the impaired expansion of $CD8^+$ T cells (Kim et al., 2020).

Conclusions and future directions

Exogenous NO plays a multifaceted and pivotal role in cancer treatment and immune modulation. Its direct influence extends to various aspects, affecting the immunogenicity of cancer cells as well as the functions, activation, and differentiation of different immune cell subtypes. These intricate interactions have a significant impact on the immune system's complex balance between stimulation and suppression. The utilization of exogenous sources of NO has unveiled its potential to augment the immunogenicity of cancer cells. It has been shown to expand the populations of activated dendritic cells and M1-polarized macrophages, while also influencing the differentiation of $CD4^+$ T cells into diverse subtypes. This proimmunogenic aspect underscores NO's capacity to bolster the immune system's ability to recognize and combat cancer. However, NO also exhibits immunosuppressive effects that must be carefully considered. These effects include the expansion of regulatory T cells (Tregs), inhibition of cytotoxic T lymphocyte differentiation, proliferation, and migration, and the expansion of myeloid-derived suppressor cells (MDSCs). These immunosuppressive mechanisms highlight the nuanced role NO plays in shaping the immune response.

Recent research endeavors have proposed innovative strategies to harness the immune-stimulatory aspects of exogenous NO selectively. These strategies involve precise control over NO dosage through advanced drug delivery systems. Additionally, the exploration of combination therapy approaches, combining NO donors with other chemotherapeutic or immune-modulatory agents, offers exciting prospects for optimizing cancer immunotherapy. Despite these promising developments, a significant challenge in utilizing NO more broadly in antitumor immunotherapy remains. Current NO delivery systems often lack stability and targeting precision. Although efforts have led to the development of stimuli-responsive NO delivery systems that prevent the spontaneous release of NO during storage and circulation in the bloodstream, there is still a pressing need for further advancements in this field to enhance specificity and effectiveness.

In summary, our comprehension of NO's effects on the immune system has expanded rapidly in recent years, unveiling exciting opportunities to leverage its therapeutic potential. With ongoing efforts to optimize dosing regimens, explore combination therapies, and refine controlled NO donors and delivery systems, we are moving closer to realizing the therapeutic potential of nitric oxide in cancer treatment and immunotherapy.

References

Ahmed, A., & Tait, S. W. G. (2020). Targeting immunogenic cell death in cancer. *Molecular Oncology, 14*(12), 2994–3006. https://doi.org/10.1002/1878-0261.12851

Alimoradi, H., Greish, K., Gamble, A. B., & Giles, G. I. (2019). Controlled delivery of nitric oxide for cancer therapy. *Pharmaceutical Nanotechnology, 7*(4), 279–303. https://doi.org/10.2174/2211738507666190429111306

Altun, A., Temiz, T. K., Balcı, E., Polat, Z. A., & Turan, M. (2013). Effects of tyrosine kinase inhibitor E7080 and eNOS inhibitor L-NIO on colorectal cancer alone and in combination. *Chinese Journal of Cancer Research = Chung-kuo yen cheng yen chiu, 25*(5), 572–584. https://doi.org/10.3978/j.issn.1000-9604.2013.10.10

Anderson, N. M., & Simon, M. C. (2020). The tumor microenvironment. *Current Biology, 30*(16), R921–R925. https://doi.org/10.1016/j.cub.2020.06.081

Baghban, R., Roshangar, L., Jahanban-Esfahlan, R., Seidi, K., Ebrahimi-Kalan, A., Jaymand, M., Kolahian, S., Javaheri, T., & Zare, P. (2020). Tumor microenvironment complexity and therapeutic implications at a glance. *Cell Communication and Signaling, 18*(1), 59. https://doi.org/10.1186/s12964-020-0530-4

Bonavida, B., Baritaki, S., Huerta-Yepez, S., Vega, M. I., Chatterjee, D., & Yeung, K. (2008). Novel therapeutic applications of nitric oxide donors in cancer: Roles in chemo- and immunosensitization to apoptosis and inhibition of metastases. *Nitric Oxide, 19*(2), 152–157. https://doi.org/10.1016/j.niox.2008.04.018

Bonavida, B. (2020). Sensitizing activities of nitric oxide donors for cancer resistance to anticancer therapeutic drugs. *Biochemical Pharmacology, 176*, 113913. https://doi.org/10.1016/j.bcp.2020.113913

Brüne, B., Dehne, N., Grossmann, N., Jung, M., Namgaladze, D., Schmid, T., von Knethen, A., & Weigert, A. (2013). Redox control of inflammation in macrophages. *Antioxidants and Redox Signaling, 19*(6), 595–637. https://doi.org/10.1089/ars.2012.4785

Bui, I., Baritaki, S., Libra, M., Zaravinos, A., & Bonavida, B. (2023). Cancer resistance is mediated by the upregulation of several anti-apoptotic gene products via the inducible nitric oxide synthase/nitric oxide pathway: Therapeutic implications. *Antioxidants and Redox Signaling, 39*(13–15), 853–889. https://doi.org/10.1089/ars.2023.0250

Chen, G., Cai, Y., Li, B., Lin, M., Wang, X., Wang, Z., & Shuai, X. (2022). Theranostic nanosystem mediating cascade catalytic reactions for effective immunotherapy of highly immunosuppressive and poorly penetrable pancreatic tumor. *Science China Chemistry, 65*(7), 1383–1400. https://doi.org/10.1007/s11426-022-1262-x

Chen, J., Wei, Y., Yang, W., Huang, Q., Chen, Y., Zeng, K., & Chen, J. (2022). IL-6: The link between inflammation, immunity and breast cancer. *Frontiers in Oncology, 12*. https://doi.org/10.3389/fonc.2022.903800

Chen, Q., Huo, D., Cheng, H., Lyu, Z., Zhu, C., Guan, B., & Xia, Y. (2019). Near-infrared-triggered release of Ca^{2+} ions for potential application in combination cancer therapy. *Advanced Healthcare Materials, 8*(6). https://doi.org/10.1002/adhm.201801113

Datta, M., Coussens, L. M., Nishikawa, H., Hodi, F. S., & Jain, R. K. (2019). Reprogramming the tumor microenvironment to improve immunotherapy: Emerging strategies and combination therapies. *American Society of Clinical Oncology Educational Book, 39*, 165–174. https://doi.org/10.1200/EDBK_237987

De Boo, S., Kopecka, J., Brusa, D., Gazzano, E., Matera, L., Ghigo, D., Bosia, A., & Riganti, C. (2009). iNOS activity is necessary for the cytotoxic and immunogenic effects of doxorubicin in human colon cancer cells. *Molecular Cancer, 8*(1), 108. https://doi.org/10.1186/1476-4598-8-108

Di Lorenzo, A., Lin, M. I., Murata, T., Landskroner-Eiger, S., Schleicher, M., Kothiya, M., Iwakiri, Y., Yu, J., Huang, P. L., & Sessa, W. C. (2013). eNOS derived nitric oxide regulates endothelial barrier function via VE cadherin and Rho GTPases. *Journal of Cell Science*. https://doi.org/10.1242/jcs.115972

Du, L., He, H., Xiao, Z., Xiao, H., An, Y., Zhong, H., Lin, M., Meng, X., Han, S., & Shuai, X. (2022). GSH-responsive metal–organic framework for intratumoral release of NO and Ido inhibitor to enhance antitumor immunotherapy. *Small, 18*(15). https://doi.org/10.1002/smll.202107732

Dunlap, T., Abdul-Hay, S. O., Chandrasena, R. E. P., Hagos, G. K., Sinha, V., Wang, Z., Wang, H., & Thatcher, G. R. J. (2008). Nitrates and NO-NSAIDs in cancer chemoprevention and therapy: In vitro evidence querying the NO donor functionality. *Nitric Oxide, 19*(2), 115–124. https://doi.org/10.1016/j.niox.2008.04.013

Ekmekcioglu, S., Grimm, E. A., & Roszik, J. (2017). Targeting iNOS to increase efficacy of immunotherapies. *Human Vaccines & Immunotherapeutics, 13*(5), 1105–1108. https://doi.org/10.1080/21645515.2016.1276682

Emami Nejad, A., Najafgholian, S., Rostami, A., Sistani, A., Shojaeifar, S., Esparvarinha, M., Nedaeinia, R., Haghjooy Javanmard, S., Taherian, M., Ahmadlou, M., Salehi, R., Sadeghi, B., & Manian, M. (2021). The role of hypoxia in the tumor microenvironment and development of cancer stem cell: A novel approach to developing treatment. *Cancer Cell International, 21*(1), 62. https://doi.org/10.1186/s12935-020-01719-5

Fallica, A. N., Barbaraci, C., Amata, E., Pasquinucci, L., Turnaturi, R., Dichiara, M., Intagliata, S., Gariboldi, M. B., Marras, E., Orlandi, V. T., Ferroni, C., Martini, C., Rescifina, A., Gentile, D., Varchi, G., & Marrazzo, A. (2021). Nitric oxide photo-donor hybrids of ciprofloxacin and norfloxacin: A shift in activity from antimicrobial to anticancer agents. *Journal of Medicinal Chemistry, 64*(15), 11597–11613. https://doi.org/10.1021/acs.jmedchem.1c00917

Firdaus, F., Kuchakulla, M., Qureshi, R., Dulce, R. A., Soni, Y., Van Booven, D. J., Shah, K., Masterson, T., Rosete, O. J., Punnen, S., Hare, J. M., Ramasamy, R., & Arora, H. (2022). S-nitrosylation of CSF1 receptor increases the efficacy of CSF1R blockage against prostate cancer. *Cell Death & Disease, 13*(10), 859. https://doi.org/10.1038/s41419-022-05289-4

Forstermann, U., & Sessa, W. C. (2012). Nitric oxide synthases: Regulation and function. *European Heart Journal, 33*(7), 829–837. https://doi.org/10.1093/eurheartj/ehr304

Fraix, A., Parisi, C., Seggio, M., & Sortino, S. (2021). Nitric oxide photoreleasers with fluorescent reporting. *Chemistry - A European Journal, 27*(50), 12714–12725. https://doi.org/10.1002/chem.202101662

Hao, H., Yu, M., Yi, Y., Sun, S., Huang, X., Huang, C., Liu, Y., Huang, W., Wang, J., Zhao, J., & Wu, M. (2022). Mesoporous calcium peroxide-ignited NO generation for amplifying photothermal immunotherapy of breast cancer. *Chemical Engineering Journal, 437*, 135371. https://doi.org/10.1016/j.cej.2022.135371

Havunen, R., Santos, J. M., Sorsa, S., Rantapero, T., Lumen, D., Siurala, M., Airaksinen, A. J., Cervera-Carrascon, V., Tähtinen, S., Kanerva, A., & Hemminki, A. (2018). Abscopal effect in non-injected tumors achieved with cytokine-armed oncolytic adenovirus. *Molecular Therapy - Oncolytics, 11*, 109–121. https://doi.org/10.1016/j.omto.2018.10.005

He, Z., & Zhang, S. (2021). Tumor-associated macrophages and their functional transformation in the hypoxic tumor microenvironment. *Frontiers in Immunology, 12*. https://doi.org/10.3389/fimmu.2021.741305

Huang, T., Sun, X., Qi, Y., Yang, X., Fan, L., Chen, M., Yue, Y., Ge, H., Li, Y., Nie, G., Min, H., & Sun, X. (2023). Immunogenic cell death effects induced by doxorubicin improved chemo-immunotherapy via restoration of granzyme B activity. *Nano Research*. https://doi.org/10.1007/s12274-023-5581-6

Huang, X., Zhong, Y., Li, Y., Zhou, X., Yang, L., Zhao, B., Zhou, J., Qiao, H., Huang, D., Qian, H., & Chen, W. (2022). Black phosphorus-synergic nitric oxide nanogasholder spatiotemporally regulates tumor microenvironments for self-amplifying immunotherapy. *ACS Applied Materials & Interfaces, 14*(33), 37466–37477. https://doi.org/10.1021/acsami.2c10098

Huerta, S., Baay-Guzman, G., Gonzalez-Bonilla, C. R., Livingston, E. H., Huerta-Yepez, S., & Bonavida, B. (2009). In vitro and in vivo sensitization of SW620 metastatic colon cancer cells to CDDP-induced apoptosis by the nitric oxide donor DETANONOate: Involvement of AIF. *Nitric Oxide, 20*(3), 182–194. https://doi.org/10.1016/j.niox.2008.11.006

Ji, C., Si, J., Xu, Y., Zhang, W., Yang, Y., He, X., Xu, H., Mou, X., Ren, H., & Guo, H. (2021). Mitochondria-targeted and ultrasound-responsive nanoparticles for oxygen and nitric oxide codelivery to reverse immunosuppression and enhance sonodynamic therapy for immune activation. *Theranostics, 11*(17), 8587–8604. https://doi.org/10.7150/thno.62572

Jiang, W., Dong, W., Li, M., Guo, Z., Wang, Q., Liu, Y., Bi, Y., Zhou, H., & Wang, Y. (2022). Nitric oxide induces immunogenic cell death and potentiates cancer immunotherapy. *ACS Nano, 16*(3), 3881–3894. https://doi.org/10.1021/acsnano.1c09048

Jiang, W., Shen, Z., Guo, Z., Wang, Q., Li, Q., Hu, J., & Wang, Y. (2023). Overcoming oxygen heterogeneity of tumor microenvironments to boost cancer immunotherapy by oxygen-switchable ROS/RNS nanogenerators. *Nano Today, 48*, 101696. https://doi.org/10.1016/j.nantod.2022.101696

Jin, G., Gao, Z., Liu, Y., Zhao, J., Ou, H., Xu, F., & Ding, D. (2021). Polymeric nitric oxide delivery nanoplatforms for treating cancer, cardiovascular diseases, and infection. *Advanced Healthcare Materials, 10*(3). https://doi.org/10.1002/adhm.202001550

Kashfi, K., Rayyan, Y., Qiao, L. L., Williams, J. L., Chen, J., del Soldato, P., Traganos, F., & Rigas, B. (2002). Nitric oxide-donating nonsteroidal anti-inflammatory drugs inhibit the growth of various cultured human cancer cells: Evidence of a tissue type-independent effect. *Journal of Pharmacology and Experimental Therapeutics, 303*(3), 1273–1282. https://doi.org/10.1124/jpet.102.042754

Khan, F. H., Dervan, E., Bhattacharyya, D. D., McAuliffe, J. D., Miranda, K. M., & Glynn, S. A. (2020). The role of nitric oxide in cancer: Master regulator or NOt? *International Journal of Molecular Sciences, 21*(24), 9393. https://doi.org/10.3390/ijms21249393

Kim, J., Francis, D. M., Sestito, L. F., Archer, P. A., Manspeaker, M. P., O'Melia, M. J., & Thomas, S. N. (2022). Thermosensitive hydrogel releasing nitric oxide donor and anti-CTLA-4 micelles for anti-tumor immunotherapy. *Nature Communications, 13*(1), 1479. https://doi.org/10.1038/s41467-022-29121-x

Kim, J., Sestito, L. F., Im, S., Kim, W. J., & Thomas, S. N. (2020). Poly(cyclodextrin)-Polydrug nanocomplexes as synthetic oncolytic virus for locoregional melanoma chemoimmunotherapy. *Advanced Functional Materials, 30*(16). https://doi.org/10.1002/adfm.201908788

Kim, J., & Thomas, S. N. (2022). Opportunities for nitric oxide in potentiating cancer immunotherapy. *Pharmacological Reviews, 74*(4), 1146–1175. https://doi.org/10.1124/pharmrev.121.000500

Kopecka, J., Salaroglio, I. C., Perez-Ruiz, E., Sarmento-Ribeiro, A. B., Saponara, S., De Las Rivas, J., & Riganti, C. (2021). Hypoxia as a driver of resistance to immunotherapy. *Drug Resistance Updates, 59*, 100787. https://doi.org/10.1016/j.drup.2021.100787

Korde Choudhari, S., Chaudhary, M., Bagde, S., Gadbail, A. R., & Joshi, V. (2013). Nitric oxide and cancer: A review. *World Journal of Surgical Oncology, 11*(1), 118. https://doi.org/10.1186/1477-7819-11-118

Kraehenbuehl, L., Weng, C.-H., Eghbali, S., Wolchok, J. D., & Merghoub, T. (2022). Enhancing immunotherapy in cancer by targeting emerging immunomodulatory pathways. *Nature Reviews Clinical Oncology, 19*(1), 37–50. https://doi.org/10.1038/s41571-021-00552-7

Kuryk, L., Møller, A.-S. W., & Jaderberg, M. (2019). Abscopal effect when combining oncolytic adenovirus and checkpoint inhibitor in a humanized NOG mouse model of melanoma. *Journal of Medical Virology, 91*(9), 1702–1706. https://doi.org/10.1002/jmv.25501

Labadie, B. W., Bao, R., & Luke, J. J. (2019). Reimagining Ido pathway inhibition in cancer immunotherapy via downstream focus on the tryptophan–kynurenine–aryl hydrocarbon Axis. *Clinical Cancer Research, 25*(5), 1462–1471. https://doi.org/10.1158/1078-0432.CCR-18-2882

Lampson, B. L., Kendall, S. D. S., Ancrile, B. B., Morrison, M. M., Shealy, M. J., Barrientos, K. S., Crowe, M. S., Kashatus, D. F., White, R. R., Gurley, S. B., Cardona, D. M., & Counter, C. M. (2012). Targeting eNOS in pancreatic cancer. *Cancer Research, 72*(17), 4472–4482. https://doi.org/10.1158/0008-5472.CAN-12-0057

Laneri, F., Licciardello, N., Suzuki, Y., Graziano, A. C. E., Sodano, F., Fraix, A., & Sortino, S. (2022). A supramolecular nanoassembly of lenvatinib and a green light-activatable NO releaser for combined chemo-phototherapy. *Pharmaceutics, 15*(1), 96. https://doi.org/10.3390/pharmaceutics15010096

Li, C.-Y., Anuraga, G., Chang, C.-P., Weng, T.-Y., Hsu, H.-P., Ta, H. D. K., Su, P.-F., Chiu, P.-H., Yang, S.-J., Chen, F.-W., Ye, P.-H., Wang, C.-Y., & Lai, M.-D. (2023). Repurposing nitric oxide donating drugs in cancer therapy through immune modulation. *Journal of Experimental & Clinical Cancer Research, 42*(1), 22. https://doi.org/10.1186/s13046-022-02590-0

Li, Y., Qian, H., Huang, X., Zhou, X., Zhao, B., Yang, L., Chen, J., Qiao, H., Huang, D., Zhong, Y., & Chen, W. (2022a). Up-converted nano-gasholder with precise nitric oxide release remodels immunosuppressive microenvironment and potentiates tumor immunotherapy. *Nano Today, 42*, 101381. https://doi.org/10.1016/j.nantod.2022.101381

Li, Y., Yoon, B., Dey, A., Nguyen, Van Q., & Park, J. H. (2022). Recent progress in nitric oxide-generating nanomedicine for cancer therapy. *Journal of Controlled Release, 352*, 179–198. https://doi.org/10.1016/j.jconrel.2022.10.012

Li, Z., Lai, X., Fu, S., Ren, L., Cai, H., Zhang, H., Gu, Z., Ma, X., & Luo, K. (2022). Immunogenic cell death activates the tumor immune microenvironment to boost the immunotherapy efficiency. *Advanced Science, 9*(22). https://doi.org/10.1002/advs.202201734

Lin, X., Fang, Y., Jin, X., Zhang, M., & Shi, K. (2021). Modulating repolarization of tumor-associated macrophages with targeted therapeutic nanoparticles as a potential strategy for cancer therapy. *ACS Applied Bio Materials, 4*(8), 5871–5896. https://doi.org/10.1021/acsabm.1c00461

Liu, Y., Liang, X., Yin, X., Lv, J., Tang, K., Ma, J., Ji, T., Zhang, H., Dong, W., Jin, X., Chen, D., Li, Y., Zhang, S., Xie, H. Q., Zhao, B., Zhao, T., Lu, J., Hu, Z.-W., Cao, X., Qin, F. X.-F., & Huang, B. (2017). Blockade of Ido-kynurenine-AhR metabolic circuitry abrogates IFN-γ-induced immunologic dormancy of tumor-repopulating cells. *Nature Communications, 8*(1), 15207. https://doi.org/10.1038/ncomms15207

Mabrouk, N., Racoeur, C., Shan, J., Massot, A., Ghione, S., Privat, M., Dondaine, L., Ballot, E., Truntzer, C., Boidot, R., Hermetet, F., Derangère, V., Bruchard, M., Végran, F., Chouchane, L., Ghiringhelli, F., Bettaieb, A., & Paul, C. (2023). GTN enhances antitumor effects of doxorubicin in TNBC by targeting the immunosuppressive activity of PMN-MDSC. *Cancers, 15*(12), 3129. https://doi.org/10.3390/cancers15123129

Malanga, M., Seggio, M., Kirejev, V., Fraix, A., Di Bari, I., Fenyvesi, E., Ericson, M. B., & Sortino, S. (2019). A phototherapeutic fluorescent β-cyclodextrin branched polymer delivering nitric oxide. *Biomaterials Science, 7*(6), 2272–2276. https://doi.org/10.1039/C9BM00395A

McGinity, C. L., Palmieri, E. M., Somasundaram, V., Bhattacharyya, D. D., Ridnour, L. A., Cheng, R. Y. S., Ryan, A. E., Glynn, S. A., Thomas, D. D., Miranda, K. M., Anderson, S. K., Lockett, S. J., McVicar, D. W., & Wink, D. A. (2021). Nitric oxide modulates metabolic processes in the tumor immune microenvironment. *International Journal of Molecular Sciences, 22*(13), 7068. https://doi.org/10.3390/ijms22137068

Mintz, J., Vedenko, A., Rosete, O., Shah, K., Goldstein, G., Hare, J. M., Ramasamy, R., & Arora, H. (2021). Current advances of nitric oxide in cancer and anticancer therapeutics. *Vaccines, 9*(2), 94. https://doi.org/10.3390/vaccines9020094

Mishra, D., Patel, V., & Banerjee, D. (2020). Nitric oxide and S-nitrosylation in cancers: Emphasis on breast cancer. *Breast Cancer: Basic and Clinical Research, 14*. https://doi.org/10.1177/1178223419882688

Navasardyan, I., & Bonavida, B. (2021). Regulation of T Cells in cancer by nitric oxide. *Cells, 10*(10), 2655. https://doi.org/10.3390/cells10102655

Omidkhah, N., & Ghodsi, R. (2022). NO-HDAC dual inhibitors. *European Journal of Medicinal Chemistry, 227*, 113934. https://doi.org/10.1016/j.ejmech.2021.113934

Rapozzi, V., Ragno, D., Guerrini, A., Ferroni, C., Pietra, E.della, Cesselli, D., Castoria, G., Di Donato, M., Saracino, E., Benfenati, V., & Varchi, G. (2015). Androgen receptor targeted conjugate for bimodal photodynamic therapy of prostate cancer in vitro. *Bioconjugate Chemistry, 26*(8), 1662–1671. https://doi.org/10.1021/acs.bioconjchem.5b00261

Ravishankar, B., Liu, H., Shinde, R., Chaudhary, K., Xiao, W., Bradley, J., Koritzinsky, M., Madaio, M. P., & McGaha, T. L. (2015). The amino acid sensor GCN2 inhibits inflammatory responses to apoptotic cells promoting tolerance and suppressing systemic autoimmunity. *Proceedings of the National Academy of Sciences, 112*(34), 10774–10779. https://doi.org/10.1073/pnas.1504276112

Reis, A. K. C. A., Stern, A., & Monteiro, H. P. (2019). S-nitrosothiols and H2S donors: Potential chemo-therapeutic agents in cancer. *Redox Biology, 27*, 101190. https://doi.org/10.1016/j.redox.2019.101190

Ren, X., Wang, Y., Jia, L., Guo, X., He, X., Zhao, Z., Gao, Di, & Yang, Z. (2022). Intelligent nanomedicine approaches using medical gas-mediated multi-therapeutic modalities against cancer. *Journal of Biomedical Nanotechnology, 18*(1), 24–49. https://doi.org/10.1166/jbn.2022.3224

Riganti, C., Miraglia, E., Viarisio, D., Costamagna, C., Pescarmona, G., Ghigo, D., & Bosia, A. (2005). Nitric oxide reverts the resistance to doxorubicin in human colon cancer cells by inhibiting the drug efflux. *Cancer Research, 65*(2), 516–525.

Sahebnasagh, A., Saghafi, F., Negintaji, S., Hu, T., Shabani-Borujeni, M., Safdari, M., Ghaleno, H. R., Miao, L., Qi, Y., Wang, M., Liao, P., Sureda, A., Simal-Gándara, J., Nabavi, S. M., & Xiao, J. (2022). Nitric oxide and immune responses in cancer: Searching for new therapeutic strategies. *Current Medicinal Chemistry, 29*(9), 1561–1595. https://doi.org/10.2174/0929867328666210707194543

Salimian Rizi, B., Achreja, A., & Nagrath, D. (2017). Nitric oxide: The forgotten child of tumor metabolism. *Trends in Cancer, 3*(9), 659–672. https://doi.org/10.1016/j.trecan.2017.07.005

Somasundaram, V., Basudhar, D., Bharadwaj, G., No, J. H., Ridnour, L. A., Cheng, R. Y. S., Fujita, M., Thomas, D. D., Anderson, S. K., McVicar, D. W., & Wink, D. A. (2019). Molecular mechanisms of nitric oxide in cancer progression, signal transduction, and metabolism. *Antioxidants and Redox Signaling, 30*(8), 1124–1143. https://doi.org/10.1089/ars.2018.7527

Szeliga, M., & Albrecht, J. (2021). Roles of nitric oxide and polyamines in brain tumor growth. *Advances in Medical Sciences, 66*(1), 199–205. https://doi.org/10.1016/j.advms.2021.02.006

Theivendran, S., Gu, Z., Tang, J., Yang, Y., Song, H., Yang, Y., Zhang, M., Cheng, D., & Yu, C. (2022). Nanostructured organosilica nitric oxide donors intrinsically regulate macrophage polarization with antitumor effect. *ACS Nano, 16*(7), 10943–10957. https://doi.org/10.1021/acsnano.2c03348

Tran, A. N., Boyd, N. H., Walker, K., & Hjelmeland, A. B. (2017). NOS expression and NO function in glioma and implications for patient therapies. *Antioxidants and Redox Signaling, 26*(17), 986–999. https://doi.org/10.1089/ars.2016.6820

Vedenko, A., Panara, K., Goldstein, G., Ramasamy, R., & Arora, H. (2020). *Tumor microenvironment and nitric oxide: Concepts and mechanisms*. https://doi.org/10.1007/978-3-030-50224-9_10. http://link.springer.com/10.1007/978-3-030-50224-9_10

Vong, L. B., & Nagasaki, Y. (2020). Nitric oxide nano-delivery systems for cancer therapeutics: Advances and challenges. *Antioxidants, 9*(9), 791. https://doi.org/10.3390/antiox9090791

Wang, Y., Xie, H., Wu, Y., Xu, S., Li, Y., Li, J., Xu, X., Wang, S., Li, Y., & Zhang, Z. (2022). Bioinspired lipoproteins of furoxans–oxaliplatin remodel physical barriers in tumor to potentiate T-cell infiltration. *Advanced Materials, 34*(14). https://doi.org/10.1002/adma.202110614

Xie, D., Wang, Q., & Wu, G. (2022). Research progress in inducing immunogenic cell death of tumor cells. *Frontiers in Immunology, 13*. https://doi.org/10.3389/fimmu.2022.1017400

Xu, W., Liu, L. Z., Loizidou, M., Ahmed, M., & Charles, I. G. (2002). The role of nitric oxide in cancer. *Cell Research, 12*(5–6), 311–320. https://doi.org/10.1038/sj.cr.7290133

Yang, Z., Gao, Di, Guo, X., Jin, L., Zheng, J., Wang, Y., Chen, S., Zheng, X., Zeng, L., Guo, M., Zhang, X., & Tian, Z. (2020). Fighting immune cold and reprogramming immunosuppressive tumor microenvironment with red blood cell membrane-camouflaged nanobullets. *ACS Nano, 14*(12), 17442–17457. https://doi.org/10.1021/acsnano.0c07721

Yarlagadda, K., Hassani, J., Foote, I. P., & Markowitz, J. (2017). The role of nitric oxide in melanoma. *Biochimica et Biophysica Acta (BBA) - Reviews on Cancer, 1868*(2), 500–509. https://doi.org/10.1016/j.bbcan.2017.09.005

Zhang, Y., Pan, Z., Chen, C., Tan, Y., Wang, X., Wang, L., Zhang, L., Chen, Y., & He, G. (2023). Design, synthesis and anti-ovarian cancer activities of thieno[2,3-d]pyrimidine based chimeric BRD4 inhibitor/nitric oxide-donator. *European Journal of Medicinal Chemistry, 246*, 114970. https://doi.org/10.1016/j.ejmech.2022.114970

Zhao, Z., Shan, X., Zhang, H., Shi, X., Huang, P., Sun, J., He, Z., Luo, C., & Zhang, S. (2023). Nitric oxide-driven nanotherapeutics for cancer treatment. *Journal of Controlled Release, 362*, 151–169. https://doi.org/10.1016/j.jconrel.2023.08.038

Zheng, P., Ding, B., Shi, R., Jiang, Z., Xu, W., Li, G., Ding, J., & Chen, X. (2021). A multichannel Ca 2+ nanomodulator for multilevel mitochondrial destruction-mediated cancer therapy. *Advanced Materials, 33*(15). https://doi.org/10.1002/adma.202007426

CHAPTER

9

Control of nitric oxide synthase 2: Role of NRF2-regulated distal enhancer

Eros Di Giorgio, Valentina Rapozzi and Luigi E. Xodo

Department of Medicine, Laboratory of Biochemistry, University of Udine, Udine, Italy

Abbreviations

AP-1 Activator protein 1
BH4 (6R)-5,6,7,8-tetrahydrobiopterin
CaM Calmodulin
DAF-FM DA 4-Amino-5-methylamino-2′,7′-difluorofluorescein diacetate
DEGs Differentially expressed genes
ENCODE Encyclopedia of DNA Elements
FAD Flavin adenine dinucleotide
FMN Flavin mononucleotide
GMPc Cyclic guanosine monophosphate
GTP Guanosine triphosphate
H3K27ac Histone 3 acetylated at lysine 27
H3K4me1 Histone 3 monomethylated at lysine 4
IFN-γ Type II interferon
IL-1 Interleukin-1
IL-1R1 Interleukin-1 receptor type I
JAKs Janus family kinases
KRAS Kirsten Ras gene
LPS Lipopolysaccharides
NADPH Nicotinamide adenine dinucleotide phosphate reduced
NF-kB nuclear factor kappa-light-chain-enhancer of activated B cells
NO Nitric oxide
NOS1 Neuronal nitric oxide synthase (also called nNOS)
NOS2 Inducible nitric oxide synthase (also called iNOS)
NOS3 Endothelialnitric oxide synthase (also called eNOS)
PDAC Pancreatic ductal adenocarcinoma cells
sGC Soluble Guanylyl cyclase

Therapeutic Applications of Nitric Oxide in Cancer and Inflammatory Disorders
https://doi.org/10.1016/B978-0-443-22354-9.00009-7

STAT1 Signal transducer and activator of transcription 1
TLR4 Toll-like receptor 4
TNF-α Tumor necrosis factor alpha
TRAF6 TNF receptor-associated factor 6

Introduction

Inducible nitric oxide synthase (NOS2) is an enzyme that plays a crucial role in the production of nitric oxide (NO) in cells (Alderton et al., 2001). NO is a signaling molecule with multiple physiological functions that is involved in immune response, neurotransmission, and vasodilation and may contribute to carcinogenesis (Bredt, 2009; Levine et al., 2012). There are three isoforms of nitric oxide synthase: neuronal NOS (nNOS or NOS1), endothelial NOS (eNOS or NOS3), and inducible NOS (iNOS or NOS2) (Andrew & Mayer, 1999; Levine et al., 2012; Michel & Feron, 1997). While NOS1 and NOS3 are constitutively expressed in certain tissues, NOS2 is usually not present at significant levels under normal conditions. Its expression is induced in response to various stimuli, including inflammatory cytokines such as interferon-gamma (INF-γ) and tumor necrosis factor-alpha (TNF-α), bacterial lipopolysaccharides (LPSs), and other microbial products (Kone et al., 2003; Sharma et al., 2007).

After induction, NOS2 produces significant amounts of nitric oxide over a prolonged period of time compared to the other NOS isoforms (MacMicking et al., 1997). This sustained production of NO is often associated with immune responses and inflammation. The NO produced by NOS2 serves as a potent antimicrobial agent and helps the immune system to fight infection (Chakravortty & Hensel, 2003). NOS2 is frequently expressed in immune cells, including macrophages and neutrophils, as part of the body's defense against pathogens. Due to its involvement in immune responses and inflammation, NOS2 has become a target for therapeutic intervention in conditions such as septic shock, autoimmune diseases, and chronic inflammation (Sharma et al., 2007). Interfering with NOS2 activity is emerging as a potential strategy for the treatment of inflammatory conditions, although careful consideration is required as NO has both protective and deleterious effects.

The role of NOS2 in carcinogenesis is complex and may vary depending on the type of cancer, stage of cancer development, and the surrounding microenvironment (Vannini et al., 2015). Some studies suggest that elevated levels of NOS2 and nitric oxide may promote cancer cell growth, invasion, and the formation of new blood vessels to support tumor growth (angiogenesis) (Lala & Orucevic, 1998; Nathan, 1997; Wink et al., 1998). However, there are also studies that link NO to antitumor immune responses (MacMicking et al., 1997). Overall, NO plays a dual role in cancer and the mechanisms leading to these two activities are not yet fully understood (Wink et al., 2008; Choudhari et al., 2013; Xu et al., 2002). In this chapter, we will mainly focus on how NOS2 is regulated in cancer cells, especially in pancreatic cancer cells. We will discuss how NOS2 is regulated by NRF2, which binds to a distal enhancer in the NOS2 locus, and how the dysregulation of NOS2 expression can be exploited in combination therapies against pancreatic ductal adenocarcinoma cells (PDACs).

NOS2 structure and function

NOS2 is a 131 kDa mammalian protein consisting of 1153 amino acids that forms a homodimeric enzyme that catalyzes the conversion of L-arginine to L-citrulline and NO (uniProt, P35228 NOS2_HUMAN). Each subunit is characterized by two main domains: (i) a carboxy-terminal reductase domain, containing flavin adenine dinucleotide (FAD) and flavin mononucleotide (FMN) as cofactors, with a structure similar to that of P450 reductase; and (ii) an N-terminal catalytic oxygenase domain containing a heme group and (6R) 5,6,7,8-tetrahydro-L-biopterin (BH4) (Fig. 9.1A) (Alderton et al., 2005). The conversion of L-arginine to L-citrulline and NO occurs via an electron transport chain from nicotinamide adenine dinucleotide phosphate (NADPH) to FAD, FMN, and heme/O_2 (Fig. 9.1A–C). At the interface between the two subunits, a zinc ion is coordinated to two Cys-containing motifs (Hemmens et al., 2000; Li et al., 1999; Raman et al., 1998). The zinc ion is required for the binding of BH4 and L-arginine to the oxygenase domains of the NOS2 enzyme. Reduction equivalents from the two-electron donor NADPH in the reductase domain are transferred to the heme in the oxygenase domain so that Fe^{3+} is reduced to Fe^{2+}, that is, to a reduced state that allows the binding of O_2 to heme and its activation (Giroud et al., 2010). It is likely that BH_4 is also involved in the electron transport to the heme. The reductase and oxygenase domains are connected by a Ca^{2+}—calmodulin (CaM) linker (Abu-Soud & Stuehr, 1993), which facilitates electron flow to the oxygenase domain. Relatively low concentrations of intracellular calcium (~40 nM) are necessary for CaM to bind to the NOS2 enzyme. As the typical cytoplasmic calcium concentration is around 100 nM, regulation by calcium is not relevant for NOS2, whereas it is essential for the other NOS isoforms. The activated oxygen bound to heme is used to oxidize L-arginine in a two-step reaction (Fig. 9.1D). First, L-arginine is converted to N_{ω}-hydroxy-L-arginine, and then N^{ω}-hydroxy-L-arginine is converted to L-citrulline and NO (Noble et al., 1999; Stuehr et al., 2001). All NOS enzymes are only able to produce NO if they are present in the dimeric form. This is because the monomeric form is unable to bind to L-arginine and BH4 and produce NO. In the monomeric form, electrons therefore flow from NADPH to FAD, FMN, and molecular oxygen, which is reduced to superoxide ($O_2{\bullet}^-$) (Förstermann & Sessa, 2012).

The amount of NO that can be produced by the NOS enzymes varies between the three isoenzymes. While NOS1 and NOS3 only produce nanomolar NO concentrations for short periods of time (seconds to minutes), the inducible NOS2 produces larger amounts of NO in the micromolar range and for longer periods of time: hours or even days (Heinecke et al., 2014).

NO exerts its biological effects via canonical and noncanonical pathways. The canonical pathway involves the activation of soluble guanylate cyclase (sGC) by NO, which catalyzes the conversion of GTP to cGMP: a molecule that, in turn, activates cGMP-dependent kinases that transduce cell signaling through protein phosphorylation (Arnold et al., 1997; Heinrich et al., 2013).

Mechanistically, NO binds to the heme in the N-terminal domain of sGC, triggering a conformational change that activates the catalytic site in the C-terminal domain of the enzyme (Kang et al., 2019). The direct binding of NO to the heme iron is called nitrosylation. Nitrosylation reactions regulate the activity of many other proteins (Arnold et al., 1997) (Fig. 9.2). While sGC, for example, is activated by nitrosylation, cytochrome *c* oxidase

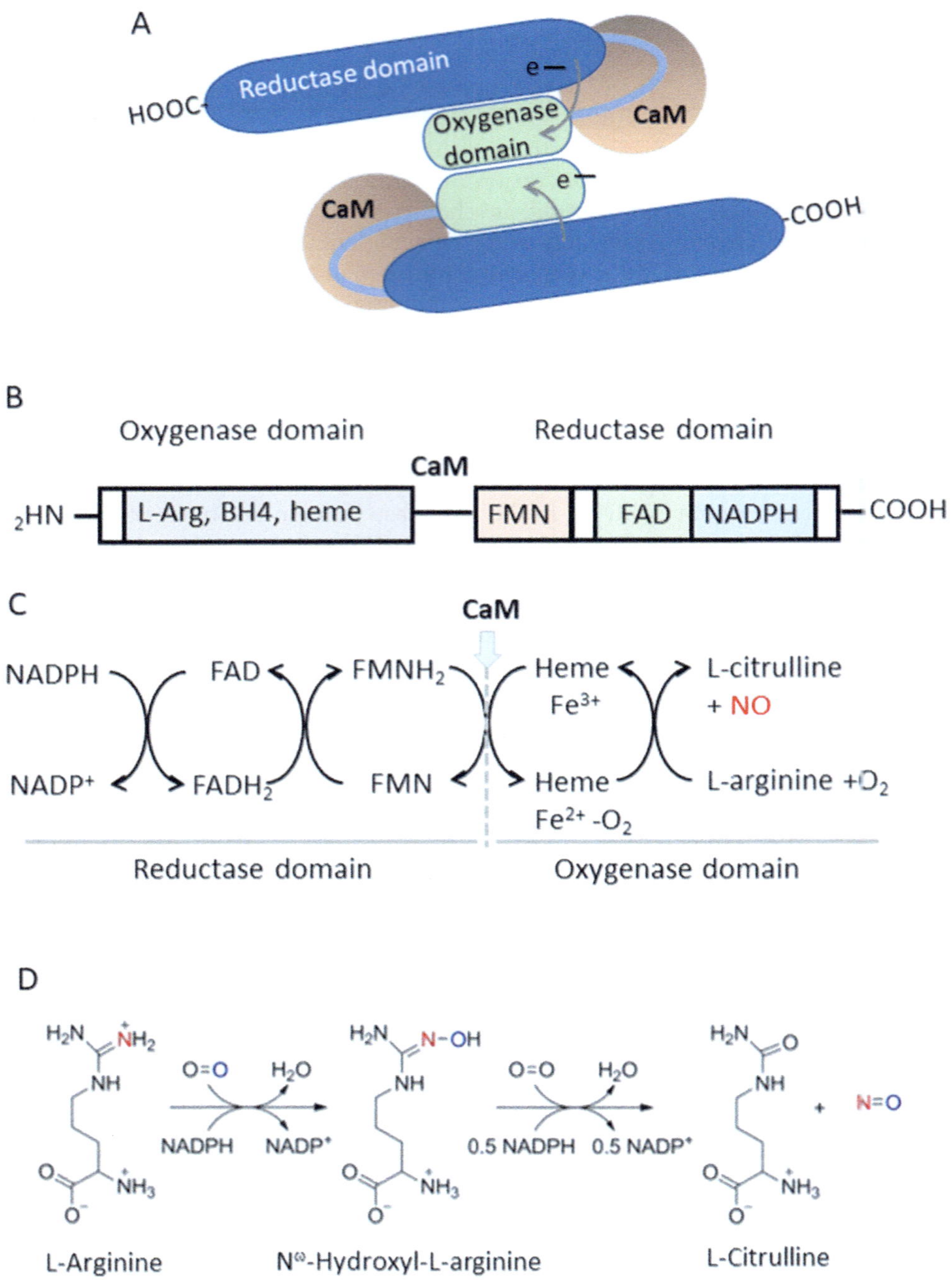

FIGURE 9.1 **Nitric oxide synthase 2: structure and function.** (A, B) Schematic structure of NOS2. Each subunit contains the heme moiety, a substrate binding site, a binding site for H4B, which functions in electron transfer, and a tetracoordinated zinc ion, which is important for structural integrity. Each subunit provides two of the cysteine ligands for zinc ion; (C, D) As a dimer, NOS2 catalyzes the transformation of L-arginine to L-citrulline and NO. This occurs through a complex reaction involving both the reductase and oxygenase domains of NOS2. L-arginine, protoporphyrin IX and Fe (heme) and BH4 are bound to the oxygenase domain, while FMN, FAD, and NADPH are bound to the reductase domain. Calmodulin (CaM) is noncovalently bound to interface between the two subunits of the enzyme. The binding of calmodulin is crucial for the activation of NOS2. The binding of CaM stabilizes the active conformation of the enzyme, facilitating the production of nitric oxide in response to specific physiological signals or stimuli.

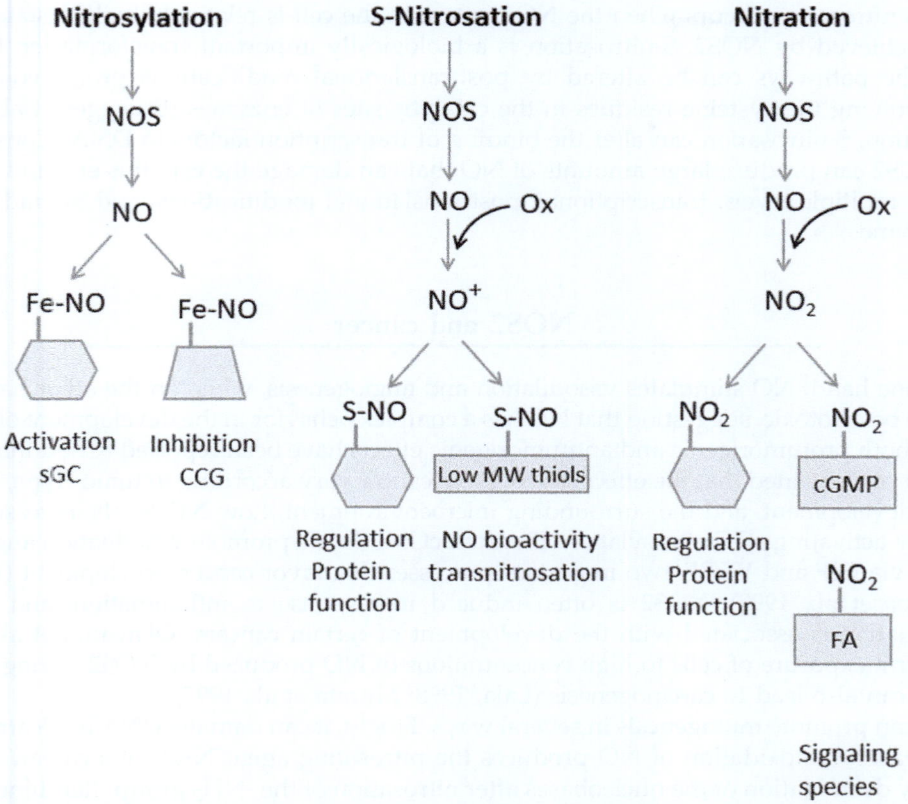

FIGURE 9.2 Chemistry of NO signaling. NO produced by NOS enzymes can activate different signaling pathways. Nitrosylation is the direct binding of NO to transition metals, heme-Fe. This reaction is at the basis of the canonical NO-induced pathway that activates sGC and inhibits cytochrome *c* oxidase (CCO). Fe–S clusters of electron transport chain and other proteins can also be nitrosylated. S-nitrosation refers to the covalent binding of NO^+, an oxidized form of NO, to thiols to form S-nitrosothiols (S–NO). This modification regulates the activity of a multitude of proteins. Low molecular weight molecules such as Cys and glutathione can also be S-nitrosated. With the term "nitration," we refer to the reaction of nitrogen dioxide with tyrosine, a reaction that alters the function of proteins.

(CCO) and thus the electron transport chain is inhibited by nitrosylation (Lundberg et al., 2022). In addition to nitrosylation of sGC and CCO, NO can also react with iron–sulfur clusters contained in certain proteins, resulting in nitrosylation of the iron and degradation of the cluster, another effect on the electron transport chain that explains how NO exerts its toxic effect (Crack et al., 2011).

The noncanonical NO pathway is based on S-nitrosylation, a reaction in which the nitrosyl cation (NO^+) binds covalently to protein thiols and forms nitrosothiols, $SH + NO^+ \rightarrow -SNO$ (Heinrich et al., 2013) (Fig. 9.2). The reaction of the nitrosyl cation with cysteine or glutathione produces low molecular weight nitroso compounds that behave as NO donors and are themselves bioactive as NO. In fact, as NO, they are potent vasodilators (Ignarro et al.,

1980). S-nitrosation occurs when the NO content in the cell is relatively high: a state that is easily achieved by NOS2. S-nitrosation is a biologically important transformation because metabolic pathways can be altered by posttranslational modifications promoted by NO and involving the cysteine residues in the catalytic sites of enzymes (Bruegger et al., 2018). In addition, S-nitrosation can alter the binding of transcription factors to DNA. Considering that NOS2 can produce large amounts of NO that can damage the cell, this enzyme is regulated at multiple levels: transcriptional, posttranslational modifications, and degradation of the enzyme.

NOS2 and cancer

On one hand, NO stimulates vasodilation and angiogenesis, while on the other hand, NO can also be cytotoxic, suggesting that NO has a complex behavior in the development of cancer. In fact, both protumorigenic and antitumorigenic effects have been reported (Choudhari et al., 2013). It can be stated that the effects of NOS2 in cancer vary according to tumor type, stage of cancer development, and the surrounding microenvironment. Low NO levels increase cGMP levels by activating sGC nitrosylation. Elevated cGMP levels promote vasodilation and angiogenesis via HIF and VEGF: two important processes that favor cancer development (Papapetropoulos et al., 1997). NOS2 is often induced in response to inflammation, and chronic inflammation is associated with the development of certain cancers (Okayama et al., 2013). Long-term exposure of cells to high concentrations of NO produced by NOS2 during inflammation can also lead to carcinogenesis (Lala, 1998; Murata et al., 1997).

NO can promote mutagenesis in several ways. Firstly, it can damage DNA by deaminating nucleobases. Autoxidation of NO produces the nitrosating agent N_2O_3, which can damage DNA by deamination of the nucleobases after nitrosation of the -NH_2 group. Xanthine or uracil is formed from the base pair G:C during deamination (Caufield et al., 1998). Thirdly, NO can inactivate DNA repair proteins by S-nitration of thiols on the protein surface (Graziewicz et al., 1996; Laval & Wink, 1994) (Fig. 9.2). NO can also promote carcinogenesis by inactivating the tumor suppressor oncoprotein p53 through nitration of tyrosine residues, which causes a reduction in the DNA-binding capacity of the protein (Calmels et al., 1997; Forrester et al., 1996). As already mentioned, NO can have a cytostatic/cytotoxic effect. By inhibiting proteasomal degradation, NO can promote the accumulation of p53 (Glockzin et al., 1999). The functions of p53 include the transcriptional activation of genes that block the cell cycle (p21 and cyclin G) and the activation of apoptosis (BAX, FAS). High NO concentrations can therefore have both cytostatic (interruption of the cell cycle) and cytotoxic (apoptosis) effects (Ambs et al., 1997; Lala & Chakraborty, 2001).

Transcription factors activating NOS2 expression

The human *NOS2* gene is located on chromosome 17q11.2-q12 and, according to the ENSEMBL database, codes for two transcripts: NOS2_001 (NM_000625.4) and NOS2_201 (NM_153292.1). NOS2_001 comprises 27 exons and encodes the primary NOS2 protein with a length of 1153 amino acids (Xu et al., 1996). In contrast, NOS2_201, a splice variant of NOS2_001 spanning 28 exons, encodes a protein 1114 amino acids in length whose expression level remains unclear. The regulation of NOS2 expression is complex and context-dependent. Several extracellular stimuli activate signaling pathways that converge at the NOS2 promoter. In addition, epigenetic modifications and chromatin remodeling can influence NOS2 expression. Several studies have suggested nuclear factor kappa B (NF-kB) as the most important upstream regulator of NOS2. NF-kB consists of five members—p50, p52, RelA (p65), c-Rel, and RelB—which form homo- and heterodimers that can bind to gene promoters and enhancers to modulate gene expression (Gilmore & Gilmore, 2006; Xu et al., 1996). The pivotal p65/p50 heterodimer is of central importance for the transcriptional regulation of many genes, including *NOS2* (Hayden & Ghosh, 2008; Xie et al., 1994). Activation of NOS2 transcription often follows a general mechanism triggered by proinflammatory cytokines such as interleukin-1 (IL-1), TNF-α, IFN-γ, and LPS (Fig. 9.3). These cytokines bind to specific cell surface receptors and activate kinases that phosphorylate various intracellular proteins. Phosphorylation leads to the activation of certain transcription factors including

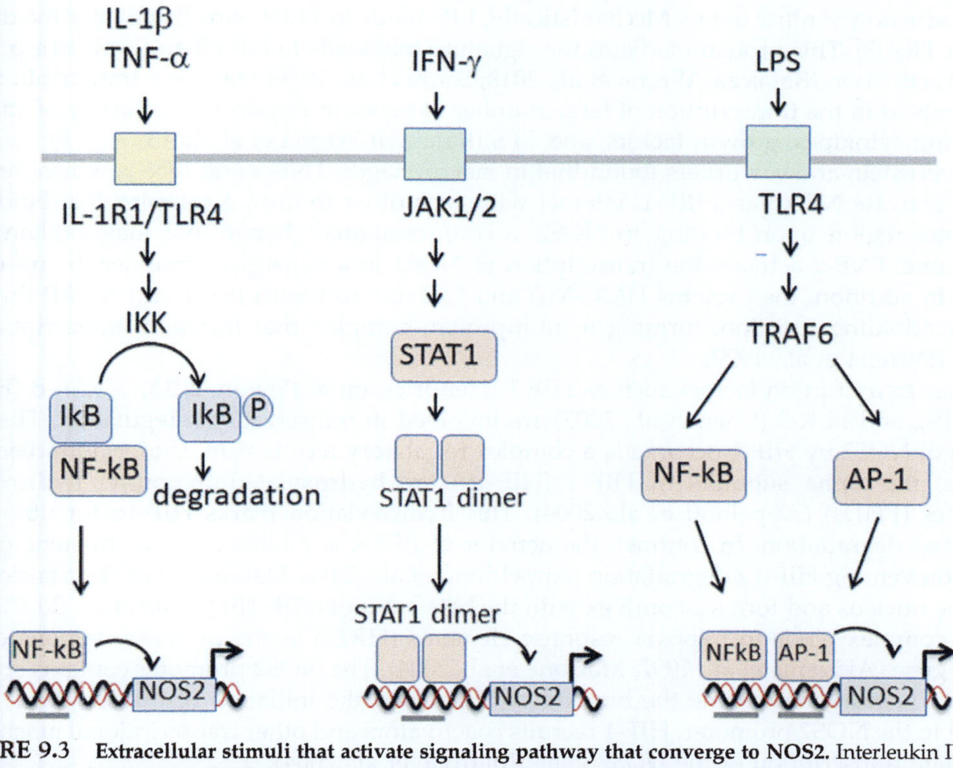

FIGURE 9.3 Extracellular stimuli that activate signaling pathway that converge to NOS2. Interleukin IL-1 and tumor necrosis factor-alpha (TNF-a) stimulate a signaling pathway that leads to the activation of NF-kB. Interferon (IFN-γ) activates the JAK-STA1 signaling pathway, which stimulates NOS_2 transcription via STAT1. Lipopolysaccharides (LPS) can also activate NOS2 via NF-kB and AP-1.

NF-κB, STAT-1, NF-IL6, JAK, AP-1, and ELK-3 (Arias-Salvatierra et al., 2011; Lowenstein & Padalko, 2004; Ratajczak-Wrona et al., 2013; Vannini et al., 2015). Subsequently, these active factors migrate into the nucleus, bind to the promoter or enhancer region of the NOS2 gene, and induce gene expression.

As shown in Fig. 9.3, IL-1 and TNF-α bind to the interleukin-1 receptor 1 (IL-1R1) and the toll-like receptor 4 (TLR4), respectively, and activate the IKK kinase. Activated IKK promotes the phosphorylation of IkB, the inhibitor of NF-kB. Phosphorylated IkB separates from NF-kB, is ubiquitinated and degraded by the proteosome. Free NF-kB (p50/p65) translocates from the cytoplasm to the nucleus, where it binds to the kB element in the 5′-flanking region of NOS2 and activates its transcription (Xie et al., 1994).

In addition to IL-1, the IFN-γ signaling pathway also leads to the activation of NOS2. Upon binding to its receptor, IFN-γ activates the JAK-STAT signaling pathway, which leads to the formation of IRF-1 and STAT1 homodimers, two transcription factors essential for NOS2 transcription (Kamijo et al., 1994; Meraz et al., 1996). In addition, LPSs, large molecules found in the outer membrane of gram-negative bacteria, are potent inducers of immune responses and can activate various immune cells, including macrophages. When macrophages recognize LPS, they can initiate a signaling cascade that leads to the activation of NOS2 and the production of nitric oxide. Mechanistically, LPS binds to TLR4, which activates the downstream TRAF6. This protein mediates the signaling that leads to NF-kB and activator protein (AP-1) activation (Ratajczal-Wrona et al., 2013; Saura et al., 1999). AP-1 is a transcription factor involved in the transcription of large number of genes in response to a variety of stimuli, including cytokines, growth factors, and LPS (Ratajczal-Wrona et al., 2013).

Lowernstein and coworkers found that in macrophages, TNF-α and IFN-γ, which, respectively, activate NF-kB and IRF-1, interact with each other to form a complex that binds the promoter region upon binding to NOS2, a conformational change that may explain why IFN-γ and TNF-α activate the transcription of NOS2 in a synergistic manner (Saura et al., 1999). In addition, the proteins HMG-Y(I) and CBP interact with IRF-1 and NF-kB through their scaffolding function, forming a multiprotein complex that increases transcription of NOS2 (Perrella et al., 1999).

Other transcription factors such as HIF-1 (Hendrickson & Poyton, 2015), CREB-B (Hou e al., 2019), and ELK-3 (Chen et al., 2003) are involved in transcriptional regulation. The activation of NOS2 by HIF-1 occurs via a complex regulatory mechanism. Under normoxic conditions, the alpha subunits of HIF-1 (HIF-1α) are hydroxylated by prolyl hydroxylase enzymes (PHDs) (Appelhoff et al., 2004). This hydroxylation marks HIF-1α for ubiquitin-mediated degradation. In contrast, the activity of PHDs is inhibited under hypoxic conditions, preventing HIF-1α degradation (Appeldhoff et al., 2004). Stabilized HIF-1α translocates into the nucleus and forms a complex with the beta subunit (HIF-1β) (Ziello et al., 2007). This HIF-1 complex binds to hypoxia response elements (HREs) in the promoter regions of the *NOS2* gene (Akiyama et al., 1997; Matrone et al., 2004). The NOS2 promoter contains specific DNA sequences that enable the binding of HIF-1 and the initiation of transcription. Once bound to the NOS2 promoter, HIF-1 recruits coactivators and other transcriptional machinery to initiate transcription of the *NOS2* gene (Matrone et al., 2004).

NRF2 controls the expression of NOS2

In a recent study, we found that the transcription factor NRF2, a key regulator of redox homeostasis in cells, has the ability to rewire metabolism in pancreatic cancer cells under stress conditions (Di Giorgio et al., 2023). Together with other researchers, we have also found that NRF2 expression is stimulated by oncogenic KRAS (Ferino, Rapozzi, & Xodo, 2020; Ferino, Nicoletto, et al., 2020; Tao et al., 2014; Yang et al., 2021). A plot of KRAS and NRF2 mRNA levels from 175 Cancer Genome Atlas (TCGA) samples shows that an increase in *KRAS* expression leads to an increase in NRF2 expression (Fig. 9.4A). In addition, the survival probability of PDAC patients with high KRAS and NRF2 levels is lower than that of patients with low KRAS and NRF2 levels (Fig. 9.4B).

Putting these findings together, it can be concluded that KRAS controls metabolism, or at least part of it, via NRF2. A particularly interesting activity among the metabolic pathways regulated by NRF2 is that related to the catabolism of arginine. Since arginine serves as a substrate for NOS enzymes, we investigated whether NRF2, and consequently KRAS, exerts control over NOS2 and NO production in PDACs. To address this question, we used CRISPR/Cas9 to suppress the *NRF2* gene in PDAC cells: Panc-1 and MIA PaCa-2, and obtained a clone of each cell line designated $NRF2^{-/-}$ (Di Giorgio et al., 2023). To investigate the effects of NRF2 silencing on the transcriptome, we performed RNA-seq analysis on Panc-1 $NRF2^{-/-}$ cells (GEO: GSE217965). We found that NRF2 silencing induced the downregulation of 1888 genes and the upregulation of 666 genes, according to a threshold $|\log 2\ FC \geq 1|$, $P < .05$ (Di Giorgio et al., 2023). Using bioinformatic analysis, we found that the suppression of NRF2 promotes a metabolic rewiring that causes cells to lose their anaerobic properties in favor of aerobic metabolism, with arginine playing a prominent role (Di Giorgio et al., 2023). Gene ontology (GO) analysis of differentially expressed genes (DEGs) in $NRF2^{-/-}$ cells compared to wild-type (WT) Panc-1 cells revealed that four GO-enriched terms affect the *NOS2* gene (Di Giorgio et al., 2024). One of these is the HIF-1 signaling pathway, which is considered to be the main driver of adaptation of cellular metabolism to oxygen consumption and oxidative stress. Of the DEGs associated with the HIF-1 signaling pathway, five are upregulated and 15 are downregulated. *NOS2* is one of the upregulated DEGs in $NRF2^{-/-}$ Panc-1 cells, as it has a $\log_2 FC = 6.84$ (FC = fold change) and a P-value $= 5.53 \times 10^{-6.}$ Interestingly, *NOS1* was instead strongly downregulated with a $\log_2 FC = -5.77$ and a P-value $= 9.21 \times 10^{-3}$, while *NOS3* was not affected by NRF2 silencing as it showed similar levels in $NRF2^{-/-}$ and WT cells ($\log_2 FC = 0.348$, $P = 0\ .225$) (Di Giorgio et al., 2024). Analysis of DEGs using GENEMANIA, a bioinformatic tool for predicting the function of a specific group of genes, showed that the upregulated DEGs of the KEGG-HIF pathway – NOS2, PRKCB, IGF1R, CYBB, AKT3 – are predicted to interact with the genes encoding the subunits of the enzymatic complex that converts GTP to GMPc, the second messenger of the classical NO-mediated signaling pathway (Di Giorgio et al., 2024). The downregulated DEGs of the HIF signaling pathway encode enzymes involved in anaerobic metabolism: glycolysis, pentose phosphate pathway (PPP), glutathione cycle, consistent with the fact that suppression of NRF2 renders Panc-1 cells dependent on aerobic metabolism (Di Giorgio et al.,

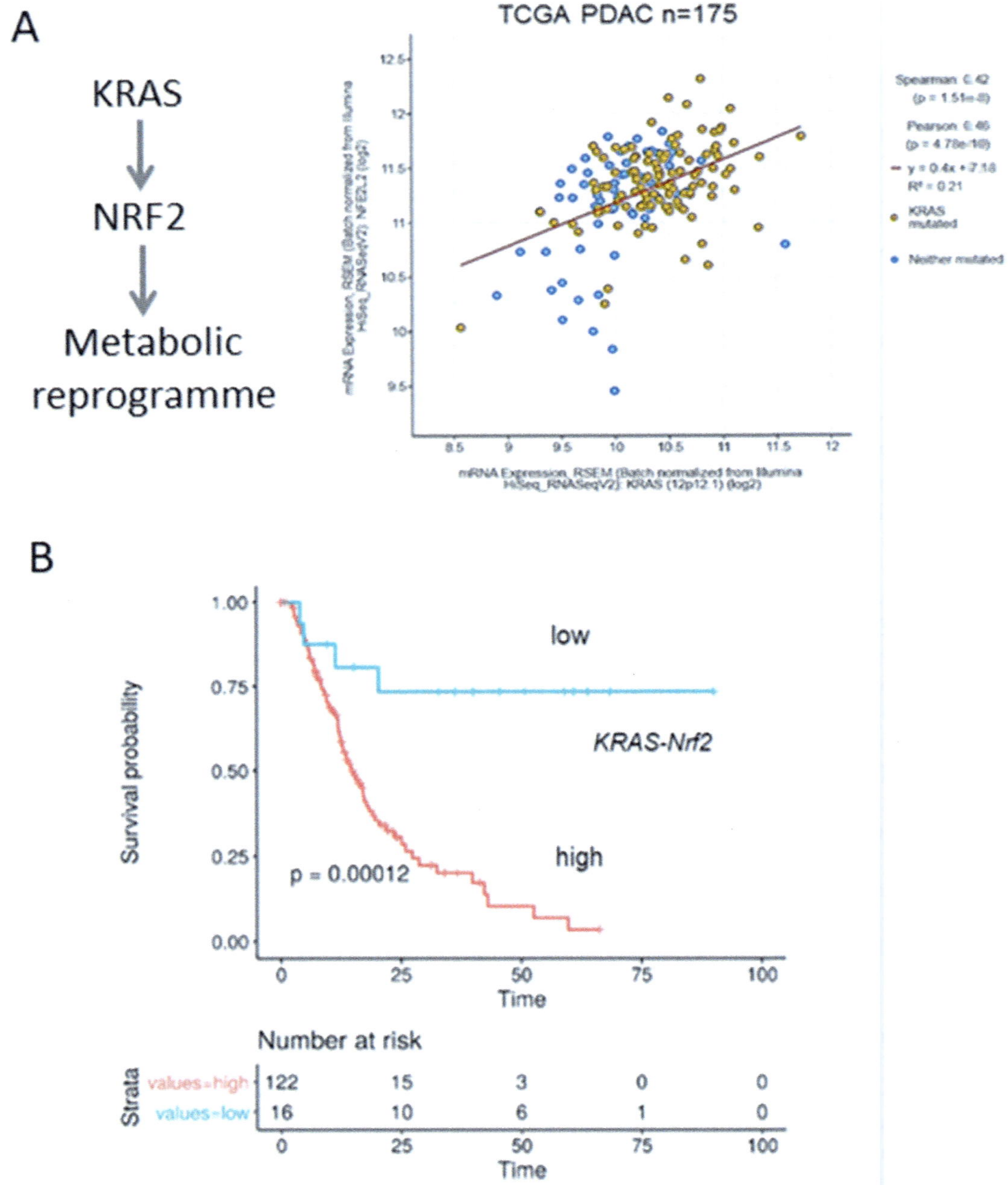

FIGURE 9.4 The KRAS-NRF2 axis. (A) The KRAS-NRF2 scheme and the correlation diagram between NRF2 and *KRAS* mRNA levels in 175 PDAC samples from The Cancer Genome Atlas (TCGA). The graph shows that increasing KRAS expression leads to an increase in NRF2 mRNA; (B) survival probability of PDAC patients with high (above third quartile) and low KRAS and NRF2 levels.

2024). To provide further evidence for a link between NRF2 and NOS2/NO, we showed by GSEA that the genes downregulated in breast cancer cells treated with DETA-NONOate, an NO donor (Pervin et al., 2001), were also downregulated in NRF2$^{-/-}$ Panc-1 cells compared to WT-Panc-1 cells. Moreover, the genes that are upregulated in NOS2$^{-/-}$ mice were also upregulated in NRF2$^{+/+}$ Panc-1 cells. Overall, the GSEA analysis shows that the genes involved in the response to an increase in nitric oxide, as detected in breast cancer and NOS2-deficient mice lungs, are also activated in NRF2$^{-/-}$ Panc-1 cells.

Another important observation is that glycolytic Panc-1 cells, which mainly use glucose for ATP and biomass production, have low levels of NOS2/NO, suggesting that their metabolism is not dependent on arginine. In contrast, NRF2$^{-/-}$ Panc-1 cells exhibited approximately fourfold higher NOS2 levels, as these cells utilize an aerobic metabolism and have an altered metabolism of arginine (Di Giorgio et al., 2024). Measurement of NO with DAF-FM DA, a nonfluorescent molecule that reacts with NO to form a fluorescent benzotriazole, shows that NRF2$^{-/-}$ Panc-1 cells contain about fourfold more NO than WT cells at low glucose levels and 2.5-fold more NO at high glucose levels (Di Giorgio et al., 2024). When NRF2 is rescued in NRF2$^{-/-}$ Panc-1 cells, the amount of NO decreases to the levels observed in WT cells. The rescue experiments show that NOS2/NO in pancreatic cancer cells are indeed under the control of NRF2 or, in a more integrated manner, the KRAS-NRF2 axis.

NRF2 binds to a distal enhancer of NOS2

In view of these results, we asked how NRF2 controls *NOS2*. Does NRF2 bind directly to the regulatory elements of the *NOS2* gene or does NRF2 exert an indirect effect on *NOS2* expression by altering the metabolism of arginine? An interesting answer to this query was obtained by a reanalysis of NRF2 ChIP-seq data from IMR90 (ENCSR197WGI) and HepG2 (ENCSR488EES) cells available at ENCODE (Encyclopedia of DNA Elements). The results showed a distinct ChIP-seq peak indicating the interaction of NRF2 with a distal enhancer of the *NOS2* gene, 22 kb downstream of TSS (Fig. 9.5A). It should be remembered that enhancers are regulatory DNA sequences that can modulate the expression of target genes over long distances (Panigrahi & O'Malley, 2021). The NOS2 enhancer region is 269 nucleotides long and harbors two putative antioxidant response elements (AREs) recognized by NRF2 (Suzuki et al., 2013) (Fig. 9.5B). Although the ChIP-seq analysis was not performed in pancreatic cancer cells, it showed that NRF2 regulates NOS2 expression by directly binding to the NOS2 locus.

Direct evidence that NRF2 binds to *NOS2* in Panc-1 cells was obtained through ChIP immunoprecipitation experiments (Di Giorgio et al., 2024). These experiments showed that NRF2 binds to the NOS2 genomic region where histone H3 is epigenetically modified either by acetylation or methylation (site 7, Fig. 9.5C) (Di Giorgio et al., 2024). Acetylation of histone H3 at lysine 27, H3K27ac, is a well-known histone modification associated with active gene transcription (Kurdistani et al., 2004; Struhl, 1998). When this modification occurs at enhancer regions, it is often linked to an increased transcriptional activity of nearby genes. The presence of H3K27ac at enhancer regions indicates an active state and is associated with the recruitment of the transcriptional machinery to promote gene expression (Kang et al., 2021;

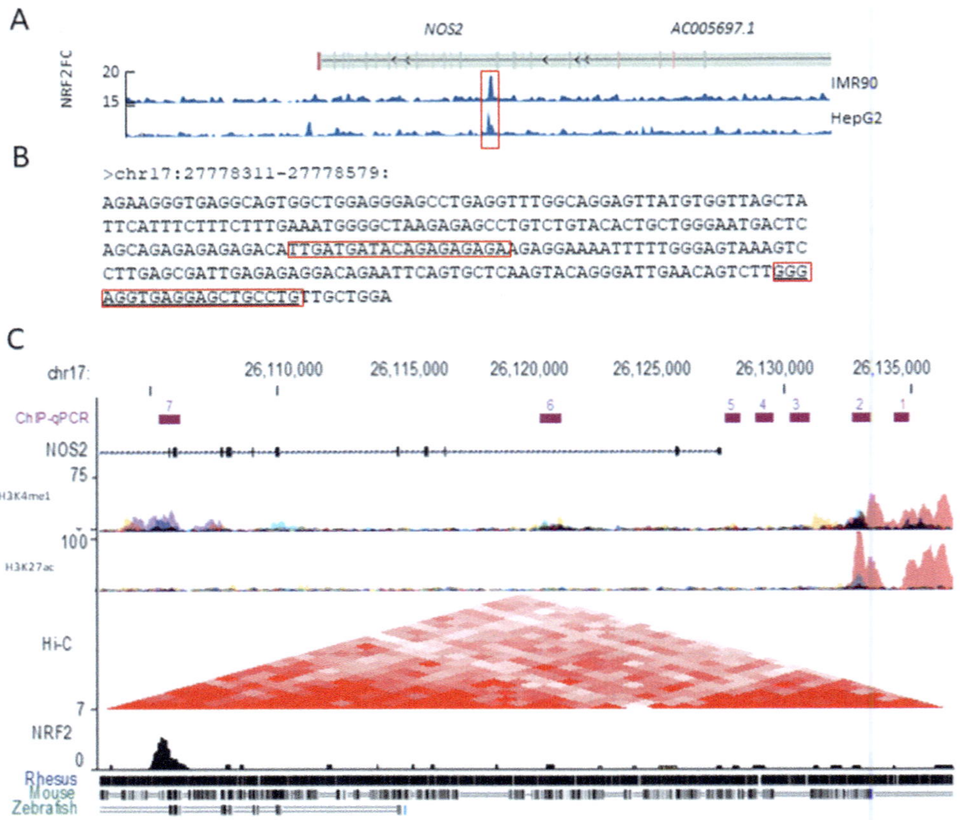

FIGURE 9.5 NRF2 binds to a distal NOS2 enhancer in a poised state in Panc-1 cells. (A) NRF2 ChIP-seq signals expressed as fold-change (FC) with respect to input at the NOS2 locus. Data retrieved from Encode; the enriched peaks are highlighted; (B) sequence of the NOS2 distal enhancer bound by NRF2. Two antioxidant response elements (ARE) domains recognized by NRF2 are shown in the box; (C) genomic organization and Hi-C data of *NOS2* locus. The amplified regions investigated in qPCR are indicated (1–7). The levels of H3K4me1 and H3K27ac obtained from Encode (cell lines: GM12878, H1-hESC, HSMM, HUVEC, K562, NHEK, NHLF), Hi-C from IMR90 cells (http://3dgenome.fsm.northwestern.edu/view.php) are shown. *Adapted from figure 8 of Ref. Di Giorgio, E., Cortolezzis, Y., Gualandi, N., Agostini, F., Rapozzi, V., & Xodo L. E. (2014). NRF2 interacts with distal enhancer and inhibits nitric oxide synthase 2 expression in KRAS-driven pancreatic cancer cells.* Biochimica et Biophysica Acta (BBA) - Molecular Cell Research, 1871, *119606–119619.*

Yao et al., 2020). Monomethylation of histone H3 at lysine 4 (H3K4me1) is also associated with enhancer regions (Yao et al., 2020). H3K4me1 marks poised or active enhancers and is often found at regulatory regions that modulate gene expression. Together, these histone modifications are part of the histone code, a complex language that cells use to regulate gene expression (Zentner et al., 2011). The presence of H3K27ac and H3K4me1 at specific genomic loci is indicative of regions that are actively involved in the regulation of gene expression, particularly through the modulation of enhancer activity (Zentner et al., 2011).

The DNA region in NOS2 bound by NRF2 was found H3K27ac negative (H3K27ac$^-$) and H3K4me1 positive (H3K4me1$^+$) in Panc-1 cells (Fig. 9.5C). H3K27ac$^-$ means the absence or low levels of acetylation at the lysine 27 of histone H3. Since histone acetylation is generally associated with open chromatin and gene transcription, H3K27ac$^-$ suggests a state between activation and repression that is little or no transcribed. Instead, H3K4me1$^+$ indicates the presence of monomethylation at the lysine 4 of H3. H3K4me1$^+$ alone does not necessarily indicate active transcription, but is rather a marker for potential enhancer activity. In conclusion, the combination of histone modifications, H3K27ac$^-$ and H3K4me1$^+$, indicates a poised, nonfully active state of the enhancer, ready to be fully activated under specific cellular conditions or stimuli (for instance hypoxia, vide infra) (Di Giorgio et al., 2024).

As PDAC spheroids can better reproduce the structure and biological features of tumors than 2D cellular models, we asked if NRF2 controls the activity of the distal NOS2 enhancer and NOS2 transcription also in 3D tumor spheroids. H3K27ac ChIP-seq of WT and NRF2$^{-/-}$ Panc-1 spheroids showed that significant acetylation was found in more than 13500 genes. In particular, 569 genes were found associated to acetylation in NRF2$^{-/-}$ spheroids, while 734 in WT NRF2$^{+/+}$ spheroids. The genes acetylated exclusively in NRF2$^{-/-}$ cells belong to arginine metabolism, epithelial-mesenchymal transition (EMT), and nitric oxide related pathways. To confirm the control of the NOS2 enhancer by NRF2, we plotted the H3K27ac signal around the *NOS2* locus and found significant increased acetylation in NRF2$^{-/-}$ spheroids compared to WT, at the distal enhancer (Fig. 9.5C). This increase in acetylation spreads up to the regulative element of NOS2 located 5 kb upstream the TSS (Di Giorgio et al., 2024). The data confirmed also in 3D Panc-1 spheroids that a NOS2 enhancer is controlled by NRF2.

However, in addition to NRF2, another factor plays a role in the control of NOS2 transcription: hypoxia (Di Giorgio et al., 2024). Our study has shown that inhibition of NRF2 leads to the release of the distal enhancer, allowing activating proteins to bind to the enhancer and activate NOS2 transcription. The chromatin structure at the NOS2 locus is only partially open, as only monomethylating of histone H3 at lysine 4 (H3K4me1$^+$) is present. Under hypoxic conditions, however, histone H3 is also acetylated at lysine 27 (H3K27ac$^+$), so that the chromatin is further opened. In addition, hypoxia increases the level of HIF-1, which binds to the HRE domains of NOS2, and thereby further stimulates transcription. As shown in Fig. 9.6A, we propose a three-step mechanism for the full expression of NOS2: (i) liberation of the enhancer from NRF2; (ii) acetylation of H3K27 leading to chromatin decompaction at the NOS2 locus; (iii) binding of hypoxia-regulated transcription factors to the NOS2 promoter (Matrone et al., 2004). Fig. 9.6B shows a cartoon illustrating the interaction between the enhancer and the NRF2/activator proteins, which activates NOS2 transcription.

Epigenetic mechanisms for the control of *NOS2* have been previously reported. For instance, de Andrés et al. showed that the abnormal expression of *NOS2* in osteoarthritic human chondrocytes correlates with changes in DNA methylation in the promoter and enhancers of *NOS2* (de Andrés et al., 2013). The authors reported that an NF-kB enhancer region at −5.8 kb from the transcription starting site was demethylated in osteoarthritic compared to control chondrocytes. This finding showed that the demethylation of the NF-kB enhancer is necessary for the transactivation of *NOS2*. A similar epigenetic mechanism was observed in human macrophages for the control of *NOS2* (Ross et al., 2014). While mouse macrophages express significant levels of *NOS2*, human macrophages express little NOS2 and NO. This different behavior is due to the fact that the human NOS2 gene is highly

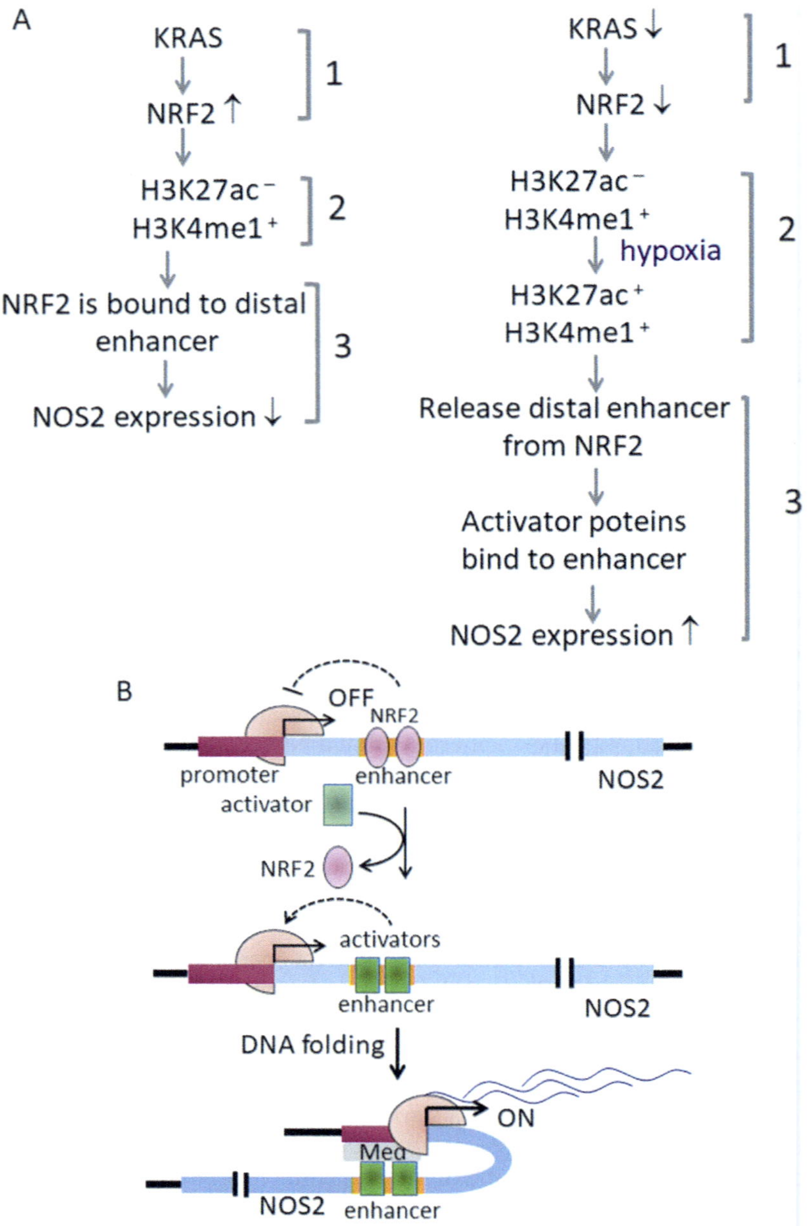

FIGURE 9.6 A three-step mechanism of NOS2 gene expression. (A, left) The diagram shows the normal state in which KRAS and NRF2 are expressed in PDAC cells (step 1). The chromatin in the enhancer region is characterized by H3K27ac$^-$ and H3K4me1$^+$, a poised chromatin conformation that does not favor NOS2 transcription (step 2). NRF2 binds to the distal enhancer and blocks NOS2 transcription (step 3); (A, right) the diagram on the right shows what happens when KRAS is suppressed by anti-KRAS molecules and hypoxia conditions. The suppression of KRAS also leads to the suppression of NRF2 (step 1). Hypoxia stimulates acetylation of H3K27, leading to chromatin decompaction characterized by H3K27ac+ and H3K4me1+ at the NOS2 locus (step 2). Open chromatin favors the release of NRF2 from the distal enhancer and the binding of activator proteins. Transcription of NOS2 is then activated; (B) proposed mechanism of repression of the NOS2 gene by NRF2 in Panc-1 cells. NRF2 binds to a distal enhancer and inactivates it. When the NRF2 level decreases, the enhancer is released and allows the binding of activator proteins that promote transcription. "Med" stands for mediator, a protein complex involved in gene expression in eukaryotic cells. *Image adapted from Ref. Di Giorgio, E., Cortolezzis, Y., Gualandi, N., Agostini, F., Rapozzi, V., & Xodo, L. E. (2024). NRF2 interacts with distal enhancer and inhibits nitric oxide synthase 2 expression in KRAS-driven pancreatic cancer cells.* Biochimica et Biophysica Acta (BBA) - Molecular Cell Research, 1871, *119606–119619.*

methylated in the region of the transcription start site, while the mouse *NOS2* gene is rather unmethylated at the CpGs in the 5′-flanking region. The analysis of chromatin accessibility and histone modification indicates that the chromatin has a close conformation at the *NOS2* locus in the human macrophages, while it is in the open conformation in the mouse macrophages. This study provides a clear contribution on how epigenetic modifications control gene expression (Ross et al., 2014).

The demonstration that cytokines strongly induce the expression of NOS2 in rodent endothelial cells but not in human endothelial cells led Dreger et al. to find that NOS2 in HUVAC cells (human umbilical vein endothelial cells) is a potential target for the histone methyltransferase EZH2, which mediates the trimethylation of histone H3 at lysine 27 (H3K27me3) (Dreger et al., 2016). The study indicates that EZH2 regulates gene expression through the epigenetic machinery. Finally, another mechanism for chromatin accessibility of the NOS2 promoter, somewhat surprising, was reported to involve the NLRC4/caspase 1axis in human macrophages (Buzzo et al., 2017). A substrate for caspase 1 is PARP-1, a protein normally associated to apoptosis but also able to induce chromatin changes necessary for gene expression (Hottiger, 2015). Normally PARP-1 maintains chromatin in the condensed state decreasing gene expression. The cleavage of PARP-1 by caspase 1, under stress conditions, determines its exclusion from chromatin promoting a local decondensation, allowing the binding to DNA of transcription factors (Erener et al., 2012).

PDAC-spheroids formation depends on NOS2: Therapeutic implications

The investigation into the biological relevance of NOS2 in the proliferation of Panc-1 cells was undertaken in our laboratory (Di Giorgio et al., 2024). Our experiments encompassed both NRF2$^{-/-}$ cells, characterized by elevated NOS2 levels, and WT Panc-1 cells expressing lower NOS2 levels. Initially, we confirmed that esiNOS2, a combination of siRNA specific to NOS2, effectively suppressed NOS2 in both NRF2$^{-/-}$ and WT Panc-1 cells. Subsequently, we assessed the 3D spheroid formation capacity of these cells following esiNOS2 treatment (Di Giorgio et al., 2024). Interestingly, the suppression of NOS2 had a minimal impact on WT cells, while NOS2$^{-/-}$ cells exhibited a substantial response to esiNOS2. In NRF2$^{-/-}$ cells, esiNOS2 significantly decreased both the 3D spheroid area and the efficiency of spheroid formation. These findings align with the crucial role of NOS2 in the metabolism of NRF2$^{-/-}$ cells, bearing significant therapeutic implications.

As mentioned earlier, NO serves as a vital regulator of cellular metabolism, activating the canonical signaling pathway via cGMP and the noncanonical signaling pathway via posttranslational nitrosylation, S-nitrosation, and tyrosine nitration. In this context, Zhou et al. discovered that NO S-nitrosates coenzyme A (CoA) to nitroso-CoA (SNO-CoA), a source of −SNO in protein nitrosation (Zhou et al., 2019). They reported that nitrosation of pyruvate kinase PKM2 leads to the inhibition of glycolysis. Consistent with the literature (Fu et al., 2019), the suppression of NRF2 in Panc-1 cells, causing an increase in NO, results in the inhibition of glycolysis (Di Giorgio et al., 2024).

Concluding remarks

In the context of our decade-long research, we consistently demonstrated the essential role of KRAS in the proliferation of PDAC cells. Inhibiting this oncogene leads to a profound inhibition of cell growth, activating apoptosis and ferroptosis (Di Giorgio et al., 2022; Ferino, Rapozzi, & Xodo, 2020; Ferino, Nicoletto, et al., 2020; MIglietta et al., 2017). Notably, KRAS stimulates the expression of NRF2, most likely through the MAPK and AKT pathways (Di Giorgio et al., 2023). Consequently, inhibiting KRAS results in the concurrent inhibition of NRF2. This implies a metabolic shift in PDAC cells treated with anti-KRAS molecules, transitioning from anaerobic (glycolysis) to aerobic metabolism, where arginine becomes critical for fueling phosphocreatine, polyamines, and the NOS2 pathways (Di Giorgio et al., 2023, 2024). Therefore, anti-KRAS-treated PDAC cells overcome metabolic stress by redirecting their metabolism toward an aerobic state, underscoring the importance of arginine in this process.

Combination therapies using anti-KRAS molecules and NOS2 inhibitors or inhibitors of arginine metabolism such as homoarginine are likely to be more effective than monotherapies. In our laboratory, we are investigating combination therapies in both 3D spheroids and mouse models. Further work is needed to determine the extent to which NOS2 inhibition can affect the metabolism and viability of PDAC cells and synergize with conventional therapies (gemcitabine, FOLFIRINOX) or KRAS inhibitors.

References

Abu-Soud, H. M., & Stuehr, D. J. (1993). Nitric oxide synthases reveal a role for calmodulin in controlling electron transfer. *Proceedings of the National Academy of Sciences of the United States of America, 90*(22), 10769–10772. https://doi.org/10.1073/pnas.90.22.10769

Akiyama, K., Suzuki, H., Grant, P., & Bing, R. J. (1997). Oxidation products of nitric oxide, NO_2 and NO_3, in plasma after experimental myocardial infarction. *Journal of Molecular and Cellular Cardiology, 29*, 1–9.

Alderton, W. K., Cooper, C. E., & Knowles, R. G. (2001). Nitric oxide synthases: Structure, function and inhibition. *Biochemical Journal, 357*(3), 593–615. https://doi.org/10.1042/0264-6021:3570593

Alderton, W. K., Angell, A. D. R., Craig, C., Dawson, J., Garvey, E., Moncada, S., Monkhouse, J., Rees, D., Russell, L. J., Russell, R. J., Schwartz, S., Waslidge, N., & Knowles, R. G. (2005). GW274150 and GW273629 are potent and highly selective inhibitors of inducible nitric oxide synthase in vitro and in vivo. *British Journal of Pharmacology, 145*(3), 301–312. https://doi.org/10.1038/sj.bjp.0706168

Ambs, S., Hussain, S. P., & Harris, C. C. (1997). Interactive effects of nitric oxideand the p53 tumor suppressor gene in carcinogenesis and tumor progression. *The FASEB Journal, 11*, 443–448.

Andrew, P. J., & Mayer, B. (1999). Enzymatic function of nitric oxide synthases. *Cardiovascular Research, 43*(3), 521–531. https://doi.org/10.1016/S0008-6363(99)00115-7

Appelhoff, R. J., Tian, Y.-M., Raval, R. R., Turley, H., Harris, A. L., Pugh, C. W., Ratcliffe, P. J., & Gleadle, J. M. (2004). Differential function of the prolyl hydroxylases PHD1, PHD2, and PHD3 in the regulation of hypoxia-inducible factor. *Journal of Biological Chemistry, 279*, 38458–38465.

Arias-Salvatierra, D., Silbergeld, E. K., Acosta-Saavedra, L. C., & Calderon-Aranda, E. S. (2011). Role of nitric oxide produced by iNOS through NF-κB pathway in migration of cerebellar granule neurons induced by Lipopolysaccharide. *Cellular Signalling, 23*, 425–435.

Arnold, W. P., Mittal, C. K., Katsuki, S., & Murad, F. (1997). Nitric oxide activates guanylate cyclase and increases guanosine 3′:5′-cyclic monophosphate levels in various tissue preparations. *Proceedings of the National Academy of Sciences, 74*, 3203–3320.

Bredt, D. S. (2009). Endogenous nitric oxide synthesis: Biological functions and pathophysiology. *Free Radical Research, 31*(6), 577–596. https://doi.org/10.1080/10715769900301161

Bruegger, J. J., Smith, B. C., Wynia-Smith, S. L., & Marletta, M. A. (2018). Comparative and integrative metabolomics reveal that S-nitrosation inhibits physiologically relevant metabolic enzymes. *Journal of Biological Chemistry, 293*, 6282–6296.

Buzzo, C. L., Medina, T., Branco, L. M., Lage, S. L., Ferreira, L. C., Amarante-Mendes, G. P., Hottiger, M. O., De Carvalho, D. D., & Bortoluci, K. R. (2017). Epigenetic regulation of nitric oxide synthase 2, inducible (*Nos2)* by NLRC4 inflammasomes involves PARP1 cleavage. *Scientific Reports, 7*, 41686–41698.

Calmels, S., Hainaut, P., & Ohshima, H. (1997). Nitric oxide induces conformational and functional modifications of wild-type p53 tumour suppressor protein. *Cancer Research, 57*, 3365–3369.

Caulfield, J. L., Wishnok, J. S., & Tannenbaum, S. R. (1998). Nitric oxide-induced deamination of cytosine and guanine in deoxynucleosides and oligonucleotides. *Journal of Biological Chemistry, 273*, 12689–12695.

Chakravortty, D., & Hensel, M. (2003). Inducible nitric oxide synthase and control of intracellular bacterial pathogens. *Microbes and Infection, 5*(7), 621–627. https://doi.org/10.1016/s1286-4579(03)00096-0

Chen, Y.-H., Layne, M. D., Chung, S. W., Ejima, K., Baron, R. M., Yet, S.-F., & Perrella, M. A. (2003). Elk-3 is a transcriptional repressor of nitric-oxide synthase 2. *Journal of Biological Chemistry, 278*, 39572–39577.

Choudhari, K., Chaudhary, S., Minal, B., Sachin, G., Amol, R., & Joshi, V. (2013). Nitric oxide and cancer: A review. *World Journal of Surgical Oncology, 11*(1), 118–129. https://doi.org/10.1186/1477-7819-11-118

Crack, J. C., Smith, L. J., Stapleton, M. R., Peck, J., Watmough, N. J., Buttner, M. J., Buxton, R. S., Green, J., Oganesyan, V. S., & Thomson, A. J. (2011). Le Brun NE mechanistic insight into the nitrosylation of the [4Fe–4S] cluster of WhiB-like proteins. *Journal of the American Chemical Society*, 133, 1112–1121.

de Andrés, M. C., Imagawa, K., Hashimoto, K., Gonzalez, A., Roach, H. I., Goldring, M. B., & Oreffo, R. O. (2013). Loss of methylation in CpG sites in the NF-κB enhancer elements of inducible nitric oxide synthase is responsible for gene induction in human articular chondrocytes. *Arthritis and Rheumatism, 65*, 732–742.

Di Giorgio, E., Ferino, A., Choudhary, H., Löffler, P. M. G., D'Este, F., Rapozzi, V., Tikhomirov, A., Shchekotikhin, A., Vogel, S., & Xodo, L. E. (2022). Photosensitization of pancreatic cancer cells by cationic alkyl-porphyrins in free form or engrafted into POPC liposomes: The relationship between delivery mode and mechanism of cell death. *Journal of Photochemistry and Photobiology B, 231*, 112449–112464.

Di Giorgio, E., Cortolezzis, Y., Gualandi, N., Agostini, F., Rapozzi, V., & Xodo, L. E. (2024). NRF2 interacts with distal enhancer and inhibits nitric oxide synthase 2 expression in KRAS-driven pancreatic cancer cells. *Biochimica et Biophysica Acta (BBA) - Molecular Cell Research, 1871*, 119606–119619.

Di Giorgio, E., Choudhary, H., Ferino, A., Cortolezzis, Y., Dalla, E., D'Este, F., Comelli, M., Rapozzi, V., & Xodo, L. E. (2023). Suppression of the KRAS-NRF2 axis shifts arginine into the phosphocreatine energy system in pancreatic cancer cells. *iScience*. https://doi.org/10.1016/j.isci.2023.108566

Dreger, H., Ludwig, A., Weller, A., Baumann, G., Stangl, V., & Stangl, K. (2016). Epigenetic suppression of iNOS expression in human endothelial cells: A potential role of Ezh2-mediated H3K27me3. *Genomics, 107*, 145–149.

Erener, S., Pétrilli, V., Kassner, I., Minotti, R., Castillo, R., Santoro, R., Hassa, P. O., Tschopp, J., & Hottiger, M. O. (2012). Inflammasome-activated caspase 7 cleaves PARP1 to enhance the expression of a subset of NF-κB target genes. *Molecular Cell, 46*, 200–211.

Ferino, A., Nicoletto, G., D'Este, F., Zorzet, S., Lago, S., Richter, S. N., Tikhomirov, A., Shchekotikhin, A., & Xodo, L. E. (2020). Photodynamic therapy for *ras*-driven cancers: Targeting G-quadruplex RNA structures with bifunctional alkyl-modified porphyrins. *Journal of Medicinal Chemistry, 63*, 1245–1260.

Ferino, A., Rapozzi, V., & Xodo, L. E. (2020). The ROS-KRAS-Nrf2 axis in the control of the redox homeostasis and the intersection with survival-apoptosis pathways: Implications for photodynamic therapy. *Journal of Photochemistry and Photobiology B, 202*, 111672–111681.

Forrester, K., Ambs, S., Lupold, S. E., Kapust, R. B., Spillare, E. A., Weinberg, W. C., Felley-Bosco, E., Wang, X. W., Geller, D. A., Tzeng, E., Billiar, T. R., & Harris, C. C. (1996). Nitric oxide induced p53 accumulation and regulation of inducible nitric oxide synthase expression by wild type p53. *Proceedings of the National Academy of Sciences of the United States of America, 93*, 2442–2447.

Förstermann, U., & Sessa, W. C. (2012). Nitric oxide synthases: Regulation and function. *European Heart Journal, 33*(7), 829–837. https://doi.org/10.1093/eurheartj/ehr304

Fu, J., Xiong, Z., Huang, C., Li, J., Yang, W., Han, Y., Paiboonrungruan, C., Major, M. B., Chen, K.-N., Kang, X., & Chen, X. (2019). Hyperactivity of the transcription factor Nrf2 causes metabolic reprogramming in mouse esophagus. *Journal of Biological Chemistry, 294*, 327–340.

Gilmore, T. D., & Gilmore, T. D. (2006). Introduction to NFkB: Players, pathways, perspectives. *Oncogene, 25*, 6680–6684.

Giroud, C., Moreau, M., Mattioli, T. A., Balland, V., Boucher, J. L., Xu-Li, Y., Stuehr, D. J., & Santolini, J. (2010). Role of arginine guanidinium moiety in nitric-oxide synthase mechanism of oxygen activation. *Journal of Biological Chemistry, 285*(10), 7233–7245. https://doi.org/10.1074/jbc.M109.038240

Glockzin, S., von Knethen, A., Scheffner, M., & Brüne, B. (1999). Activation of cell death program by nitric oxide involves inhibition of the proteosome. *Journal of Biological Chemistry, 274*, 19581–19586.

Graziewicz, M., Wink, D. A., & Laval, F. (1996). Nitric oxide inhibits DNA ligase activity: Potential mechanisms for NO-mediated DNA damage. *Carcinogenesis, 17*, 2501–2505.

Hayden, M. S., & Ghosh, S. (2008). Shared principles in NF-kappaB signalling. *Cell, 132*, 344–362.

Heinecke, J. L., Ridnour, L. A., Cheng, R. Y. S., Switzer, C. H., Lizardo, M. M., Khanna, C., Glynn, S. A., Hussain, S. P., Young, H. A., Ambs, S., & Wink, D. A. (2014). Tumor microenvironment-based feed-forward regulation of NOS2 in breast cancer progression. *Proceedings of the National Academy of Sciences of the United States of America, 111*(17), 6323–6328. https://doi.org/10.1073/pnas.1401799111

Heinrich, T. A., da, S., Roberto, S., Miranda, K. M., Switzer, C. H., Wink, D. A., & Fukuto, J. M. (2013). Biological nitric oxide signalling: Chemistry and terminology. *British Journal of Pharmacology, 169*(7), 1417–1429. https://doi.org/10.1111/bph.12217

Hemmens, B., Goessler, W., Schmidt, K., & Mayer, B. (2000). Role of bound zinc in dimer stabilization but not enzyme activity of neuronal nitric-oxide synthase. *Journal of Biological Chemistry, 275*(46), 35786–35791. https://doi.org/10.1074/jbc.m005976200

Hendrickson, M., & Poyton, R. (2015). Crosstalk between nitric oxide and hypoxia-inducible factor signaling pathways: An update. *Research and Reports in Biochemistry, 5*, 147–161.

Hottiger, M. O. (2015). Poly(ADP-ribose) polymerase inhibitor therapeutic effect: Are we just scratching the surface? *Expert Opinion on Therapeutic Targets, 19*, 1149–1152.

Hou, L., Li, B., Ding, D., Kang, L., & Wang, X. (2019). CREB-B acts as a key mediator of NPF/NO pathway involved in phase-related locomotor plasticity in locusts. *PLoS Genetics, 15*.

Ignarro, L. J., Edwards, J. C., Gruetter, D. Y., Barry, B. K., & Gruetter, C. A. (1980). Possible involvement of S-nitrosothiols in the activation of guanylate cyclase by nitrosocompounds. *FEBS Letters, 110*, 275–278.

Kamijo, R., Harada, H., Matsuyama, T., Bosland, M., Gerecitano, J., Shapiro, D., Le, J., Koh, S. I., Kimura, T., & Green, S. J. (1994). Requirement for transcription factor IRF-1 in NO synthase induction in macrophages. *Science, 263*, 1612–1615.

Kang, Y., Liu, R., Wu, J. X., et al. (2019). Structural insights into the mechanism of human soluble guanylate cyclase. *Nature, 574*, 206–210.

Kang, Y., Kim, Y. W., Kang, J., & Kim, A. (2021). Histone H3K4me1 and H3K27ac play roles in nucleosome eviction and eRNA transcription, respectively, at enhancers. *The FASEB Journal, 35*, e21781–e21795.

Kone, B. C., Kuncewicz, T., Zhang, W., & Yu, Z. Y. (2003). Protein interactions with nitric oxide synthases: Controlling the right time, the right place, and the right amount of nitric oxide. *American Journal of Physiology - Renal Physiology, 285*(2), F178–F190. https://doi.org/10.1152/ajprenal.00048.2003

Kurdistani, S. K., Tavazoie, S., & Grunstein, M. (2004). Mapping global histone acetylation patterns to gene expression. *Cell, 117*, 721–733.

Lala, P. K., & Chakraborty, C. (2001). Role of nitric oxide in carcinogenesis and tumour progression. *The Lancet Oncology, 2*, 149–156.

Lala, P. K., & Orucevic, A. (1998). Role of nitric oxide in tumor progression: Lessons from experimental tumors. *Cancer and Metastasis Reviews, 17*(1), 91–106. https://doi.org/10.1023/A:1005960822365

Lala, P. K. (1998). Significance of nitric oxide in carcinogenesis, tumourprogression and cancer therapy. *Cancer and Metastasis Reviews, 17*, 1–6.

Laval, F., & Wink, D. A. (1994). Inhibition of nitric oxide of the repair proteinO6 -methylguanine-DNA-methyltransferase. *Carcinogenesis, 15*, 443–447.

Levine, A. B., Punihaole, D., & Levine, T. B. (2012). Characterization of the role of nitric oxide and its clinical applications. *Cardiology, 122*(1), 55–68. https://doi.org/10.1159/000338150

Li, H., Raman, C. S., Glaser, C. B., Blasko, E., Young, T. A., Parkinson, J. F., Whitlow, M., & Poulos, T. L. (1999). Crystal structures of zinc-free and -bound heme domain of human inducible nitric-oxide synthase. Implications for dimer stability and comparison with endothelial nitric-oxide synthase. *Journal of Biological Chemistry, 274*(30), 21276–21284. https://doi.org/10.1074/jbc.274.30.21276

Lowenstein, C. J., & Padalko, E. (2004). iNOS at a glance. *Journal of Cell Science, 117*, 2865–2867.

Lundberg, J. O., & Weitzberg, E. (2022). Nitric oxide signalimng in health and disease. *Cell, 185*, 2853–2878.

MacMicking, J., Xie, Q. W., & Nathan, C. (1997). Nitric oxide and macrophage function. *Annual Review of Immunology, 15*, 323–350. https://doi.org/10.1146/annurev.immunol.15.1.323

Matrone, C., Pignataro, G., Molinaro, P., Irace, C., Scorziello, A., Di Renzo, G. F., & Annunziato, L. (2004). HIF-1alpha reveals a binding activity to the promoter of iNOS gene after permanent middle cerebral artery occlusion. *Journal of Neurochemistry, 90*, 368–378.

Meraz, M. A., White, J. M., Sheehan, K. C., Bach, E. A., Rodig, S. J., Dighe, A. S., Kaplan, D. H., Riley, J. K., Greenlund, A. C., & Campbell, D. (1996). Targeted disruption of the Stat1 gene in mice reveals unexpected physiologic specificity in the JAK-STAT signaling pathway. *Cell, 84*, 431–442.

Michel, T., & Feron, O. (1997). Nitric oxide synthases: Which, where, how, and why? *Journal of Clinical Investigation, 100*(9), 2146–2152. https://doi.org/10.1172/JCI119750

Miglietta, G., Cogoi, S., Marinello, J., Capranico, G., Tikhomirov, A. S., Shchekotikhin, A., & Xodo, L. E. (2017). RNA G-quadruplexes in kirsten ras (KRAS) oncogene as targets for small molecules inhibiting translation. *Journal of Medicinal Chemistry, 60*, 9448–9461.

Murata, J.-I., Tada, M., Iggo, R. D., Sawamura, Y., Shinohe, Y., & Abe, H. (1997). Nitric oxide as a carcinogen: Analysis by yeast functional assay of inactivating p53 mutations induced by nitric oxide. *Mutation Research, 379*, 211–218.

Nathan, C. (1997). Inducible nitric oxide synthase: What difference does it make? *Journal of Clinical Investigation, 100*(10), 2417–2423. https://doi.org/10.1172/JCI119782

Noble, M. A., Munro, A. W., Rivers, S. L., Robledo, L., Daff, S. N., Yellowlees, L. J., Shimizu, T., Sagami, I., Guillemette, J. G., & Chapman, S. K. (1999). Potentiometric analysis of the flavin cofactors of neuronal nitric oxide synthase. *Biochemistry, 38*(50), 16413–16418. https://doi.org/10.1021/bi992150w

Okayama, H., Saito, M., Oue, N., Weiss, J. M., Stauffer, J., Takenoshita, S., Wiltrout, R. H., Hussain, S. P., & Harris, C. C. (2013). NOS2 enhances KRAS-induced lung carcinogenesis, inflammation and microRNA-21 expression. *International Journal of Cancer, 132*, 9–18.

Panigrahi, A., & O'Malley, B. W. (2021). Mechanisms of enhancer action: The known and the unknown. *Genome Biology, 22*, 108–138.

Papapetropoulos, A., Garcia-Cardena, G., Madri, J. A., & Sessa, W. C. (1997). Nitric oxide production contributes to the angiogenic properties of vascular endothelian growth factor in human endothelial cells. *Journal of Clinical Investigation, 100*, 3131–3139.

Perrella, M. A., Pellacani, A., Wiesel, P., Chin, M. T., Foster, L. C., Ibanez, M., Hsieh, C. M., Reeves, R., Yet, S. F., & Lee, M. E. (1999). High mobility group-I(Y) protein facilitates nuclear factor-kappaB binding and transactivation of the inducible nitric-oxide synthase promoter/enhancer. *Journal of Biological Chemistry, 274*, 9045–9052.

Pervin, S., Singh, R., & Chaudhuri, G. (2001). Nitric oxide-induced cytostasis and cell cycle arrest of a human breast cancer cell line (MDA-MB-231): Potential role of cyclin D1. *Proceedings of the National Academy of Sciences of the United States of America, 98*, 3583–3588.

Raman, C. S., Li, H., Martásek, P., Král, V., Masters, B. S. S., & Poulos, T. L. (1998). Crystal structure of constitutive endothelial nitric oxide synthase: A paradigm for pterin function involving a novel metal center. *Cell, 95*(7), 939–950. https://doi.org/10.1016/S0092-8674(00)81718-3

Ratajczak-Wrona, W., Jablonska, E., Garley, M., Jablonski, J., Radziwon, P., & Iwaniuk, A. (2013). Role of AP-1 family proteins in regulation of inducible nitric oxide synthase (iNOS) in human neutrophils. *Journal of Immunotoxicology, 10*, 32–39.

Ross, T. J., Kremens, K., Powers, L. S., Brink, B., Knutson, T., Domann, F. E., Philibert, R. A., Milhem, M. M., & Monick, M. M. (2014). Epigenetic silencing of the human NOS2 gene: Rethinking the role of nitric oxide in human macrophage inflammatory responses. *Journal of Immunology, 192*, 2326–2338.

Saura, M., Zaragoza, C., Bao, C., McMillan, A., & Lowenstein, C. J. (1999). Interaction of interferon regulatory factor-1 and nuclear factor kappaB during activation of inducible nitric oxide synthase transcription. *Journal of Molecular Biology, 289*, 459–471.

Sharma, J. N., Al-Omran, A., & Parvathy, S. S. (2007). Role of nitric oxide in inflammatory diseases. *Inflammopharmacology, 15*(6), 252–259. https://doi.org/10.1007/s10787-007-0013-x

Struhl, K. (1998). Histone acetylation and transcriptional regulatory mechanisms. *Genes and Development, 12*, 599–606.

Stuehr, D., Pou, S., & Rosen, G. M. (2001). Oxygen reduction by nitric-oxide synthases. *Journal of Biological Chemistry, 276*(18), 14533–14536. https://doi.org/10.1074/jbc.R100011200

Suzuki, T., Motohashi, H., & Yamamoto, M. (2013). Toward clinical application of the Keap1-Nrf2 pathway. *Trends in Pharmacological Sciences, 34*, 340–346.

Tao, S., Wang, S., Moghaddam, S. J., Ooi, A., Chapman, E., Wong, P. K., & Zhang, D. D. (2014). Oncogenic KRAS confers chemoresistance by upregulating NRF2. *Cancer Research, 74*, 7430–7441. Erratum in: Cancer Res. 2019; 79:1015.

Vannini, F., Kashfi, K., & Nath, N. (2015). The dual role of iNOS in cancer. *Redox Biology, 6*, 334–343. https://doi.org/10.1016/j.redox.2015.08.009

Wink, D. A., Vodovotz, Y., Laval, J., Laval, F., Dewhirst, M. W., & Mitchell, J. B. (1998). The multifaceted roles of nitric oxide in cancer. *Carcinogenesis, 19*(5), 711–721. https://doi.org/10.1093/carcin/19.5.711

Wink, D. A., Ridnour, L. A., Hussain, S. P., & Harris, C. C. (2008). The reemergence of nitric oxide and cancer. *Nitric Oxide, 19*(2), 65–67. https://doi.org/10.1016/j.niox.2008.05.003

Xie, Q. W., Kashiwabara, Y., & Nathan, C. (1994). Role of transcription factor NF-kappaB/Relin induction of nitric oxide synthase. *Journal of Biological Chemistry, 269*, 4705–4708.

Xu, W., Charles, I. G., Liu, L., Moncada, S., & Emson, P. (1996). Molecular cloning and structural organization of the human inducible nitric oxide synthase gene (NOS2). *Biochemical and Biophysical Research Communications, 219*, 784–788.

Xu, W., Liu, L. Z., Loizidou, M., Ahmed, M., & Charles, I. G. (2002). The role of nitric oxide in cancer. *Cell Research, 12*(5–6), 311–320. https://doi.org/10.1038/sj.cr.7290133

Yang, L., Shen, C., Estrada-Bernal, A., Robb, R., Chatterjee, M., Sebastian, N., Webb, A., Mo, X., Chen, W., Krishnan, S., & Williams, T. M. (2021). Oncogenic KRAS drives radioresistance through upregulation of NRF2-53BP1-mediated non-homologous end-joining repair. *Nucleic Acids Research, 49*, 11067–11082.

Yao, J., Chen, J., Li, L. Y., & Wu, M. (2020). Epigenetic plasticity of enhancers in cancer. *Transcription, 11*, 26–36.

Zentner, G. E., Tesar, P. J., & Scacheri, P. C. (2011). Epigenetic signatures distinguish multiple classes of enhancers with distinct cellular functions. *Genome Research, 21*, 1273–1283.

Ziello, J. E., Jovin, I. S., & Huang, Y. (2007). Hypoxia-Inducible Factor (HIF)-1 regulatory pathway and its potential for therapeutic intervention in malignancy and ischemia. *Yale Journal of Biology and Medicine, 80*, 51–60.

Zhou, H.-L., Zhang, R., Anand, P., Stomberski, C. T., Qian, Z., Hausladen, A., Wang, L., Rhee, E. P., Parikh, S. M., Karumanchi, S. A., & Stamler, J. S. (2019). Metabolic reprogramming by the S-nitroso-CoA reductase system protects against kidney injury. *Nature*, 56596–57100.

CHAPTER

10

Nitric oxide in tumor biology: From stemness to metabolic reprogramming

Antonio Rodríguez-Ariza

Instituto Maimónides de Investigación Biomédica de Córdoba (IMIBIC), Córdoba, Spain

Abbreviations

2DG 2-deoxy-glucose
APC adenomatous polyposis coli
CRC colorectal cancer
CSC cancer stem cell
CSNO S-nitroso-L-cysteine
eNOS endothelial nitric oxide synthase
ER estrogen receptor
GSNOR S-nitrosoglutathione reductase
GSNOR S-nitrosoglutathione
NO nitric oxide
NOS nitric oxide synthase
PD-L1 programmed cell death ligand-1
PD1 programmed cell death protein 1
PDOs patient-derived organoids
PDXs patient-derived xenografts
SNO S-nitrosothiols
TCGA The Cancer Genome Atlas

Introduction

Nitric oxide (NO) plays an important and complex role in cancer biology, performing a contradictory dual function that significantly impacts tumorigenesis. On one hand, low NO levels have been shown to promote tumor progression, while conversely, elevated

Therapeutic Applications of Nitric Oxide in Cancer and Inflammatory Disorders
https://doi.org/10.1016/B978-0-443-22354-9.00010-3

concentrations of NO demonstrate potent inhibitory effects on cancer development (Khan et al., 2020; Miranda et al., 2021). This dual nature adds complexity to the development of potential NO-based therapeutic strategies against cancer, which can be broadly categorized into two distinct approaches. In the first approach, NO-based therapies seek to exploit the antitumor properties of this simple molecule. Consequently, therapies aimed at increasing NO levels or sensitizing cancer cells to its antitumor effects offer a promising avenue for intervention (Sinha, 2023; Huang et al., 2017). On the contrary, anticancer therapies directed against NO production pursue to block the protumor effects of this molecule. This strategy may involve the use of specific inhibitors targeting nitric oxide synthase (NOS) enzymes or modulating other molecular pathways related with NO-mediated signaling (Chung et al., 2021; Reddy et al., 2023).

Although the antitumor mechanisms of NO are relatively well known, the intricate mechanisms underlying the tumorigenic activity of this molecule are less understood. Thus, NO appears to play a complex role in promoting various processes that facilitate tumor progression, including DNA damage contributing to oncogenesis (Burney et al., 1999), the stimulation of angiogenesis (Morbidelli et al., 2004), the creation of an immunosuppressive microenvironment within the tumor (PeÑarando et al., 2019), the facilitation of tumor cell invasiveness and metastasis (Reddy et al., 2023), and the promotion of cancer cell survival and resistance to therapy (Hays & Bonavida, 2019). As new and intriguing aspects of tumor biology are uncovered, it becomes increasingly apparent that NO also plays a role in these phenomena. Hence, in the past decade, the concept of cancer stemness, which denotes the stem cell–like characteristics exhibited by certain cancer cells, has gained recognition for its significant involvement in various aspects of carcinogenesis (Prasad et al., 2020). In addition, recent research has significantly advanced our understanding of why tumors develop metabolic phenotypes that endow cancer cells with the capacities to adapt and persist (Faubert et al., 2020). Notably, emerging experimental evidence underscores the fundamental role of NO in these processes that profoundly impact on the ability of tumors to progress and resist therapy.

Role of NO in the acquisition of stem characteristics and therapy resistance in breast cancer

Tumors have traditionally been considered to follow a stochastic model whereby each tumor cell has the same probability of being the cell of origin and where the sequential accumulation of mutations favoring tumor initiation and progression will occur (Knudson, 2001). However, there is increasing evidence that only a subpopulation of tumor cells, known as cancer stem cells (CSCs), possess self-renewal capacity, are more resistant to therapy and are responsible for initiating and maintaining tumor growth, which ultimately results in tumor dissemination and metastasis (Huang et al., 2020; Nassar & Blanpain, 2016). Although this hierarchical model was proposed 30 years ago (Lapidot et al., 1994), experimental evidence consistently indicates that cancer is not simply the result of randomly mutating stem cells. Thus, it has been shown that there is high cellular plasticity, whereby differentiated tumor cells can reacquire stem characteristics, and rather than a constant population of stem cells in the tumor, there is a dynamic CSC population in response to environmental conditions and therapy (Batlle & Clevers, 2017; La Porta & Zapperi, 2017; Li & Laterra, 2012;

Mani et al., 2008). Importantly, studies conducted over the past decade have revealed that NO signaling also significantly influences the growth and tumorigenic capabilities of CSCs (Gao et al., 2022; López-Sánchez et al., 2021; Peñarando et al., 2018; Wang et al., 2018).

An important mechanism by which NO influences target protein function is through the attachment of a nitroso moiety to reactive thiol groups at specific cysteine residues, resulting in the creation of S-nitrosothiols (SNO) in a process commonly referred to as S-nitrosylation (Stomberski et al., 2019). In this context, we reported that disruption of enzyme-mediated denitrosylation processes, which are vital for maintaining SNO homeostasis, confers survival advantages to estrogen receptor-positive (ER^+) breast cancer cells (Cañas et al., 2012). Remarkably, our data indicated that the impairment of SNO homeostasis not only induced the phosphorylation of ER-alpha at Thr311, resulting in an almost exclusively nuclear localization of this receptor, but also caused the expansion of a CSC-like subpopulation in ER^+ breast cancer cells (Fig. 10.1). We, therefore, hypothesized that NO may also play a relevant role in the acquisition of stem characteristics in ER^+ breast cancer and the development of resistance against hormonal therapies. Accordingly, in a subsequent study (López-Sánchez et al., 2021), we demonstrated that NO depletion not only inhibited mammosphere formation and decreased CSCs subpopulations in ER+ breast cancer cells, but also increased the antitumoral efficacy of antihormonal treatment with tamoxifen (Fig. 10.2). Moreover, in this study, we also reported that low NOS2 expression was significantly associated with a higher metastasis-free survival in ER^+ breast cancer patients treated with tamoxifen; however, this difference in metastasis-free survival was not observed in ER+ breast cancer patients treated with chemotherapy. Therefore, survival data also supported that a higher NOS2 expression provides an adaptive advantage for ER-positive tumors treated with tamoxifen, resulting in lower patient metastasis-free survival rates (López-Sánchez et al., 2021).

Nitric oxide depletion decreases stemness and augments the efficacy of hormonal therapy with tamoxifen in estrogen receptor positive (ER+) breast cancer cells.

Role of NO in the biology of intestinal cancer cells

Colorectal cancer (CRC) is another tumor where CSCs are known to play a pivotal role (van der Heijden & Vermeulen, 2019). Intestinal stem cells are located at the bottom of the intestinal crypts, and the alteration of the Wnt pathway through loss or mutation of the *APC* gene in these stem cells could explain the origin of CRC (Barker et al., 2009). In this regard, and to investigate the potential role of NO in the initiation of intestinal tumors, we first analyzed murine syngeneic models and unexpectedly found that the endothelial isoform of NOS (eNOS) was upregulated in the altered intestinal crypts of mice harboring *Apc* mutations, which we also confirmed at the transcriptional level (Peñarando et al., 2018). Moreover, in human clinical samples, those tumors with mesenchymal or stem-like gene signatures also showed a higher eNOS expression (Fig. 10.3). Importantly, we confirmed that selective removal of NO in CRC cancer cells not only decreased the expression of stem markers and their ability to form colonospheres, but also directly impacted their ability to generate tumors in immunosuppressed mice (Peñarando et al., 2018). Thus, with this study, we established the important role of NO in the generation and maintenance of tumor stem cells in CRC, and suggested some therapeutic opportunities, especially for those mesenchymal colorectal tumors with the worst prognosis.

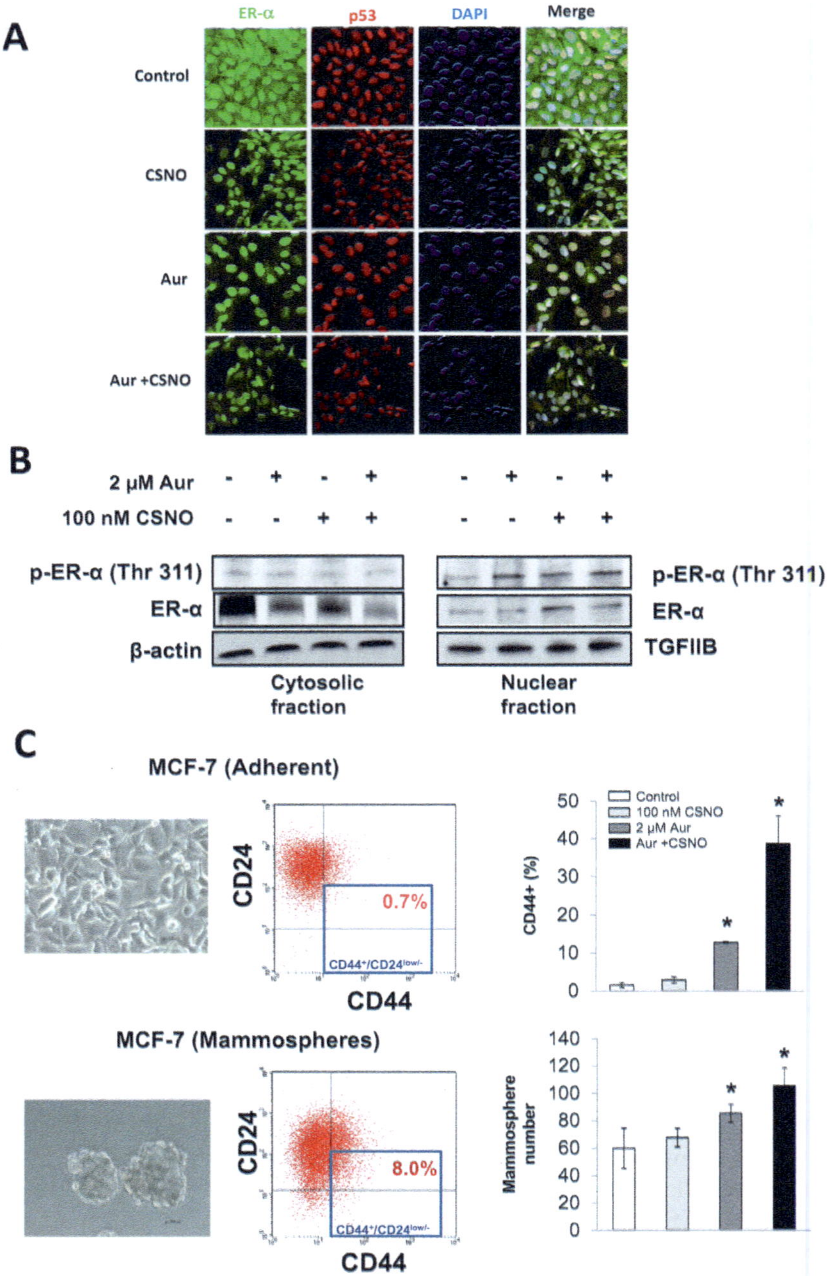

FIGURE 10.1 Impairment of S-nitrosothiol homeostasis alters the subcellular localization of estrogen receptor (ER) and expands a cancer stem cell–like subpopulation in ER-positive breast cancer cells. The impairment of SNO homeostasis in breast cancer cells was achieved with the highly specific thioredoxin reductase inhibitor auranofin (Aur) and/or exposure to S-nitroso-L-cysteine (CSNO). The impairment of SNO homeostasis not only induced the phosphorylation of ER-alpha at Thr311, resulting in an almost exclusively nuclear localization of this receptor (A, B) but also caused the expansion of a CSC-like subpopulation in ER + breast cancer cells (C). *Reproduced from Cañas, A., López-Sánchez, L. M., Valverde-Estepa, A., Hernández, V., Fuentes, E., Muñoz-Castañeda, J. R., López-Pedrera, C., De La Haba-Rodríguez, J. R., Aranda, E., Rodríguez-Ariza, A. (2012). Maintenance of S-nitrosothiol homeostasis plays an important role in growth suppression of estrogen receptor-positive breast tumors.* Breast Cancer Research, *14(6), R153, with permission of BioMed Central.*

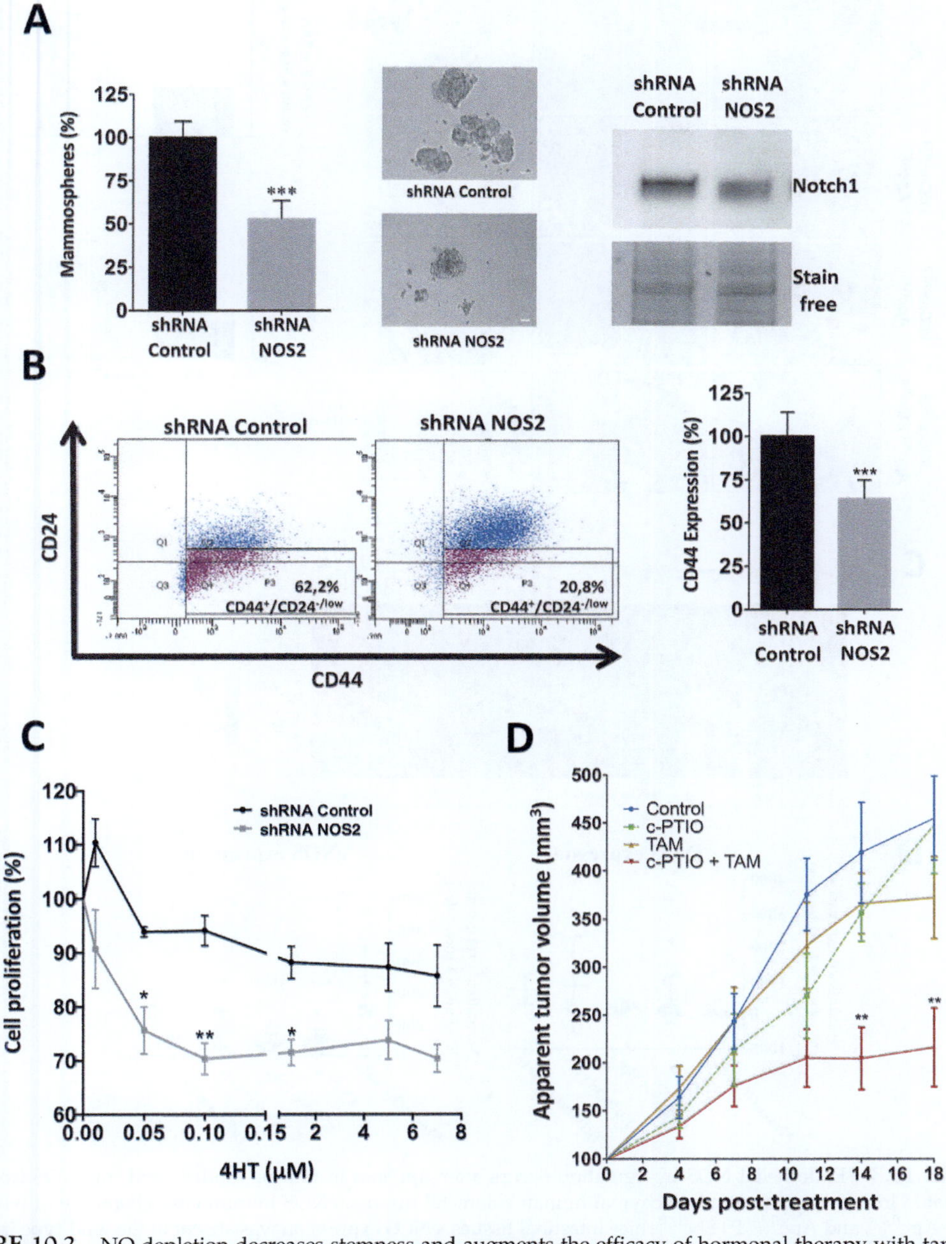

FIGURE 10.2 NO depletion decreases stemness and augments the efficacy of hormonal therapy with tamoxifen in estrogen receptor positive (ER+) breast cancer cells. NOS2 deficiency inhibits mammosphere formation and the expression of CSCs-specific markers in ER+ breast cancer cells (A, B). NOS2 silencing in vitro (C) or NO depletion in vivo (D) increases the antitumoral efficacy of antihormonal treatment with tamoxifen in ER+ breast cancer. *Reproduced from López-Sánchez, L. M., Mena, R., Guil-Luna, S., Mantrana, A., Peñarando, J., Toledano-Fonseca, M., Conde, F., De la Haba-Rodríguez, J. R., Aranda, E., Rodríguez-Ariza, A. (2021). Nitric oxide-targeted therapy inhibits stemness and increases the efficacy of tamoxifen in estrogen receptor-positive breast cancer cells.* Lab Invest, *101(3), 292–303, with permission of Elsevier.*

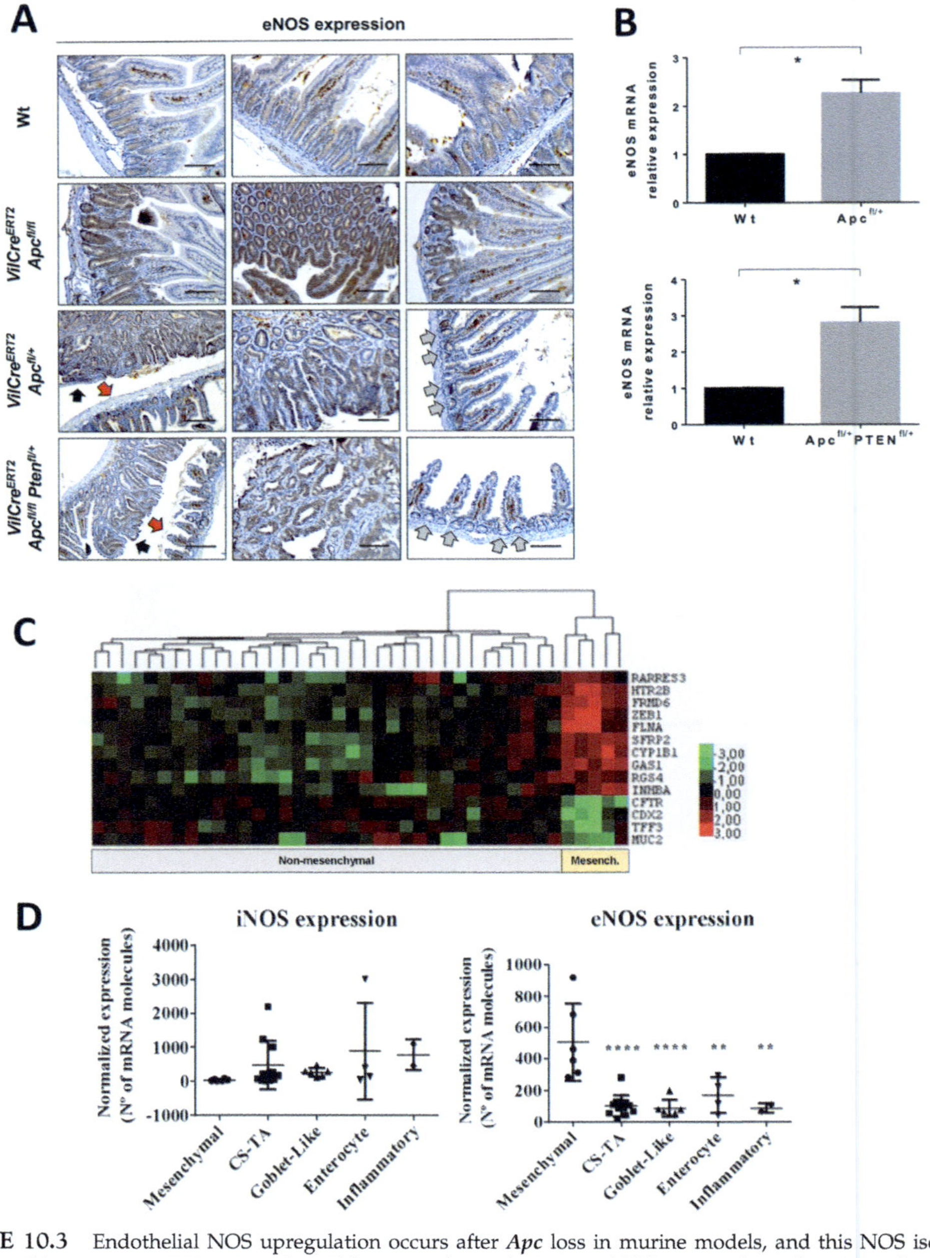

FIGURE 10.3 Endothelial NOS upregulation occurs after ***Apc*** loss in murine models, and this NOS isoform is upregulated in poor-prognosis mesenchymal human colorectal tumors. eNOS immunohistochemistry of wild-type, $Apc^{fl/fl}$, $Apc^{fl/+}$, and $Apc^{fl/+}$ $PTEN^{fl/fl}$ mice intestinal tissues. eNOS expression was absent in the wild-type intestine, whereas $Apc^{fl/fl}$ sections showed intense immunostaining in the hyperproliferative crypts. In the $Apc^{fl/+}$ and $Apc^{fl/+}$ $PTEN^{fl/+}$ mice models, hyperproliferative areas resulted in upregulation of eNOS (*black arrows*), whereas normal intestinal tissue did not show expression of this NOS isoform (*red arrow*) (A). RT-qPCR analysis of eNOS expression from wild-type, $Apc^{fl/+}$ and $Apc^{fl/+}$ $PTEN^{fl/+}$ tissue (B). Unsupervised classification of the mesenchymal subgroup of CRC tumors and nonmesenchymal subtypes. (C). iNOS and eNOS mRNA expression in human CRC subtypes (D). *Reproduced from Peñarando, J., López-Sánchez, L. M., Mena, R., Guil-Luna, S., Conde, F., Hernández, V., Toledano, M., Gudiño, V., Raponi, M., Billard, C., Villar, C., Díaz, C., Gómez-Barbadillo, J., De la Haba-Rodríguez, J., Myant, K., Aranda, E., Rodríguez-Ariza, A. (2018). A role for endothelial nitric oxide synthase in intestinal stem cell proliferation and mesenchymal colorectal cancer.* BMC Biology, 16*(1), 3, with permission of BioMed Central.*

Metabolic shift underlies tumor progression and immune evasion in S-nitrosoglutathione reductase-deficient cancer

As mentioned before, the role of NO in cancer is very complex, with a distinctive aspect of this complexity being the elaborate network of interactions between the tumor and the immune cells. Although these interactions are diverse and context dependent, experimental evidence supports that NO can be considered as an immunosuppressive mediator (PeÑarando et al., 2019). Thus, NO participates in the different strategies used by tumors to escape immune-mediated recognition, not only including the acquisition of stem-like characteristics, but also contributing to metabolic reprogramming of both tumor cells and tumor-infiltrating immune cells (López-Sánchez et al., 2020). Importantly, protein S-nitrosylation appears as a common mechanism by which NO signaling exerts these protumorigenic effects in the tumor microenvironment (López-Sánchez et al., 2020; PeÑarando et al., 2019).

In this regard, S-nitrosoglutathione reductase (GSNOR) is a key enzyme regulating SNO homeostasis, and consequently protein S-nitrosylation, by metabolizing S-nitrosoglutathione (GSNO), which is the main reservoir of NO in cells. Thus, high levels of GSNO and S-nitrosylated proteins are observed in GSNOR-deficient mice (Wei et al., 2010). Importantly, the deficiency or inhibition of this enzyme has unveiled novel mechanisms that underlie the role of NO in the initiation and progression of tumors (Barnett & Buxton, 2017; Cañas et al., 2016; He et al., 2022; Mena-Osuna et al., 2023; Rizza et al., 2023; Salvatore Rizza et al., 2020; Rizza & Filomeni, 2017; Wei et al., 2010). Moreover, this research strongly supports the suggested tumor suppressor role of this denitrosylating enzyme (Rizza & Filomeni, 2016). For instance, we demonstrated that inhibition of this enzyme in HER2-overexpressing breast cancer increases protein S-nitrosylation and reduces the efficacy of therapeutic treatment against HER2. Furthermore, low expression of the *ADH5* gene, which encodes GSNOR, was associated with lower survival rates in patients with HER2 tumors, but not in the other breast cancer subtypes (Cañas et al., 2016).

We also investigated whether this denitrosylase enzyme also plays a relevant role in the pathogenesis of CRC (Mena-Osuna et al., 2023). We first found that low GSNOR expression was associated with worse prognostic factors in CRC. Also, like in breast cancer, the worst-prognosis mesenchymal CMS4 was the molecular subtype characterized by a significant lower expression of this denitrosylase. Moreover, survival analyses using TCGA data confirmed that lower ADH5 expression was significantly associated with a lower survival rate in CRC patients. However, it is still unclear which factors are responsible for the poorer prognosis in GSNOR-negative CRC tumors. In this context, it is known that NO can directly facilitate metabolic reprogramming in the tumor microenvironment by promoting aerobic glycolysis in tumor cells and lactate secretion into the tumor milieu (López-Sánchez et al., 2020). This acidification of the tumor microenvironment promotes immune escape by recruiting and activating immunosuppressive immune cells, while it is unfavorable to the infiltration of cytotoxic effector cells (Fig. 10.4). Therefore, we decided to investigate whether this phenomenon occurs in GSNOR-negative tumors. As shown in Fig. 10.5, we found that GSNOR-negative tumors were characterized by an immunosuppressive microenvironment. Specifically, GSNOR-negative tumors showed significant downregulation of $CD8^+$ T cell—related genes. We further confirmed that GSNOR-negative tumors are characterized by less infiltration of cytotoxic or $CD8^+$ T lymphocytes. Moreover, GSNOR-negative tumors were characterized by downregulation of genes related to cytotoxic immune activity, and

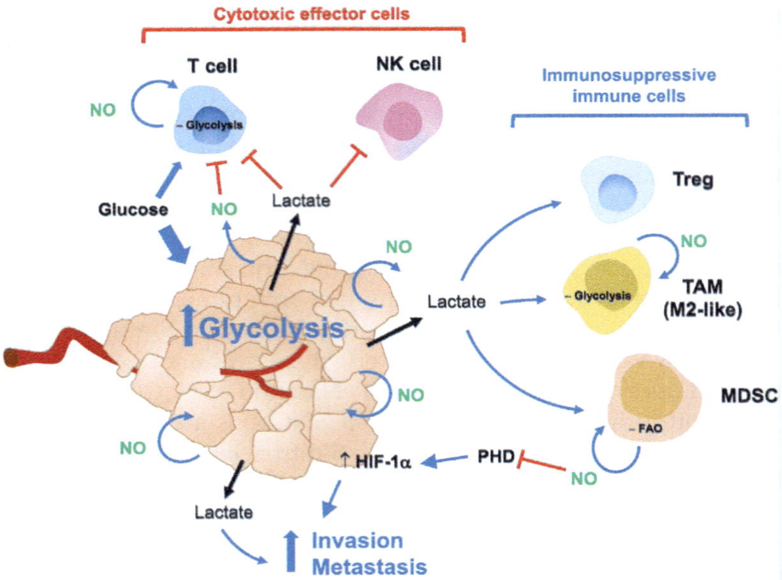

FIGURE 10.4 Nitric oxide and the metabolic switch in the tumor microenvironment. Nitric oxide (NO) has the capacity to induce metabolic reprogramming within the tumor microenvironment enhancing aerobic glycolysis in tumor cells and augmenting lactate secretion into the tumor milieu. This acidification of the tumor microenvironment promotes immune escape by recruiting and activating immunosuppressive immune cells, such as myeloid-derived suppressor cells (MDSCs), regulatory T cells (Tregs), and tumor-associated macrophages (TAMs), while being unfavorable to infiltrating cytotoxic effector cells, such as natural killer (NK) cells and T lymphocytes (T cells). NO also inhibits the prolyl hydroxylase (PHD)—mediated degradation of hypoxia-inducible factor-1α (HIF-1α), causing a hypoxic response in the tumor microenvironment. Activating or inhibiting effects of NO are shown as blue or red arrows, respectively. *Reproduced from López-Sánchez, L. M., Aranda, E., Rodríguez-Ariza, A. (2020). Nitric oxide and tumour metabolic reprogramming.* Biochemical Pharmacology, 176, *113769, with permission of Elsevier.*

induction of genes related to immunosuppression (Fig. 10.5). On the other hand, through quantitative proteomics, we validated that those CRC tumors with low GSNOR expression were not only characterized by alteration in immune-related pathways but also in metabolism, and particularly in mitochondrial energy metabolism.

To further investigate the mechanisms responsible for the increased aggressiveness of GSNOR-negative tumors, we used CRISPR to generate GSNOR knockout (GSNOR-KO) CRC cells. GSNOR deficiency was found to significantly increase the tumorigenic and tumor-initiating capacity of CRC cells. This was evident through the higher ability of GSNOR-KO cells to form tumorospheres and tumor spheroids, along with a substantial increase in their capacity to initiate tumors, as demonstrated by the limiting-dilution assay. Moreover, and in agreement with their increased PD-L1 expression, by using humanized mice models, we confirmed their resistance to anti-PD-1 immunotherapy (Mena-Osuna et al., 2023). Remarkably, GSNOR-KO cells were characterized by an alteration in their mitochondrial network, with smaller and lighter mitochondria, and an altered bioenergetic metabolism, with increased lactate secretion, lower oxidative phosphorylation rates and a higher dependence on aerobic glycolysis, a metabolic alteration related to increased tumor aggressiveness (Fig. 10.6). This greater dependence on the

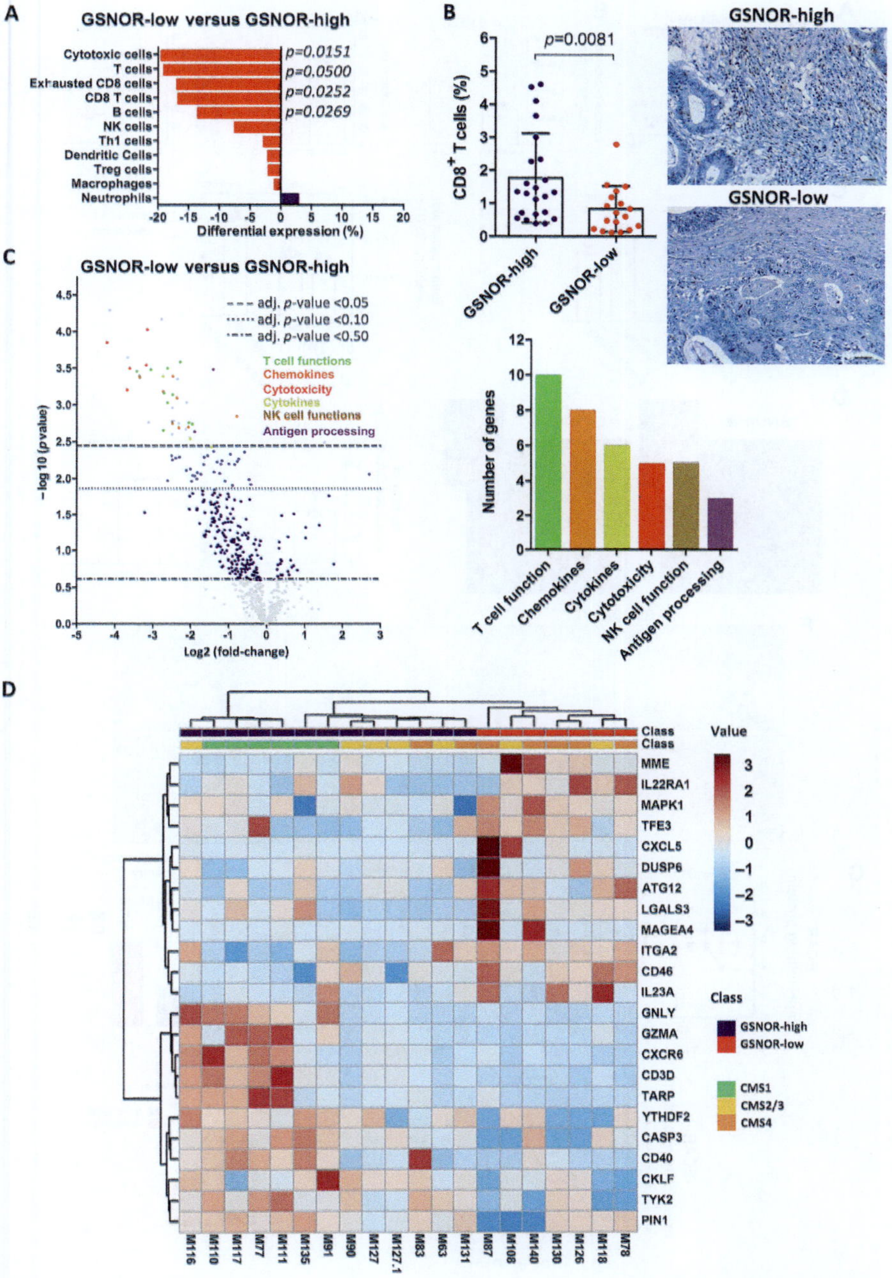

FIGURE 10.5 Low GSNOR expression is associated with an immunosuppressive microenvironment and cytotoxic T cell exclusion in clinical CRC tumors. GSNOR-negative tumors showed significant downregulation of CD8$^+$ T cell—related genes (A), less infiltration of cytotoxic or CD8$^+$ T lymphocytes (B) and downregulation of genes related to cytotoxic immune activity, along with induction of genes related to immunosuppression (C, D). *Reproduced from Mena-Osuna, R., Mantrana, A., Guil-Luna, S., Sánchez-Montero, M. T., Navarrete-Sirvent, C., Morales-Ruiz, T., Rivas-Crespo, A., Toledano-Fonseca, M., García-Ortíz, M. V., García-Jurado, G., Gómez-España, M. A., González-Fernández, R., Villar, C., Medina-Fernández, F. J., Villalba, J. M., Aranda, E., Rodríguez-Ariza, A. (2013). Metabolic shift underlies tumor progression and immune evasion in S-nitrosoglutathione reductase-deficient cancer.* Journal of Pathology, 260(3), 261–275, *with permission of John Wiley and Sons.*

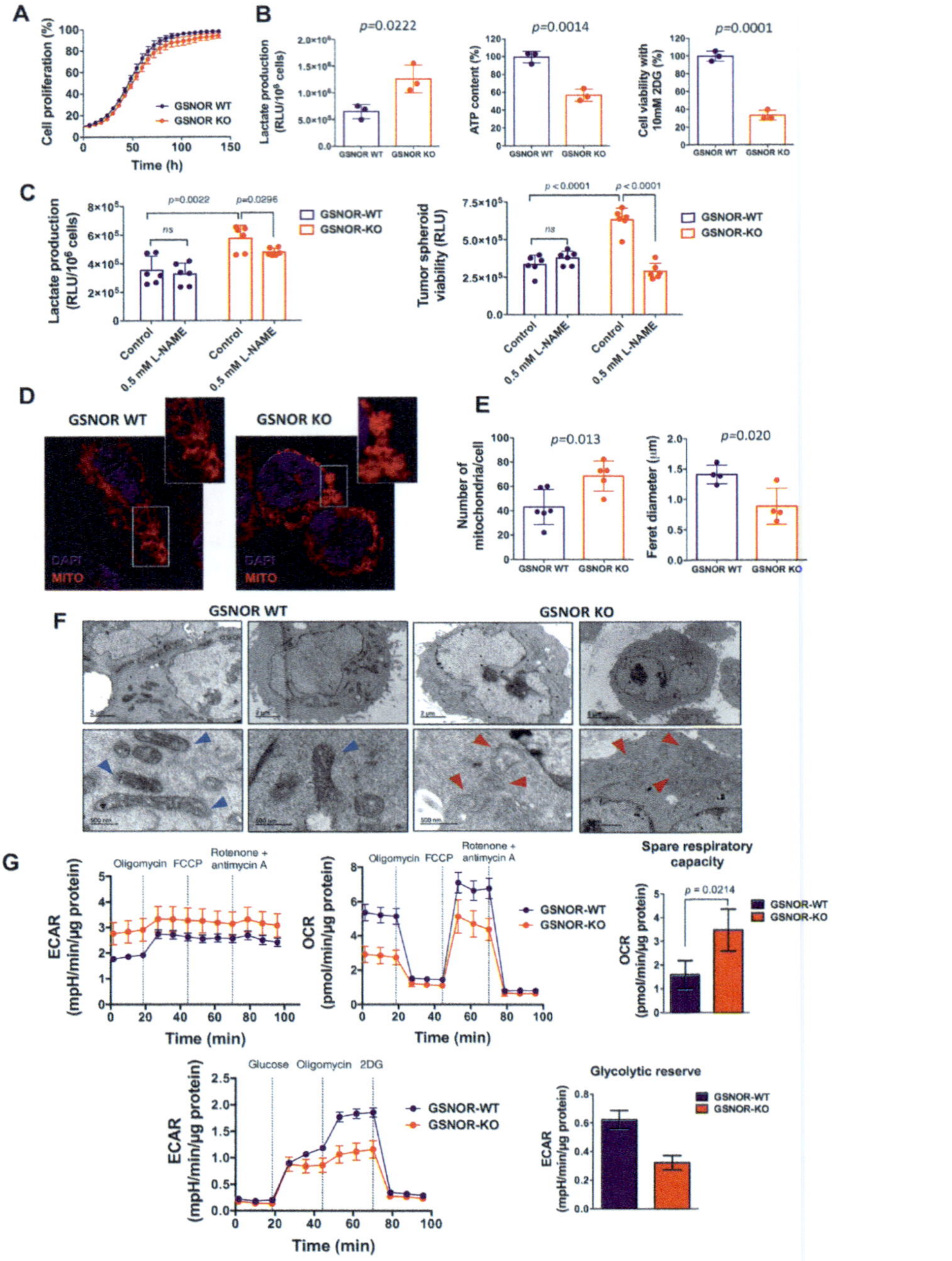

FIGURE 10.6 GSNOR deficiency is associated with metabolic disturbances and changes in the mitochondrial network. Cell proliferation rates of GSNOR-WT and GSNOR-KO cells (A). Lactate production, basal ATP content, and sensitivity to glycolysis inhibitor 2-deoxy-D-glucose (2DG) in GSNOR-WT and GSNOR-KO cells (B). Effect of

glycolytic pathway conferred to GSNOR-deficient cells a greater sensitivity to the glycolysis inhibitor 2-deoxy-glucose (2DG). Significantly, we found in relevant preclinical models, such as patient-derived organoids (PDOs) and patient-derived xenografts (PDXs), that GSNOR-deficient tumors are more vulnerable to this metabolic inhibitor (Fig. 10.7). Therefore, our study supports that metabolic rewiring induced by GSNOR deficiency constitutes an important mechanism contributing to the acquisition of aggressive and immune evasive characteristics in CRC. Furthermore, the deficiency of this denitrosylase enzyme in colorectal tumors confers metabolic vulnerabilities that can be therapeutically exploited (Mena-Osuna et al., 2023).

Conclusions

The research summarized here underscores relevant mechanisms by which NO participates in the tumorigenic process. These mechanisms range from augmenting stem cell characteristics to rewiring metabolism. Significantly, these NO-mediated effects underlie crucial aspects of cancer, such as therapeutic resistance and immune evasion, rendering them highly relevant in the context of therapy. Therefore, employing anti-NO therapies or strategies that exploit the metabolic vulnerabilities linked to abnormal NO signaling can be considered significant therapeutic avenues for treating cancer patients. Undoubtedly, future research will continue to unveil new therapeutic opportunities arising from the multifaceted role of this seemingly simple molecule in cancer.

inhibition of NO synthesis by NOS with L-NAME in lactate production and tumor spheroid formation capacity of GSNOR-WT and GSNOR-KO cells (C). Representative confocal microscopy images of mitochondrial network in GSNOR-WT and GSNOR-KO cells (D) and corresponding quantitative analysis with mean number of mitochondria per cell and mean mitochondrial diameter (E). Representative electron microscopy images, with smaller and lighter mitochondria with unidentifiable cristae morphology in GSNOR-KO cells (*red arrowheads*) compared with their parental counterparts (*blue arrowheads*) (F). Scale bars, 2 μm and 500 nm, for upper and lower images, respectively. Mitochondrial respiratory and glycolytic profiles of GSNOR-WT and GSNOR-KO cells were obtained by measuring in real time the oxygen consumption (OCR) and extracellular acidification rates (ECARs) using a Seahorse Extracellular Flux Analyzer (G). *Reproduced from Mena-Osuna, R., Mantrana, A., Guil-Luna, S., Sánchez-Montero, M. T., Navarrete-Sirvent, C., Morales-Ruiz, T., Rivas-Crespo, A., Toledano-Fonseca, M., García-Ortíz, M. V., García-Jurado, G., Gómez-España, M. A., González-Fernández, R., Villar, C., Medina-Fernández, F. J., Villalba, J. M., Aranda, E., Rodríguez-Ariza, A. (2023). Metabolic shift underlies tumor progression and immune evasion in S-nitrosoglutathione reductase-deficient cancer.* Journal of Pathology, 260*(3), 261–275, with permission of John Wiley and Sons.*

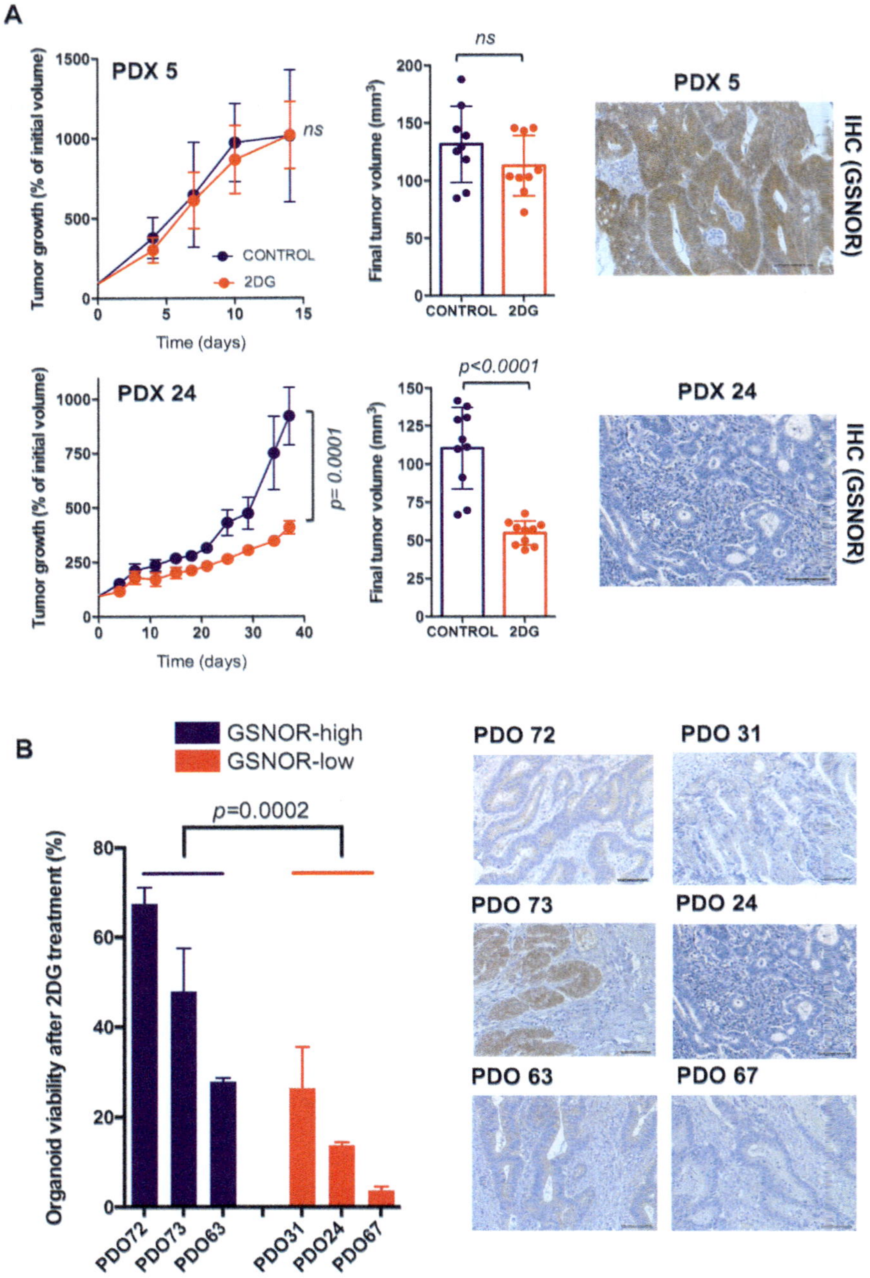

FIGURE 10.7 GSNOR-low colorectal tumors are more sensitive to glycolysis inhibitor 2DG. Effect of 2DG treatment (1 g/kg, twice a week) on growth of PDXs generated in immunocompromised mice from GSNOR-high (PDX 5) or GSNOR-low (PDX 24) tumors. GSNOR IHC analyses of tumors are shown on right (A). Effect of 2DG treatment (2.5 mM) on viability of PDOs from GSNOR-high (72, 73, 63) or GSNOR-low (24, 31, 67) tumors. GSNOR IHC analyses of corresponding tumors are shown on right (B). Mean ± SD. Scale bars, 100 μm. *Reproduced from Mena-Osuna, R., Mantrana, A., Guil-Luna, S., Sánchez-Montero, M. T., Navarrete-Sirvent, C., Morales-Ruiz, T., Rivas-Crespo, A., Toledano-Fonseca, M., García-Ortíz, M. V., García-Jurado, G., Gómez-España, M. A., González-Fernández, R., Villar, C., Medina-Fernández, F. J., Villalba, J. M., Aranda, E., Rodríguez-Ariza, A. (2023). Metabolic shift underlies tumor progression and immune evasion in S-nitrosoglutathione reductase-deficient cancer.* Journal of Pathology, 260*(3), 261–275, with permission of John Wiley and Sons.*

Acknowledgments

The research described here was supported by grants from the Consejería de Salud de la Junta de Andalucía (JA0230/09 and PI-0268–2014) and Consejería de Universidad, Investigación e Innovación of the Andalusian Government (P20-00967), and from Instituto de Salud Carlos III (PI10/00428, PI13/00553 and PI16/01508) and Ministerio de Ciencia e Innovación, Agencia Estatal de Investigación (PID2019-105256RB-I00/AEI/10.13039/501100011033), of Spanish Government. Sincere gratitude is expressed to the patients and their families who generously consented to participate in the studies described herein.

References

Barker, N., Ridgway, R. A., Van Es, J. H., Van De Wetering, M., Begthel, H., Van Den Born, M., Danenberg, E., Clarke, A. R., Sansom, O. J., & Clevers, H. (2009). Crypt stem cells as the cells-of-origin of intestinal cancer. *Nature, 457*(7229), 608–611. https://doi.org/10.1038/nature07602

Barnett, S. D., & Buxton, I. L. O. (2017). The role of S-nitrosoglutathione reductase (GSNOR) in human disease and therapy. *Critical Reviews in Biochemistry and Molecular Biology, 52*(3), 340–354. https://doi.org/10.1080/10409238.2017.1304353

Batlle, E., & Clevers, H. (2017). Cancer stem cells revisited. *Nature Medicine, 23*(10), 1124–1134. https://doi.org/10.1038/nm.4409

Burney, S., Caulfield, J. L., Niles, J. C., Wishnok, J. S., & Tannenbaum, S. R. (1999). The chemistry of DNA damage from nitric oxide and peroxynitrite. *Mutation Research - Fundamental and Molecular Mechanisms of Mutagenesis, 424*(1–2), 37–49. https://doi.org/10.1016/S0027-5107(99)00006-8

Cañas, A., López-Sánchez, L. M., Peñarando, J., Valverde, A., Conde, F., Hernández, V., Fuentes, E., López-Pedrera, C., de la Haba-Rodríguez, J. R., Aranda, E., & Rodríguez-Ariza, A. (2016). Altered S-nitrosothiol homeostasis provides a survival advantage to breast cancer cells in HER2 tumors and reduces their sensitivity to trastuzumab. *Biochimica et Biophysica Acta - Molecular Basis of Disease, 1862*(4), 601–610. https://doi.org/10.1016/j.bbadis.2016.02.005

Cañas, A., López-Sánchez, L. M., Valverde-Estepa, A., Hernández, V., Fuentes, E., Muñoz-Castañeda, J. R., López-Pedrera, C., De La Haba-Rodríguez, J. R., Aranda, E., & Rodríguez-Ariza, A. (2012). Maintenance of S-nitrosothiol homeostasis plays an important role in growth suppression of estrogen receptor-positive breast tumors. *Breast Cancer Research, 14*(6). https://doi.org/10.1186/bcr3366

Chung, A. W., Anand, K., Anselme, A. C., Chan, A. A., Gupta, N., Venta, L. A., Schwartz, M. R., Qian, W., Xu, Y., Zhang, L., Kuhn, J., Patel, T., Rodriguez, A. A., Belcheva, A., Darcourt, J., Ensor, J., Bernicker, E., Pan, P. Y., Chen, S. H., … Chang, J. C. (2021). A phase 1/2 clinical trial of the nitric oxide synthase inhibitor L-NMMA and taxane for treating chemoresistant triple-negative breast cancer. *Science Translational Medicine, 13*(624). https://doi.org/10.1126/scitranslmed.abj5070

Faubert, B., Solmonson, A., & DeBerardinis, R. J. (2020). Metabolic reprogramming and cancer progression. *Science, 368*(6487). https://doi.org/10.1126/science.aaw5473

Gao, W., Wang, Y., Yu, S., Wang, Z., Ma, T., Chan, A. M. L., Chiu, P. K. F., Ng, C. F., Wu, D., & Chan, F. L. (2022). Endothelial nitric oxide synthase (eNOS)-NO signaling axis functions to promote the growth of prostate cancer stem-like cells. *Stem Cell Research and Therapy, 13*(1). https://doi.org/10.1186/s13287-022-02864-6

Hays, E., & Bonavida, B. (2019). Nitric oxide-mediated enhancement and reversal of resistance of anticancer therapies. *Antioxidants, 8*(9). https://doi.org/10.3390/antiox8090407

He, Q., Qu, M., Shen, T., Xu, Y., Luo, J., Tan, D., Xu, C., Barkat, M. Q., Zeng, L. H., & Wu, X. (2022). Suppression of VEGFD expression by S-nitrosylation promotes the development of lung adenocarcinoma. *Journal of Experimental and Clinical Cancer Research, 41*(1). https://doi.org/10.1186/s13046-022-02453-8

Huang, T., Song, X., Xu, D., Tiek, D., Goenka, A., Wu, B., Sastry, N., Hu, B., & Cheng, S. Y. (2020). Stem cell programs in cancer initiation, progression, and therapy resistance. *Theranostics, 10*(19), 8721–8743. https://doi.org/10.7150/thno.41648

Huang, Z., Fu, J., & Zhang, Y. (2017). Nitric oxide donor-based cancer therapy: Advances and prospects. *Journal of Medicinal Chemistry, 60*(18), 7617–7635. https://doi.org/10.1021/acs.jmedchem.6b01672

Khan, F. H., Dervan, E., Bhattacharyya, D. D., McAuliffe, J. D., Miranda, K. M., & Glynn, S. A. (2020). The role of nitric oxide in cancer: Master regulator or not? *International Journal of Molecular Sciences, 21*(24), 1–30. https://doi.org/10.3390/ijms21249393

Knudson, A. G. (2001). Two genetic hits (more or less) to cancer. *Nature Reviews Cancer, 1*(2), 157–162. https://doi.org/10.1038/35101031

López-Sánchez, L. M., Aranda, E., & Rodríguez-Ariza, A. (2020). Nitric oxide and tumor metabolic reprogramming. *Biochemical Pharmacology, 176*. https://doi.org/10.1016/j.bcp.2019.113769

López-Sánchez, L. M., Mena, R., Guil-Luna, S., Mantrana, A., Peñarando, J., Toledano-Fonseca, M., Conde, F. De la Haba-Rodríguez, J. R., Aranda, E., & Rodríguez-Ariza, A. (2021). Nitric oxide-targeted therapy inhibits stemness and increases the efficacy of tamoxifen in estrogen receptor-positive breast cancer cells. *Laboratory Investigation, 101*(3), 292–303. https://doi.org/10.1038/s41374-020-00507-z

La Porta, C. A. M., & Zapperi, S. (2017). Complexity in cancer stem cells and tumor evolution: Toward precision medicine. *Seminars in Cancer Biology, 44*, 3–9. https://doi.org/10.1016/j.semcancer.2017.02.007

Lapidot, T., Sirard, C., Vormoor, J., Murdoch, B., Hoang, T., Caceres-Cortes, J., Minden, M., Paterson, B., Caligiuri, M. A., & Dick, J. E. (1994). A cell initiating human acute myeloid leukaemia after transplantation into SCID mice. *Nature, 367*(6464), 645–648. https://doi.org/10.1038/367645a0

Li, Y., & Laterra, J. (2012). Cancer stem cells: Distinct entities or dynamically regulated phenotypes? *Cancer Research, 72*(3), 576–580. https://doi.org/10.1158/0008-5472.can-11-3070

Mani, S. A., Guo, W., Liao, M. J., Eaton, E. N., Ayyanan, A., Zhou, A. Y., Brooks, M., Reinhard, F., Zhang, C. C., Shipitsin, M., Campbell, L. L., Polyak, K., Brisken, C., Yang, J., & Weinberg, R. A. (2008). The epithelial-mesenchymal transition generates cells with properties of stem cells. *Cell, 133*(4), 704–715. https://doi.org/10.1016/j.cell.2008.03.027

Mena-Osuna, R., Mantrana, Guil-Luna, S., Sánchez-Montero, M. T., Navarrete-Sirvent, C., Morales-Ruiz, T., Rivas-Crespo, Toledano-Fonseca, M., García-Ortíz, M. V., García-Jurado, G., Gómez-España, M. A., González-Fernández, R., Villar, C., Medina-Fernández, F. J., Villalba, J. M., Aranda, E., & Rodriguez-Ariza, A. (2023). Metabolic shift underlies tumor progression and immune evasion in S-nitrosoglutathione reductase-deficient cancer. *The Journal of Pathology, 260*(3), 261–275. https://doi.org/10.1002/path.6080

Miranda, K. M., Ridnour, L. A., McGinity, C. L., Bhattacharyya, D., & Wink, D. A. (2021). Nitric oxide and cancer: When to give and when to take away? *Inorganic Chemistry, 60*(21), 15941–15947. https://doi.org/10.1021/acs.inorgchem.1c02434

Morbidelli, L., Donnini, S., & Ziche, M. (2004). Role of nitric oxide in tumor angiogenesis. *Cancer Treatment and Research, 117*, 155–167. https://doi.org/10.1007/978-1-4419-8871-3_11

Nassar, D., & Blanpain, C. (2016). Cancer stem cells: Basic concepts and therapeutic implications. *Annual Review of Pathology: Mechanisms of Disease, 11*(1), 47–76. https://doi.org/10.1146/annurev-pathol-012615-044438

Peñarando, J., López-Sánchez, L. M., Mena, R., Guil-Luna, S., Conde, F., Hernández, V., Toledano, M., Gudiño, V., Raponi, M., Billard, C., Villar, C., Díaz, C., Gómez-Barbadillo, J., Haba-Rodríguez, J., Myant, K., & Aranda, E. (2018). Rodríguez-Ariza A. A role for endothelial nitric oxide synthase in intestinal stem cell proliferation and mesenchymal colorectal cancer. *BMC Biology, 16*.

PeÑarando, J., Aranda, E., & RodrÍguez-Ariza, A. (2019). Immunomodulatory roles of nitric oxide in cancer: Tumor microenvironment says "NO" to antitumor immune response. *Translational Research, 210*, 99–108. https://doi.org/10.1016/j.trsl.2019.03.003

Prasad, S., Ramachandran, S., Gupta, N., Kaushik, I., & Srivastava, S. K. (2020). Cancer cells stemness: A doorstep to targeted therapy. *Biochimica et Biophysica Acta, Molecular Basis of Disease, 866*(4), 165424. https://doi.org/10.1016/j.bbadis.2019.02.019

Reddy, T. P., Glynn, S. A., Billiar, T. R., Wink, D. A., & Chang, J. C. (2023). Targeting nitric oxide: Say NO to metastasis. *Clinical Cancer Research, 29*(10), 1855–1868. https://doi.org/10.1158/1078-0432.CCR-22-2791

Rizza, S., Di Leo, L., Mandatori, S., De Zio, D., & Filomeni, G. (2020). Mitophagy contributes to alpha-tocopheryl succinate toxicity in GSNOR-deficient hepatocellular carcinoma. *Biochemical Pharmacology, 176*. https://doi.org/10.1016/j.bcp.2020.113885

Rizza, S., & Filomeni, G. (2017). Chronicles of a reductase: Biochemistry, genetics and physio-pathological role of GSNOR. *Free Radical Biology and Medicine, 110*, 19–30. https://doi.org/10.1016/j.freeradbiomed.2017.05.014

Rizza, S., & Filomeni, G. (2016). Tumor suppressor roles of the denitrosylase GSNOR. *Critical Reviews in Oncogenesis, 21*(5–6), 433–445. https://doi.org/10.1615/critrevoncog.2017021074

Rizza, S., Di Leo, L., Pecorari, C., Giglio, P., Faienza, F., Montagna, C., Maiani, E., Puglia, M., Bosisio, F. M., Petersen, T. S., Lin, L., Rissler, V., Viloria, J. S., Luo, Y., Papaleo, E., De Zio, D., Blagoev, B., & Filomeni, G. (2023). GSNOR deficiency promotes tumor growth via FAK1 S-nitrosylation. *Cell Reports, 42*(1). https://doi.org/10.1016/j.celrep.2023.111997

Sinha, B. K. (2023). Can nitric oxide-based therapy Be improved for the treatment of cancers? A perspective. *International Journal of Molecular Sciences, 24*(17). https://doi.org/10.3390/ijms241713611

Stomberski, C. T., Hess, D. T., & Stamler, J. S. (2019). Protein S-nitrosylation: Determinants of specificity and enzymatic regulation of S-Nitrosothiol-Based signaling. *Antioxidants and Redox Signaling, 30*(10), 1331–1351. https://doi.org/10.1089/ars.2017.7403

van der Heijden, M., & Vermeulen, L. (2019). Stem cells in homeostasis and cancer of the gut. *Molecular Cancer, 18*(1). https://doi.org/10.1186/s12943-019-0962-x

Wang, R., Li, Y., Tsung, A., Huang, H., Du, Q., Yang, M., Deng, M., Xiong, S., Wang, X., Zhang, L., Geller, D. A., Cheng, B., & Billiar, T. R. (2018). INOS promotes CD24+CD133+ liver cancer stem cell phenotype through a TACE/ADAM17-dependent Notch signaling pathway. *Proceedings of the National Academy of Sciences of the United States of America, 115*(43), E10127–E10136. https://doi.org/10.1073/pnas.1722100115

Wei, W., Li, B., Hanes, M. A., Kakar, S., Chen, X., & Liu, L. (2010). S-nitrosylation from GSNOR deficiency impairs DNA repair and promotes hepatocarcinogenesis. *Science Translational Medicine, 2*(19), 19. https://doi.org/10.1126/scitranslmed.3000328. ra13.

CHAPTER

11

Impact of nitric oxide on hemeprotein maturation and its relevance to cancer and pulmonary diseases

Chaitali Ghosh[1], Mamta Sumi[2] and Arnab Ghosh[2]

[1]**Department of Biomedical Engineering, Lerner Research Institute, The Cleveland Clinic, Cleveland, OH, United States;** [2]**Department of Inflammation and Immunity, Lerner Research Institute, The Cleveland Clinic, Cleveland, OH, United States**

Abbreviations

AHSP alpha hemoglobin stabilizing protein
ANP atrial natriuretic peptide
ATP adenosine triphosphate
BAL broncho alveolar lavage
cAMP cyclic adenosine triphosphate
cGMP cyclic guanosine monophosphate
COPD chronic obstructive pulmonary disease
CTCs circulating tumor cells
EMT epithelial to mesenchymal transition
EVLP ex vivo lung perfusion
FLVCR1b feline leukemia virus subgroup c receptor GTN, nitroglycerin
Hb hemoglobin
HDM house dust mite
HO1 hemeoxygenase 1
IPF idiopathic pulmonary fibrosis
IRE iron response element
Mb myoglobin
MEJ myo-endothelial junction
MPO myeloperoxidase

Therapeutic Applications of Nitric Oxide in Cancer and Inflammatory Disorders
https://doi.org/10.1016/B978-0-443-22354-9.00011-5

NO nitric oxide
NOSs nitric oxide synthases
NPR natriuretic peptide receptor
NSAIDS, PAH pulmonary arterial hypertension
NSCLC nonsquamous cell lung cancer
PAS per-arnt-sim-type
PDE-i phosphodiesterase inhibitor
PRKG1/PKG protein kinase cGMP-dependent 1
ROS reactive oxygen species

Introduction

Ever since its benchmark discovery, nitric oxide (NO) has thus far been known to play important functional roles in a variety of physiological systems (Gantner et al., 2020). NO is a signal molecule and plays a critical role in the regulation of vascular tone, and within the vasculature, NO induces vasodilation (Chen et al., 2008; Furchgott & Zawadzki, 1980; Loscalzo & Jin, 2010; Palmer et al., 1987), inhibits platelet aggregation (Loscalzo, 2001; Loscalzo & Jin, 2010; Radomski et al., 1987; Samama et al., 1995; Simon et al., 1993), displays antiinflammatory properties (Ahluwalia et al., 2004), prevents neutrophil/platelet adhesion to endothelial cells (Sneddon & Vane, 1988), inhibits smooth muscle cell proliferation and migration (Napoli et al., 2013), and regulates apoptosis and endothelial cell barrier function (Gantner et al., 2020; Napoli et al., 2013). NO is produced by many cell types in the body and plays important roles in controlling the normal function of cells as well as in regulating key processes in the nervous and immune systems. In the immune system, NO generated by macrophages acts as an antimicrobial agent. High amount of NO produced by the macrophages is toxic to the bacteria or the pathogens, and it causes their destruction (Wink et al., 2011). However, too much NO has also been implicated in conditions where the immune system is overactive, e.g., in diseases such as arthritis and the so-called autoimmune diseases (Singh et al., 2000). Hence, the balance between physiological to pathophysiological levels of cellular NO can dictate many of the cell functions.

A previous study reported that NO could block heme insertion into inducible NO synthase (iNOS) (Albakri & Stuehr, 1996). In that study, iNOS protein expression was induced in the RAW 264.7 macrophage cell line by cytokine immune activation, and thus the effect involved NO that was generated by iNOS within the cells. After some of the early expressed iNOS protein had formed an active dimer, the NO it produced was able to progressively antagonize heme insertion into any subsequently generated iNOS protein, such that by 8 h postinduction, heme insertion into any new iNOS protein was essentially blocked and the iNOS accumulated as heme-free monomers (apo-iNOS) within the cells. A variety of experiments established that the NO did not affect heme availability within the cells but was rather acted to prevent heme insertion (Albakri & Stuehr, 1996). After this study, the possibility that NO has a broader role in regulating cellular heme insertion into hemeproteins is starting to get realized.

Heme proteins perform widespread and essential roles throughout biology, including electron transfer, redox transformations, small-molecule transport, metabolism, catabolism, and signal transduction (Milo et al., 2007; Paoli et al., 2002). To perform these roles, heme proteins have evolved into a number of structural classes. The most common group utilizes protoporphyrin IX bound within the protein. Members of this group include globins such as hemoglobin and myoglobin, heme–thiolate enzymes such as cytochrome P450s (CYP) and NO synthases (NOS), peroxidases, and per-arnt-sim-type (PAS) signaling proteins (Chan, 2001; Davies et al., 2008; Ghosh, Hossain, et al., 2016; Poulos, 2005; Zhu & Silverman, 2008). The steps of heme biosynthesis and how heme functions in proteins are well studied (Heinemann et al., 2008; Ponka, 1999; Poulos, 2007). However, with the exception of cytochrome *c* biogenesis (Richard-Fogal et al., 2009) and some aspects of heme acquisition and catabolism (West & Oates, 2008; Yi et al., 2009), relatively little was known about how heme is transported within eukaryotic cells and becomes inserted into soluble protein targets or how these processes might be regulated. These steps are critical, given that free heme is bioactive and potentially cytotoxic (Tsiftsoglou et al., 2006) and is normally kept at low intracellular levels (Wijayanti et al., 2004). Many of our earlier and current studies helped unveil these vital questions as discussed in the following section.

Impact of NO on hemeprotein maturation

NO is produced by many cell types in the body and plays important roles in controlling the normal function of cells as well as in regulating key processes in the nervous and immune systems. In cells, NO can impact heme availability and heme insertion in at least three ways. First, NO is known to cause a prolonged and significant loss of intracellular iron in several cell types once iNOS is expressed (Kim et al., 1995; Weiss et al., 1994). Second, the NO-induced loss of intracellular iron inhibits heme biosynthesis via an iron-response element (IRE)–based signal cascade, which blocks transcription of the rate-limiting enzyme in heme biosynthesis (D-aminolevulinate synthase) (Klausner & Rouault, 1993). Third, NO is also known to directly inhibit the mitochondrial enzyme ferrochelatase, which is responsible for iron insertion into the newly synthesized porphyrin (Furukawa et al., 1995). Possible relationships also exist between NO synthesis, heme utilization, and conditions such as anemia. In sub-Saharan Africa, children with acute malaria show elevated NO levels and decreased Hb levels (Anstey et al., 1999), indicating a possible link between increased NO production and malarial anemia (Keller et al., 2004). Relationships also exist between increased NO and the anemia associated with many chronic inflammatory conditions (John et al., 2005). While several mechanisms are likely to participate in these forms of anemia, it is possible that a downregulation in heme biosynthesis and/or heme insertion brought on by increased NO production could be an important contributing component.

NO generated by NOS enzymes (Marletta, 1994; Murad, 2006; Stuehr, 1999; Stuehr & Haque, 2019) produced by noncanonical means (Kapil et al., 2020) is a central component of the NO-sGC-cGMP signal pathway and works by activating the soluble guanylyl cyclase (sGC), an obligate heterodimer of an α-subunit (sGCα) and a β-subunits (sGCβ) (Bryan et al., 2009; Derbyshire & Marletta, 2009, 2012). Although much attention has been directed in the past two decades to study the physiological effects of NO-cGMP signaling (Sandner et al., 2017, 2019; Stasch et al., 2002; Stasch et al., 2011; Tsai & Kass, 2009), little is known about its hematologic effects and how this pathway contributes to erythropoiesis (Moras et al., 2017). The only known hematologic effects of NO-cGMP signaling reveal that it is capable of stimulating erythropoiesis at the transcriptional level both in vivo and in vitro (Ikuta et al., 2016). NO increases cGMP via sGC activation, which increases Hbγ (fetal) levels in human erythroid cells (Tupta et al., 2022), whereas inhibition of sGC prevents NO-induced increase in Hbγ gene expression (Cokic et al., 2003). Our studies suggest that even Hbβ (adult), requires sGC as it does not express in cell systems (COS-7), which lack sGC (Ghosh et al., 2019), suggesting that adult Hb also requires sGC (Ghosh et al., 2022). While NO doses can increase Mb protein levels (Rayner et al., 2009), our recent study suggests that protein levels of Mb in COS-7 improve with transfected sGC (Ghosh et al., 2022). While our earlier studies had established a role for chaperon hsp90 to be involved in the heme-maturation for adult (β) and fetal (γ) hemoglobin (Hb) (Ghosh et al., 2018) and for myoglobin (Mb), which also required an active sGC (Ghosh et al., 2019), our recent findings suggest that cellular GAPDH is involved in heme-maturation of myoglobin and all three hemoglobins (Hb and Mb) Fig. 11.1 (Tupta et al., 2022). In these studies, we established that in all these globin maturation events, GAPDH acts as a heme chaperon, allocating/providing mitochondria generated heme to the apo-Hsp90/AHSP-$Hb_{\alpha/\beta/\gamma}$/Mb complexes, where these globin heme maturations are influenced by iron provision and magnitude of expression of GAPDH, D-aminolevulinic acid, and FLVCR1b (Chiabrando et al., 2012; Tupta et al., 2022). These findings imply that hsp90 and GAPDH play key roles in globin maturations and these studies link the activation of sGC to maturation of the globins (Mb/Hb), thereby contributing to the formation of a novel NO-sGC-globin axes (Fig. 11.2).

Hsp90 complex, cellular GAPDH, and an active sGC enable hemoglobin and myoglobin to mature. These globins can then function to sustain life.

NO-sGC-globin connection stimulates globins (Hb/Mb) to mature, and this enables certain cancer cells to thrive.

NO is an uncharged radical, and hence the cell redox milieu leads its biological functions (Thomas et al., 2003). While interactions of NO with marquee hemeproteins such as soluble guanylate cyclase are known to cause activation (Ghosh & Stuehr, 2012), those with hemoglobin are known to cause scavenging of the NO (Donadee et al., 2011; Owusu et al., 2012; Premont et al., 2020). While there are several other facets of NO hemeprotein relationship, which are implicitly stated in the published literature, our earlier and current studies found that low NO doses trigger a rapid heme-insertion into immature heme-free apo-sGCβ1, resulting in mature sGC heterodimer (Dai et al., 2022; Ghosh et al., 2014, 2016), and more examples of this low NO effect were clearly lacking. Whether this low NO phenomena is part of the NO-sGC signal pathway activation and whether other hemeproteins such as globins also undergo such NO driven heme-insertion/maturation were largely unidentified. There is evidence that shows that high NO doses can block heme-insertion into hemeproteins (Waheed et al., 2010), and there is implicit literature stating that proteins such as MPO can also have their heme-dependent activities regulated by low and high NO levels, suggesting that NO may have a dual effect on hemeprotein maturation (Abu-Soud & Hazen, 2000). However, a comprehensive study demonstrating the impact of the dual role of NO on hemeprotein maturation, including applications to highlight the role of low NO as a therapy, was

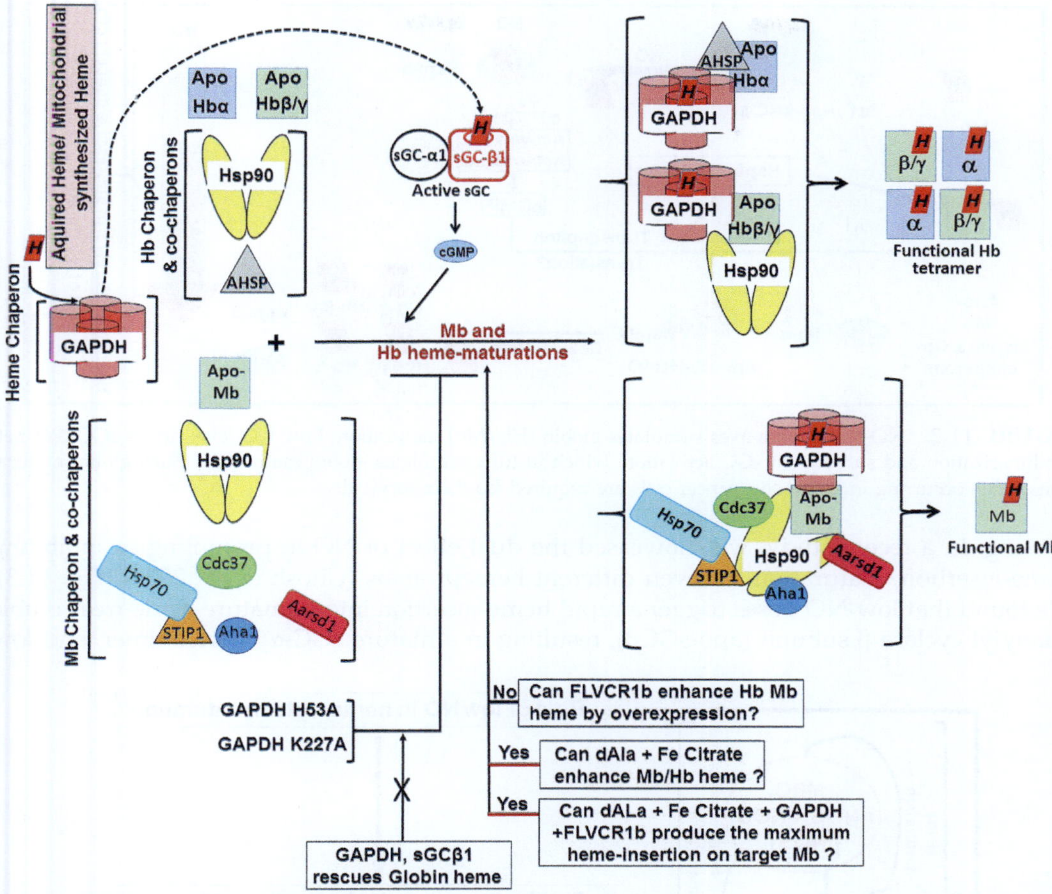

FIGURE 11.1 Hsp90 chaperon complex, an active sGC, and GAPDH are involved in globin heme-maturations. Cellular GAPDH acting as heme chaperon is involved in heme-maturation of Mb and alpha, beta (adult), and gamma (fetal) Hb (α, β, γ), along with specific chaperons/cochaperons for the globins as indicated. The apo-Hbα/apo-Hbβ/γ or the apo-Mb proteins bind to specific chaperons/cochaperons, with GAPDH and hsp90 being common chaperons to both apo-Hb or apo-Mb. The Mb and Hb heme-maturation processes require an active sGCα1β1 heterodimer making cGMP, and this induction may be transcriptional. GAPDH itself delivers heme to sGCβ1 and can regulate these processes. GAPDH binds heme and may transfer its bound heme directly to the globins or transfer its heme to hsp90/AHSP to enable heme-insertion into the corresponding client globins, followed by chaperon-assisted folding of these globins, resulting in heme-containing Hb dimers, tetramers, and Mb monomers. GAPDH mutants H53A and K227A downregulate or inhibit Hb/Mb heme-maturation, and this inhibition can be completely rescued by over-expressing GAPDH or partly by sGCβ1 overexpression. While addition of D-Ala and ferric citrate can enhance heme [1]insertion into Hb and Mb, a combination of overexpressed GAPDH, FLVCR1b, and supplementation of D-ALa plus ferric citrate gives the maximum heme-insertion into target Mb, suggesting that these factors act in a concerted way to enhance heme-insertion into Hb or Mb and are important for cellular heme homeostasis.

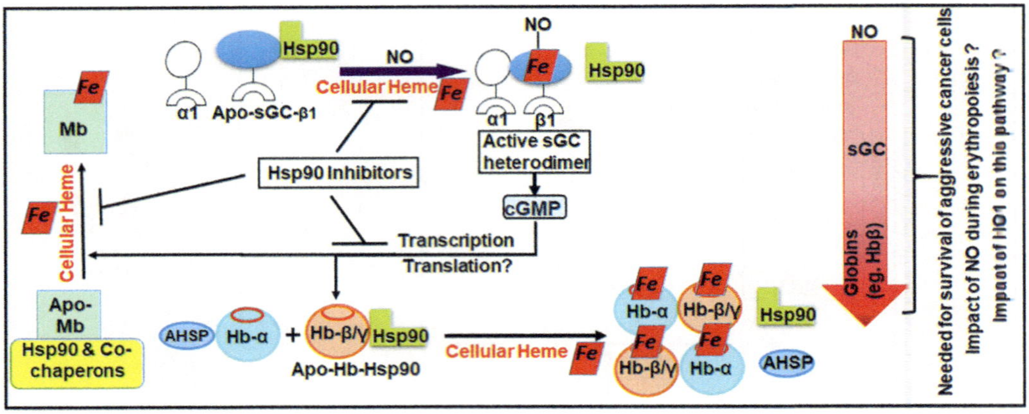

FIGURE 11.2 NO-sGC-globin axes stimulates globin (Hb/Mb) maturation. Low NO level drives sGCα1β1 heterodimerization and subsequent sGC activation, which in turn stimulates globin maturation. Such globin maturations when occurring in aggressive cancer cells are required for their survival.

missing. In a recent study, we showcased the dual effect of NO in promoting or inhibiting heme-insertion/maturation in seven different hemeproteins (Ghosh et al., 2022) (Fig. 11.3). We found that low NO doses trigger a rapid heme-insertion into immature heme-free soluble guanylyl cyclase β subunit (apo-sGCβ), resulting in a mature sGC-αβ heterodimer, and low

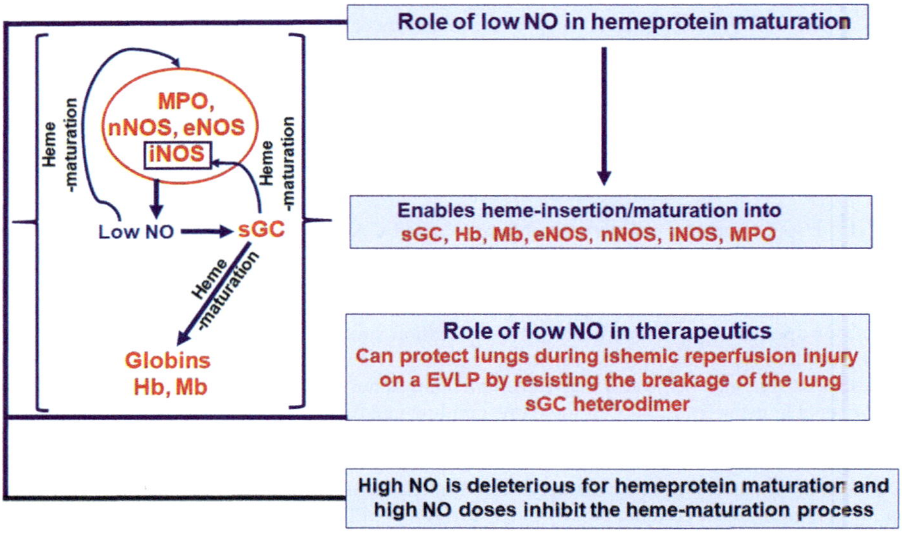

FIGURE 11.3 Impact of NO on hemeprotein maturation and its therapeutic role. While high NO levels inhibit hemeprotein maturation, low NO levels enable the heme-maturation of NOSs, sGC, Hb, Mb, and MPO. Among these hemeproteins, the components of the NO-sGC-globin axis, namely NOSs, sGC, Hb, and Mb have their heme-maturation driven by low NO levels, and an active sGC is also needed for Hb, Mb, and iNOS heme-maturation. NO at ppm levels also manifests itself as a therapy and is found to be protective against ischemia reperfusion lung injury on an ex vivo lung perfusion (EVLP) system by resisting the breakage of the lung sGC heterodimer.

NO can also trigger heme-maturation of hemoglobin and myoglobin. This low NO phenomena was not only limited to sGC and the globins but also found to occur in all three nitric oxide synthases (iNOS, nNOS, and eNOS) and myeloperoxidase (MPO). Interestingly, high NO doses were inhibitory to heme-insertion for these hemeproteins, suggesting that NO has a dose-dependent dual effect as it can act both/ways to induce or inhibit heme-maturation. Corroborating the impact of endogenous NO on sGC and globin maturations (Hb and Mb) using tissues from NOS double (n/eNOS$^{-/-}$) and triple (n/i/eNOS$^{-/-}$) knockout mice (Tsutsui et al., 2015), we found the effects of low NO to be significantly diminished in these NOS knock out mice where lung sGC was found be heme-free and the myoglobin or hemoglobin from the heart/lungs was found be low in heme, suggesting that loss of endogenous NO globally impacts the whole animal and that this impact of low NO is both essential and physiologically relevant for hemeprotein maturation (Ghosh et al., 2022). While low NO stimulated heme-insertion of globins required the presence of the NO-sGC-cGMP signal pathway, iNOS heme-maturation like the globins also required the presence of an active sGC (Fig. 11.3). Effects of low NO were also found to be protective against ischemia reperfusion injury on an ex vivo lung perfusion (EVLP) (Niikawa et al., 2019; Okamoto et al., 2019) system prior to lung transplant, which further suggests that low NO levels are also therapeutic.

Low levels of NO enable hemeprotein maturation and also have a therapeutic role.

Impact of NO on GAPDH regulation and HO1 induction may also hold significance in globin maturations. Interestingly, it is known that GAPDH mRNA is increased by NO, and as GAPDH is the heme allocator/provider (Dai et al., 2020; Tupta et al., 2022) for iNOS or sGC (Chakravarti et al., 2010; Dai et al., 2020), this may further aid sGC/iNOS heme-insertion/maturation at specific NO doses (Bereta & Bereta, 1995; Dai et al., 2020). This also suggests that an NO-GAPDH synergy may be vital for hemeprotein maturation for NOS, sGC, and the globins. NO produced endogenously (by NOSs) or given exogenously is known to induce heme-oxygenase-1 (HO1) (Durante et al., 1997), which in turn stimulates the production of CO from breakdown of cellular heme by HO1 (Fraser et al., 2011; Otterbein et al., 2016). This CO is also known to activate sGC in vascular smooth muscle cells (Nitti et al., 2020; Siow et al., 1999), thereby creating an additional NO—HO1 synergy, which generates cGMP for downstream signaling and may also aid in globin synthesis/maturation. NO is a diffusible-free radical generated primarily by NO synthases (NOSs). Endothelial cells as well as macrophages, components of hematopoietic microenvironment are potent NO producers, playing an active role in the modulation of human hematopoietic cell growth and differentiation (Čokić & Schechter, 2008). *Given our current findings of low NO doses triggering cellular heme-insertion into apo-globins or apo-sGC, it is imperative to determine what implications this would hold for RBCs production or erythropoiesis.*

NO in cancer

NO is a vital player in the progress and suppression and tumorigenesis depending on the source and concentration of NO (Mocellin et al., 2007; Thomas et al., 2008; Vahora et al., 2016). Although NO participates in normal signaling via NO-sGC-cGMP pathway (e.g., vasodilatation and neurotransmission), NO is also a cytotoxic or apoptotic molecule when produced at high concentrations by inducible nitric oxide synthase (iNOS or NOS-2) (Bian et al., 2012). Molecular mechanisms exist by which NO induces DNA damage (Nguyen et al., 1992; Thomas et al., 1998), influences DNA damage repair, and in turn modulates cell cycle arrest (Hoeijmakers, 2009; Napoli et al., 2013; Visconti et al., 2016). In other instances, NO can cause delay in cell cycle progression, allowing for aberrant DNA repair that promotes the accumulation of mutations and tumor heterogeneity (Hickok & Thomas, 2010; Napoli et al., 2013). The role of the NO-sGC-cGMP signaling in tumor progression has been extensively investigated over the past three decades, and although a general consensus regarding the role of this signaling axis in neoplastic transformation has not been reached, vital aspects of the tumor morphology may dictate differential targeting of this pathway as cGMP-dependent (NO/sGC/cGMP pathway) components may vary among different tissues (Bian et al., 2012). As depicted in Fig. 11.4, solid tumors contain two

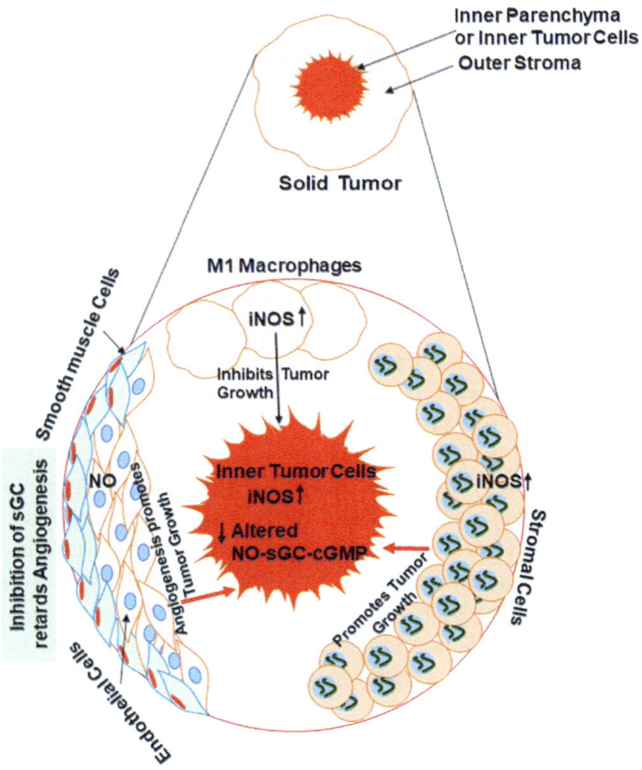

FIGURE 11.4 Representation of the NO-sGC-cGMP signal pathway within a malignant tumor. sGC activation may vary within a malignant tumor, in its outer or inner core. The NO-sGC-cGMP pathway is active in its outer core and can promote angiogenesis, which nourishes the inner tumor core. Targeting the NO-sGC-cGMP pathway for inhibition in the outer core can offer useful strategies to inhibit tumor growth. *Adapted and modified from Bian, K., Ghassemi, F., Sotolongo, A., Siu, A., Shauger, L., Kots, A., & Murad, F. (2012). NOS-2 signaling and cancer therapy.* IUBMB Life, 64*(8)*, 1521–6543, 676–683. *Wiley https://doi.org/10.1002/iub.1057.*

compartments, the inner core parenchyma, which comprises of neoplastic cells and the outer stroma (containing nonmalignant support tissue including connective tissue, blood vessels, endothelial cells, and macrophages) with divergent NO biology. NOS-2 expression resulting in high NO production in the parenchyma may contribute to the formation of an inflammatory cancer microenvironment. This high NO may result in a broken sGC heterodimer as our studies suggest (Ghosh et al., 2014, 2016), causing an altered sGC, PKG signaling, which influences differentiation or proliferation of cells in the tumor parenchyma and normal NO-sGC signaling needs to be restored to cause this inhibition of cell proliferation. In contrast, the outer stroma in the tumor microenvironment generates low to moderate levels of NO derived from the endothelial cells, and this can activate angiogenesis and epithelial-to-mesenchymal transition, promoting an aggressive phenotype. Here, in the stormal cells, the NO-sGC pathway may support angiogenesis (Morbidelli et al., 2010; Papapetropoulos et al., 1997; Pyriochou et al., 2006), and inhibition of sGC may be essential to abrogate angiogenesis as this fuels the inner parenchyma Fig. 11.4 (Bian et al., 2012).

Outer core of the malignant tumor can be targeted by drugs to inhibit the NO-sGC-cGMP pathway, thus inhibiting tumor growth.

The role of cyclic nucleotides in the regulation of cell proliferation and tumor growth was witnessed in the 1960s (Ryan & Heidrick, 1968). cAMP and cGMP were believed to act as biological antagonists in the regulation of cell growth at the beginning (Goldberg et al., 1975), and elevated cGMP levels have been associated in certain human tumor tissues (DeRubertis et al., 1976; Takemoto et al., 1982). Inhibition of proliferation by cGMP was first observed in vascular smooth muscle cells (Garg & Hassid, 1989) and then found in endothelial cells (Yang et al., 1994) and the cells from nervous system, e.g., cerebellar glial (Garg et al., 1992) or neuroblastoma cells (Murillo-Carretero et al., 2002). Use of 8-bromo-cyclic guanosine monophosphate (8-bromo-cGMP), a cell membrane–permeable cGMP analog resistant to PDE hydrolysis, significantly reduced proliferation of glioma cells in a concentration-dependent manner (Zhu et al., 2011), and elevating endogenous cGMP through NPR agonists (ANP) also abrogated tumor proliferation. Inhibiting PDE activity is another approach to elevate intracellular cGMP. PDE-5 inhibitors (PDE-5i) blunt the function of this critical recycling enzyme and block the breakdown of cGMP to its linear GMP, which enhances the NO/cGMP signaling pathway (DeRubertis et al., 1976). U87 cells treated with either a nonspecific PDE inhibitor IBMX or with a specific PDE5 inhibitor, zaprinast either alone or in combination with an NPR agonist resulted in a significant inhibition of cell proliferation (Zhu et al., 2011). Thus, glioma cell proliferation was significantly inhibited upon pharmacological restoration of sGC function through increasing intracellular cGMP levels. Studies also established an altered expression of sGC in malignant brain tumors and suggest therapeutic approaches by pharmacologically or genetically manipulating sGC/cGMP signaling (Zhu et al., 2011). Orthotopic implantation of glioma cells transfected with an active mutant form of sGC (sGC$\alpha1\beta1^{cys-105}$) in athymic mice increased the survival time by fourfold over the control (Zhu et al., 2011). Histological analysis of xenografts overexpressing sGC$\alpha1\beta1^{cys-105}$ revealed changes in cellular architecture, which resemble the morphology of normal cells. In addition, a decrease in angiogenesis also contributed to glioma inhibition by sGC/cGMP therapy, which may suggest that an active sGC$\alpha1\beta1^{cys-105}$, however, maintains high cGMP levels but may downregulate angiogenesis unlike WT sGC (Papapetropoulos et al., 1997; Pyriochou et al., 2006). Thus, these studies promoted a concept that showed that suppressed expression/function of sGC in the NO/cGMP pathway may be associated with an aggressive course of glioma, and significant inhibition of cell proliferation can be achieved by restoring normal function of the NO-sGC-cGMP axis.

Relative to normal tissues, the expression of NO-cGMP signaling components varies significantly in different human cancers (Bian et al., 2012; Mujoo et al., 2010). While NO synthase (NOS)-1 is downregulated in brain, kidney, and lung cancer, NOS-2 is upregulated in breast cancer but decreased in lung cancer. NOS-3 is higher in malignant intestine tumour, but inhibited in both brain and breast cancers (Bian et al., 2012). Even though sGC is an obligate heterodimer for function, sGC subunits are differentially expressed in cancers with sGCα1 being lower in brain, lung, and breast cancers, but higher in prostate and ovarian cancers while sGCβ1 is attenuated in brain, lung, liver, breast cancers, but elevated in lymphoma, ovarian, head, and neck cancers, which may suggest that sGC stimulators/activators as drugs can have a variable impact (Bian & Murad, 2003; Bian et al., 2012). The data regarding membrane-bound guanylyl cyclases shows that NPRs are also variable in the various cancers (Bian & Murad, 2003; Bian et al., 2012). The key molecules downstream of cGMP, PRKG1 are overexpressed in brain cancer and decreased in lung, breast, and ovarian cancers. PRKG2 is only changed in myeloid leukemia. The major cGMP-specific phosphodiesterase, PDE-5, shows broad depression in brain, lung, prostate, breast, malignant tumor of intestine, and myeloid leukaemia, suggesting that it is a good druggable target by PDE-5 inhibitors. While majority of these expression data are derived from mRNAs, actual protein data is a key requirement for these studies before a select group of patients need to follow a specific drug regimen. There are four major classes of NO-based anticancer therapies: NO donors, phosphodiesterase inhibitors (PDE-i), soluble guanylyl cyclase (sGC) activators, and immunomodulators. Of these, NO donors are well studied and are the most promising. NO donors increase NO levels or its NO isoforms (NO^-/NO^+) without the need for endogenous production (Toda et al., 2005). NO donors impact different mechanisms through its rate of NO release (slow or high), which is critical, and this can drastically influence the end result (Alimoradi et al., 2019; Megson & Webb, 2002) as our current studies suggest (Ghosh et al., 2022) and hence the use of NO as an anticancer therapy should be looked at with cautious optimism. Modulation of NO concentration has led to a number of proposed clinical uses for NO donors even outside of cancer (Grosse et al., 2020; Snyder et al., 2002). The vasculature is sensitive to the effect of NO, as the systemic veno- and vasodilator effects are useful for treating angina, acute coronary syndromes, and other cardiovascular diseases (Iachini Bellisarii et al., 2012), and in these cases, it is critical to maintain a low effective concentration of NO (pico to low micromolar range) such that the labile sGC$\alpha\beta$ heterodimer is high (Ghosh et al., 2014, 2022). In cancer therapy, NO donors are particularly helpful as chemo- and radiotherapeutic agents (Huang et al., 2017). Advanced malignancies are often characterized by incomplete vascularization, which induces localized hypoxia, which results in the induction of hypoxia-inducible factor1α (HIF-1α), priming cancer cells for survival against a variety of cellular death mechanisms induced by radio- or chemotherapy including autophagy, apoptosis, and DNA damage (Huang et al., 2017; Jing et al., 2019). NO donors attempt to reverse this effect by increasing tumor perfusion, enhancing the effect of antitumor therapy (Huang et al., 2017), and here again, NO concentrations can become the deciding factor. Multiple clinical studies have demonstrated the efficacy of nitroglycerin (GTN), for treatment of cancer. For example, in one phase II clinical trial, transdermal GTN improved outcomes for patients with advanced nonsquamous cell lung cancer (NSCLC) (Yasuda et al., 2006). Not only has GTN shown efficacy in treating, but GTN has demonstrated clinical effectiveness in liver, colorectal, and prostate cancer as well (Liu et al., 2012; Reinmuth et al., 2014). The NO release rate of GTN in these studies can again be a critical point for its clinical success. Elevated PDE-5 levels have been documented in various types of human carcinomas including prostate, pancreatic, lung, colon adenocarcinoma, breast, and bladder squamous carcinoma (Das et al., 2015; Peng et al., 2018). PDE-5i function similar to NO donors by boosting the impact of NO on tissue; however, they depend on endogenous sources of NO to maintain their effect rather than the exogenous NO from an NO donor (Chuang et al., 1998). PDE-5i are commonly used to treat erectile dysfunction (ED); however, a wide range of PDE-i have shown anticancer activity, with many tested in clinical trials (Peng et al., 2018). The primary differences between PDEs and the corresponding inhibitors that determine their functional significance are their different tissue bed distributions in addition to different regulatory feedback mechanisms and affinities for cGMP/cAMP (Peng et al., 2018). For example, thymoquinone, a natural herb with PDE-1i activity, has shown efficacy at inhibiting the growth of acute lymphoblastic lymphoma, cervical, and malignant central nervous system tumor cells, and an active clinical trial investigating the efficacy of thymoquinone to treat premalignant leukoplakia has recently been completed (CT03208790) (Farkhondeh et al., 2017; Gali-Muhtasib et al., 2006). Numerous other PDE-i have also shown similar clinical effectiveness in preclinical studies (Peng et al., 2018). Interestingly, one report also suggests that PDE-5i can prevent the progression and development of prostate cancer, although other studies have shown zero and even contradictory results (Chavez et al., 2013; Peak et al., 2016). Moreover, PDE-i may be limited by dose toxicity and systemic side effects unrelated to the primary tumor site. Furthermore, the effectiveness of PDE-i as chemotherapeutics may be linked to their ability to enhance chemotherapy. Similar to PROLI/NO, PDE-5i also demonstrated that they were able to increase the transport of doxorubicin across the blood–brain barrier in a rat brain tumor model, increasing the effect of chemotherapy (Black et al., 2008). In another study, the addition of sildenafil, a PDE-5i, with or without roflumilast, a PDE-4i, and theophylline, a methylxanthine, to lung cancer cell lines showed increased apoptosis and growth inhibition when given alongside a platinum chemotherapeutic (Domvri et al., 2017). Although numerous preclinical studies outline the potential roles for PDE-i in anticancer therapy, clinical trials have only begun to show evidence of efficacy (Peng et al., 2018; Weed et al., 2015). Large, multicenter studies are therefore needed before widespread clinical adoption of PDE-i is followed in cancer therapeutics. Activation of sGC is known to be impaired in a number of cancer cell lines including prostate, breast, and glioma, and restoration of sGC may decrease disease progression (Bian et al., 2012; Chen et al., 2002; Wen et al., 2015). sGC stimulators/activators, stimulate the sGC in a heme-dependent or independent fashion irrespective of endogenous NO (Sandner, 2018). Although current clinical uses of sGC activators are limited to pulmonary hypertension and heart failure, other promising results exist for a host of other diseases including chronic kidney disease, systemic sclerosis, COPD, and even cancer (Lee et al., 2017; Paul et al., 2018; Sandner, 2018; Tuttle et al., 2017). Efficacy of sGC activators in cancer is not well studied, although two preclinical studies suggest that sGC can mitigate platinum chemotherapeutic resistance in oropharyngeal squamous cell carcinoma (Lee et al., 2017; Tuttle et al., 2017). Nevertheless, some sGC activators were noted to be metabolites of procarcinogenic organic compounds, which could also be partly responsible for carcinogenesis (DeRubertis & Craven, 1977). Although of potential benefit for some cancers, the contradictory results of sGC activators provide further evidence of the dual-nature of NO in cancer pathogenesis. More preclinical studies are thus needed to demonstrate the efficacy of sGC activator compounds in other cancers or to make conclusions on their future potential as therapeutics.

Significance of the NO-sGC-cGMP pathway in globin maturation and relevance to cancer progression

Implications of an active NO-sGC-Globin axes are an explored area in cancer research and its related involvement in the genesis of circulating tumor cells (CTCs), which are critical for cancer cell metastasis. There are two possible roles of NO/cGMP signaling in malignant tumors. It is known that NOS-2 expression and NO overproduction may contribute to the formation of an inflammatory cancer microenvironment, which derives angiogenesis, as in case of gliomas (Fukumura et al., 2006; Bian et al., 2012). Second, sGC/cGMP signaling may influence proliferation and/or differentiation of the tumor cells. Inhibiting NOS-2 overexpression and the tumor inflammatory microenvironment, together with normalizing sGC/cGMP signaling, may be one of the favorable ways to reduce malignant tumors (Fukumura et al., 2006; Ka Bian et al., 2012). Cancer metastasis is the leading cause of cancer-related death. CTCs are cancerous cells that are shed from either primary or metastatic tumors during an intermediate stage of metastasis into the bloodstream. Over the past few years, there has been quite a bit of research in this area to the understanding that CTCs are shed from tumors and circulate in the bloodstream (Castro-Giner & Aceto, 2020; Zhong et al., 2020). CTCs flow across the bloodstream avoiding the immune system by a process called epithelial to mesenchymal transition (EMT), so that they do not get consumed by phagocytes and then can metastasize into another part of the body by extraversion (leaving the bloodstream) (Castro-Giner & Aceto, 2020). This process can occur by individual CTCs or CTC clusters. While it is becoming increasingly clear that CTCs are heterogenous at multiple levels and that only a fraction of them are capable of initiating metastasis such that the tumors spread to other parts of the body (Castro-Giner & Aceto, 2020), it also appears that CTCs adopt multiple ways to enhance their metastatic potential (Castro-Giner & Aceto, 2020). While some studies state the role of a lung derived Hbβ to be antimetastatic for neuroblastoma cells (Maman et al., 2017), other studies in CTCs suggest evidence which is contrary (Ponzetti et al., 2017; Zheng et al., 2017). For example, it was found that asymmetric Hbβ expression in a variety of tumor cells and in the CTCs increased their metastatic potential, possibly by enhancing tumor cell survival during blood-borne dissemination (Zheng et al., 2017). Here single-cell RNA-Seq profiles of CTCs from breast, prostate, and lung cancers revealed a consistent induction of Hbβ (*HBB*), but not Hbα (*HBA*). The tumor-specific origin of Hbβ was confirmed by sequence polymorphisms within human xenograft-derived CTCs in mouse models. It was determined that increased intracellular reactive oxygen species (ROS) in cultured breast CTCs triggered Hbβ induction, via activation of transcriptional regulator *KLF4*. Depletion of Hbβ in CTC-derived cultures had no significant effect on primary tumor growth, but it greatly increased apoptosis following ROS exposure and drastically reduced

CTC-derived lung metastases. These effects are reversed by the antioxidant *N*-acetyl cysteine, while overexpression of Hbβ was sufficient to suppress intracellular ROS within CTCs. These data suggest that Hbβ is essential for CTCs progression as it protects the CTCs against harmful effects of ROS and is induced by the ROS itself. This discovery also highlights the importance of the NO-sGC-cGMP axis where a globin is needed for cell survival in aggressive cancer/tumor cells. Since tumors have an inner parenchyma with dysregulated NO-sGC signaling (Fig. 11.4), such asymmetric Hbβ expression can be supported by the NO-sGC signal (cGMP) (Ghosh et al., 2022; Cokic et al., 2003) from the outer stromal cells along with ROS-mediated KLF4 activation (Zheng et al., 2017). Also even a dysregulated NO-sGC pathway as may exist in the inner parenchyma of the tumor (Fig. 11.4) can also support such Hbβ induction, as only a small amount of cGMP is essentially required for transcriptional activation of Hbβ. As there are multiple pathways of regulation to maintain the NO-sGC-globin axes, it offers encouraging check points where this pathway could be modulated by therapeutic intervention. Open-ended strategies of both inhibiting or reinforcing sGC activation should be tried to determine what factors can curtail expression of this asymmetric globin (Hbβ), which may inhibit the metastatic potential of these tumors. Interestingly our studies also revealed that hsp90 is needed for heme-maturation of Hbβ and not Hbα (Ghosh et al., 2018) (Fig. 11.1), and this would increase the selectivity of hsp90 inhibitors to block Hbβ heme-maturation (Ghosh et al., 2018). Thus, modulating the NO-sGC signaling in these CTCs or using hsp90 inhibitors to restrain Hbβ maturation can be a favorable way to stop the spread of malignant tumors.

NO in asthma and pulmonary diseases

Asthma is defined by airway inflammation and hyperresponsiveness and contributes to morbidity and mortality worldwide (Ghosh & Erzurum, 2011). The NO-sGC-cGMP signal pathway is a well-known dilation pathway in the vasculature; however, its role in lung bronchodilation was equivocal. In a study (Ghosh, Hossain, et al., 2016), we found that the NO-sGC-cGMP pathway plays a significant role in lung bronchodilation and that lung sGC becomes dysfunctional in asthma due to high levels of NO generated from iNOS induction during inflammation in the airway epithelium (Ghosh, Hossain, et al., 2016; Guo et al., 1997), which desensitizes the sGC resident in the airway smooth muscles below. Working with mouse asthma models, we found that high NO levels cause S-nitrosylation of sGCαβ, leading to generation of heme-free sGCβ with subsequent breakage in sGCαβ heterodimerization and association of heme-free sGCβ with hsp90 (Ghosh et al., 2014, 2016). This hsp90-bound sGCβ does not respond to its natural activator NO but can be activated by sGC agonists such as BAY 60-2770 (Pankey et al., 2011), which can activate the enzyme independent of heme and NO to produce bronchodilation, and such agonists are future drugs for asthma. The association of hsp90 with heme-free sGC seemed to stabilize the sGCβ subunit rather than priming it for degradation and makes this pathological sGC more drug receptive. This NO-mediated dysfunction of sGC in asthma also has a clear mechanistic trajectory. Our recent study (Sumi et al., 2023) found that lung Hbαβ present in the airway epithelium scavenges the excess NO generated in the lung epithelium, thus protecting the underlying smooth muscle sGC. In asthma, this lung Hb itself becomes heme-free (Sumi et al., 2023) and is unable to scavenge the generated airway NO, thus exposing the sGC and triggering its dysfunction in asthma.

In COPD patients, the magnitude of the NO signal is considerably less than in asthma; however importantly, the major causative agent is cigarette smoke, which induces lung inflammation and oxidative stress (Malerba et al., 2014). These again cause iNOS upregulation and subsequent NO generation, which makes sGC heme-free and BAY 58-2667/BAY 60-2270 responsive (Glynos et al., 2013; Malerba et al., 2014). To determine whether similar mechanisms of sGC dysfunction operate, we performed immunoprecipitations in the lung tissue supernatants from asthma, COPD, and PAH (pulmonary arterial hypertension) samples, which revealed that Hbα-β interactions were lost but were preserved in the control samples and the reciprocal Hbβ-hsp90 interactions were elevated in asthma, COPD, or PAH, suggesting that lung Hbαβ was low in heme in these pathologies and these were further confirmed by Olis Clarity spectral data and heme-stain (Sumi et al., 2023). In the pulmonary endothelium, Hbα is expressed by the endothelial cells and enriched at the MEJ (Straub et al., 2012). This Hbα scavenges the NO generated by its interaction with the endothelial eNOS to control the NO diffusion into the smooth muscle cells, thereby regulating the NO-sGC signaling. This scavenging is an impediment in diseases such as PAH where NO levels are very low (Alvarez et al., 2017; Klinger et al., 2013; Zhang et al., 2016) and sGC cannot be activated. However, disruption of the Hbα-eNOS interaction removes this inhibition and improves NO signaling and has been postulated as a potential therapy to improve NO levels in PAH (Alvarez et al., 2017). Here the heme on lung Hb and on sGC in PAH samples may be low since NO levels are very low in PAH due to endothelial dysfunction and scavenging of endothelial NO by Hbα resident in the MEJ (Alvarez et al., 2017; Klinger et al., 2013; Straub et al., 2012; Zhang et al., 2016), which may cause disrupted heme-maturation of these proteins (Ghosh et al., 2022). Thus, both excess NO and low NO can generate heme-free Hb or heme-free sGC, but the causative mechanisms might be similar for asthma or COPD and different for PAH. It is noteworthy here that expression of tissue Hb can have contrasting roles as Hbβ expression in tumors favors disease progression via CTCs or the function of Hbα in the pulmonary endothelium can be an impediment to vasodilation in PAH, while role of Hbαβ in the lungs is protective for sGC function.

In studies on asthma relating to airway epithelium and inflammation, chaperon hsp90 has also been directly implicated. Extracellular hsp90α (eHsp90α) is shown to mediate HDM-induced human bronchial epithelial dysfunction, suggesting that eHsp90α is a potential therapeutic target for the treatment of asthma (Dong et al., 2017). Another report (Pezzulo et al., 2019) suggests that hsp90 inhibition reverts IL13- and IL-17-induced goblet cell metaplasia in human airway epithelia. Hsp90 inhibitors have also been shown to reduce airway inflammation in mouse models of allergic asthma and another study showed that hsp90 inhibitors hampered airway relaxation (Intapad et al., 2012). Given our current understanding of iNOS heme-maturation, most of these processes of hsp90 inhibition may involve inhibition of iNOS heme-insertion, suppressing iNOS induction in the airway epithelium and reducing inflammation (Ghosh, Hossain, et al., 2016). Studies also suggest that hsp90 is overexpressed in the pulmonary arterioles of PAH patients (Wang et al., 2016) and in the idiopathic pulmonary fibrosis (IPF) lungs (Sibinska et al., 2017). Hence, small molecule inhibitors of hsp90, which target the hsp90 ATPase like 17-AAG and 17-DMAG, have been used to inhibit pulmonary vascular remodeling in PAH and lung fibrosis, respectively (Sibinska et al., 2017; Wang et al., 2016). More recently, sGC agonists (Sandner et al., 2017) are also being tried to treat age-related fibrosis. hsp90 and other chaperon machinery slows down with age (Verbeke et al., 2001), causing low NO levels, which may reduce sGC activation, thereby making such sGC agonists as the drugs of choice. Moreover, a slow hsp90 chaperon machinery may cause reduced heme-maturation causing a buildup of heme-free sGC, which can be activated by BAY 60-2770 like drugs, further encouraging the use of these agonists to treat age-related fibrosis. Hence, a clever use of sGC agonists with hsp90 inhibitors can be the therapeutics of choice in such lung diseases.

Conclusion and future directions

The discovery of multiple NO-mediated pathways within cancer has unveiled a number of novel NO-based therapies. Many of these novel therapies center around delivery of NO directly to the tumor environment, and such localized increases in NO may reverse chemotherapeutic and radiotherapeutic resistance mediated by HIF-1α, although preclinical trials have suggested the efficacy with other mechanisms as well. The current drug opinion on NO donors as anticancer drugs is limited by pharmacodynamics, pharmacokinetics, or systemic absorption and toxicity (Alimoradi et al., 2019; Pinto et al., 2019). To address these challenges, a number of NO-releasing compounds tagged to nonsteriodal antiinflammatory drugs (NSAIDs) given along with chemotherapeutics drugs such as doxorubicin (Alimoradi et al., 2019; Chegaev et al., 2011; Kashfi & Rigas, 2005), or to PDE-i bound to specific antibodies (immunotherapy) such as $CD24^+$ (a widely expressed hepatocellular marker) (Sun et al., 2019) have been developed. These compounds have shown efficacy in chemotherapy by increasing the cytotoxicity within the tumor (Alimoradi et al., 2019; Chegaev et al., 2011; Kashfi & Rigas, 2005) or by increasing the specificity of NO delivery in immunotherapies, thereby enhancing the pharmacodynamic properties of NO as an anticancer agent. Thus, the future of NO therapy may lie upon clinicians finding the right mechanism to deliver NO alongside chemo-, radio-, or immunotherapies.

Based on our studies of low NO-driven hemeprotein maturations, at this stage, we cannot rule out additional mechanisms where low NO levels can itself directly act on the apo-globins or apo-NOSs to directly cause heme downloading using the GAPDH-Hsp90 heme transfer cascade (Tupta et al., 2022). This way low NO can act at many levels to cause prolonged heme-maturation in select proteins. The impact of low NO can be visualized as being tumor-promoting (NO-sGC-hlobin axes); however, the applications of low NO can also be put to beneficial therapeutic use. For example, low NO treatment may improve human lung function on the EVLP prior to lung transplant as shown by our study (Ghosh et al., 2022). Moreover, our findings on low NO should encourage the use of sGC stimulators such as BAY 41-2272, which can produce increased cGMP following low NO treatment to cause an elevated sGC heterodimer. Moreover, since sGC maturation in fetal airway smooth muscle cells (human ASMCs) is essential for lung alveolar development in neonates (Britt et al., 2015), similar applications of low NO treatment with drugs, which synergize with NO, can be aimed to improve fetal sGC maturation in neonates, or low NO treatment of asthma HASMCs, where the lung sGC is dysfunctional (Ghosh et al., 2021), should also be pursued in future studies for better therapeutic intervention. Our present finding can thus amplify the therapeutic use of low NO levels and will help widen the scope of research in these directions.

On the functional significance of lung Hb, we need to establish the heme status of this lung Hb in a greater number of COPD, asthma, or severe asthma patient samples and compare it with age-, race-, and gender-matched healthy controls. These can be tested in the bronchial brushings collected from the BAL (bronchoalveolar lavage) fluid. Establishing the prevalence in the generation of heme-free Hb in human asthma or COPD from the Hb heme spectral data and then drawing correlations with sGC activation toward heme-independent sGC activator, BAY 60-2770 will be a great progressive step in this direction. This may in turn indicate the extent of disease progression whereby both occurrence of heme-free Hb and heme-free sGC may correlate linearly as suggested by our recent study (Sumi et al., 2023).

Acknowledgments

This work was supported by National Institute of Health Grants R56HL139564 and R01HL150049 (A.G.)

References

Abu-Soud, H. M., & Hazen, S. L. (2000). Nitric oxide modulates the catalytic activity of myeloperoxidase. *Journal of Biological Chemistry, 275*(8), 5425–5430. https://doi.org/10.1074/jbc.275.8.5425

Ahluwalia, A., Foster, P., Scotland, R. S., McLean, P. G., Mathur, A., Perretti, M., Moncada, S., & Hobbs, A. J. (2004). Antiinflammatory activity of soluble guanylate cyclase: cGMP-dependent down-regulation of P-selectin expression and leukocyte recruitment. *Proceedings of the National Academy of Sciences of the United States of America, 101*(5), 1386–1391. https://doi.org/10.1073/pnas.0304264101

Albakri, Q. A., & Stuehr, D. J. (1996). Intracellular assembly of inducible NO synthase is limited by nitric oxide-mediated changes in heme insertion and availability. *Journal of Biological Chemistry, 271*(10), 5414–5421. https://doi.org/10.1074/jbc.271.10.5414

Alimoradi, H., Greish, K., Gamble, A. B., & Giles, G. I. (2019). Controlled delivery of nitric oxide for cancer therapy. *Pharmaceutical Nanotechnology, 7*(4), 279–303. https://doi.org/10.2174/2211738507666190429111306

Alvarez, R. A., Miller, M. P., Hahn, S. A., Galley, J. C., Bauer, E., Bachman, T., Hu, J., Sembrat, J., Goncharov, D., Mora, A. L., Rojas, M., Goncharova, E., & Straub, A. C. (2017). Targeting pulmonary endothelial hemoglobin a improves nitric oxide signaling and reverses pulmonary artery endothelial dysfunction. *American Journal of Respiratory Cell and Molecular Biology, 57*(6), 733–744. https://doi.org/10.1165/rcmb.2016-0418OC

Anstey, N. M., Granger, D. L., Hassanali, M. Y., Mwaikambo, E. D., Duffy, P. E., & Weinberg, J. B. (1999). Nitric oxide, malaria, and anemia: Inverse relationship between nitric oxide production and hemoglobin concentration in asymptomatic, malaria- exposed children. *The American Journal of Tropical Medicine and Hygiene, 61*(2), 249–252. https://doi.org/10.4269/ajtmh.1999.61.249

Bereta, J., & Bereta, M. (1995). Stimulation of glyceraldehyde-3-phosphate dehydrogenase mRNA levels by endogenous nitric oxide in cytokine-activated endothelium. *Biochemical and Biophysical Research Communications, 217*(1), 363–369. https://doi.org/10.1006/bbrc.1995.2785

Bian, K., & Murad, F. (2003). Nitric oxide (NO)–biogeneration, regulation, and relevance to human diseases. *Frontiers in Bioscience, 8*, 264–278.

Bian, K., Ghassemi, F., Sotolongo, A., Siu, A., Shauger, L., Kots, A., & Murad, F. (2012). NOS-2 signaling and cancer therapy. *IUBMB Life, 64*(8), 676–683. https://doi.org/10.1002/iub.1057

Black, K. L., Yin, D., Ong, J. M., Hu, J., Konda, B. M., Wang, X., Ko, M. H. K., Bayan, J. A., Sacapano, M. R., Espinoza, A., Irvin, D. K., & Shu, Y. (2008). PDE5 inhibitors enhance tumor permeability and efficacy of chemotherapy in a rat brain tumor model. *Brain Research, 1230*, 290–302. https://doi.org/10.1016/j.brainres.2008.06.122

Britt, R. D., Thompson, M. A., Kuipers, I., Stewart, A., Vogel, E. R., Thu, J., Martin, R. J., Pabelick, C. M., & Prakash, Y. S. (2015). Soluble guanylate cyclase modulators blunt hyperoxia effects on calcium responses of developing human airway smooth muscle. *American Journal of Physiology - Lung Cellular and Molecular Physiology, 309*(6), L537–L542. https://doi.org/10.1152/ajplung.00232.2015

Bryan, N. S., Bian, K., & Murad, F. (2009). Discovery of the nitric oxide signaling pathway and targets for drug development. *Frontiers in Bioscience, 14*(1), 1–18. https://doi.org/10.2741/3228

Čokić, V. P., & Schechter, A. N. (2008). Effects of nitric oxide on red blood cell development and phenotype. *Current Topics in Developmental Biology, 82*, 169–215. https://doi.org/10.1016/S0070-2153(07)00007-5

Castro-Giner, F., & Aceto, N. (2020). Tracking cancer progression: From circulating tumor cells to metastasis. *Genome Medicine, 12*(1). https://doi.org/10.1186/s13073-020-00728-3

Chakravarti, R., Aulak, K. S., Fox, P. L., & Stuehr, D. J. (2010). GAPDH regulates cellular heme insertion into inducible nitric oxide synthase. *Proceedings of the National Academy of Sciences of the United States of America, 107*(42), 18004–18009. https://doi.org/10.1073/pnas.1008133107

Chan, M. K. (2001). Recent advances in heme-protein sensors. *Current Opinion in Chemical Biology, 5*(2), 216–222. https://doi.org/10.1016/S1367-5931(00)00193-9

Chavez, A. H., Scott Coffield, K., Hasan Rajab, M., & Jo, C. (2013). Incidence rate of prostate cancer in men treated for erectile dysfunction with phosphodiesterase type 5 inhibitors: Retrospective analysis. *Asian Journal of Andrology, 15*(2), 246–248. https://doi.org/10.1038/aja.2012.162

Chegaev, K., Riganti, C., Lazzarato, L., Rolando, B., Guglielmo, S., Campia, I., Fruttero, R., Bosia, A., & Gasco, A. (2011). Nitric oxide donor doxorubicins accumulate into doxorubicin-resistant human colon cancer cells inducing cytotoxicity. *ACS Medicinal Chemistry Letters, 2*(7), 494–497. https://doi.org/10.1021/ml100302t

Chen, K., Pittman, R. N., & Popel, A. S. (2008). Nitric oxide in the vasculature: Where does it come from and where does it go? A quantitative perspective. *Antioxidants and Redox Signaling, 10*(7), 1185–1198. https://doi.org/10.1089/ars.2007.1959

Chen, Z. J., Vetter, M., Che, D., Liu, S., Tsai, M. L., & Chang, C. H. (2002). The bradykinin/soluble guanylate cyclase signaling pathway is impaired in androgen-independent prostate cancer cells. *Cancer Letters, 177*(2), 181–187. https://doi.org/10.1016/S0304-3835(01)00788-1

Chiabrando, D., Marro, S., Mercurio, S., Giorgi, C., Petrillo, S., Vinchi, F., Fiorito, V., Fagoonee, S., Camporeale, A., Turco, E., Merlo, G. R., Silengo, L., Altruda, F., Pinton, P., & Tolosano, E. (2012). The mitochondrial heme exporter FLVCR1b mediates erythroid differentiation. *Journal of Clinical Investigation, 122*(12), 4569–4579. https://doi.org/10.1172/jci62422

Chuang, A. T., Strauss, J. D., Murphy, R. A., & Steers, W. D. (1998). Sildenafil, a type-5 CGMP phosphodiesterase inhibitor, specifically amplifies endogenous cGMP-dependent relaxation in rabbit corpus cavernosum smooth muscle in vitro. *The Journal of Urology, 160*(1), 257–261. https://doi.org/10.1016/S0022-5347(01)63100-8

Cokic, V. P., Smith, R. D., Beleslin-Cokic, B. B., Njoroge, J. M., Miller, J. L., Gladwin, M. T., & Schechter, A. N. (2003). Hydroxyurea induces fetal hemoglobin by the nitric oxide-dependent activation of soluble guanylyl cyclase. *Journal of Clinical Investigation, 111*(2), 231–239. https://doi.org/10.1172/JCI200316672

Dai, Y., Faul, E. M., Ghosh, A., & Stuehr, D. J. (2022). NO rapidly mobilizes cellular heme to trigger assembly of its own receptor. *Proceedings of the National Academy of Sciences of the United States of America, 119*(4). https://doi.org/10.1073/pnas.2115774119

Dai, Y., Sweeny, E. A., Schlanger, S., Ghosh, A., & Stuehr, D. J. (2020). GAPDH delivers heme to soluble guanylyl cyclase. *Journal of Biological Chemistry, 295*(24), 8145–8154. https://doi.org/10.1074/JBC.RA120.013802

Das, A., Durrant, D., Salloum, F. N., Xi, L., & Kukreja, R. C. (2015). PDE5 inhibitors as therapeutics for heart disease, diabetes and cancer. *Pharmacology and Therapeutics, 147*, 12–21. https://doi.org/10.1016/j.pharmthera.2014.10.003

Davies, M. J., Hawkins, C. L., Pattison, D. I., & Rees, M. D. (2008). Mammalian heme peroxidases: From molecular mechanisms to health implications. *Antioxidants and Redox Signaling, 10*(7), 1199–1234. https://doi.org/10.1089/ars.2007.1927

Derbyshire, E. R., & Marletta, M. A. (2009). Biochemistry of soluble guanylate cyclase. *Handbook of Experimental Pharmacology, 191*, 17–31. https://doi.org/10.1007/978-3-540-68964-5_2

Derbyshire, E. R., & Marletta, M. A. (2012). Structure and regulation of soluble guanylate cyclase. *Annual Review of Biochemistry, 81*, 533–559. https://doi.org/10.1146/annurev-biochem-050410-100030

DeRubertis, F. R., Chayoth, R., & Field, J. B. (1976). The content and metabolism of cyclic adenosine 3′, 5′ monophosphate and cyclic guanosine 3′, 5′ monophosphate in adenocarcinoma of the human colon. *Journal of Clinical Investigation, 57*(3), 641–649. https://doi.org/10.1172/JCI108320

DeRubertis, F. R., & Craven, P. A. (1977). Stimulation of soluble guanylate cyclase activity and cellular accumulation of cyclic guanosine 3′, 5′-monophosphate by the carcinogen 4-nitroquinoline 1-oxide: Brief communication 2. *Journal of the National Cancer Institute, 59*(6), 1741–1745. https://doi.org/10.1093/jnci/59.6.1741

Domvri, K., Zarogoulidis, K., Zogas, N., Zarogoulidis, P., Petanidis, S., Porpodis, K., Kioseoglou, E., & Hohenforst-Schmidt, W. (2017). Potential synergistic effect of phosphodiesterase inhibitors with chemotherapy in lung cancer. *Journal of Cancer, 8*(18), 3648–3656. https://doi.org/10.7150/jca.21783

Donadee, C., Raat, N. J. H., Kanias, T., Tejero, J., Lee, J. S., Kelley, E. E., Zhao, X., Liu, C., Reynolds, H., Azarov, I., Frizzell, S., Meyer, E. M., Donnenberg, A. D., Qu, L., Triulzi, D., Kim-Shapiro, D. B., & Gladwin, M. T. (2011). Nitric oxide scavenging by red blood cell microparticles and cell-free hemoglobin as a mechanism for the red cell storage lesion. *Circulation, 124*(4), 465–476. https://doi.org/10.1161/CIRCULATIONAHA.110.008698

Dong, H.m., Le, Y.q., Wang, Y.h., Zhao, H.j., Huang, C.w., Hu, Y.h., Luo, L.s., Wan, X., Wei, Y.l., Chu, Z.q., Li, W., & Cai, S.x. (2017). Extracellular heat shock protein 90α mediates HDM-induced bronchial epithelial barrier dysfunction by activating RhoA/MLC signaling. *Respiratory Research, 18*(1). https://doi.org/10.1186/s12931-017-0593-y

Durante, W., Kroll, M. H., Christodoulides, N., Peyton, K. J., & Schafer, A. I. (1997). Nitric oxide induces heme oxygenase-1 gene expression and carbon monoxide production in vascular smooth muscle cells. *Circulation Research, 80*(4), 557–564. https://doi.org/10.1161/01.RES.80.4.557

Farkhondeh, T., Samarghandian, S., Hozeifi, S., & Azimi-Nezhad, M. (2017). Therapeutic effects of thymoquinone for the treatment of central nervous system tumors: A review. *Biomedicine and Pharmacotherapy, 96*, 1440–1444. https://doi.org/10.1016/j.biopha.2017.12.013

Fraser, S. T., Midwinter, R. G., Berger, B. S., & Stocker, R. (2011). Heme oxygenase-1: A critical link between iron metabolism, erythropoiesis, and development. *Advances in Hematology, 2011*. https://doi.org/10.1155/2011/473709

Fukumura, D., Kashiwagi, S., & Jain, R. K. (2006). The role of nitric oxide in tumour progression. *Nature Reviews Cancer, 6*(7), 521–534. https://doi.org/10.1038/nrc1910

Furchgott, R. F., & Zawadzki, J. V. (1980). The obligatory role of endothelial cells in the relaxation of arterial smooth muscle by acetylcholine. *Nature, 288*(5789), 373–376. https://doi.org/10.1038/288373a0

Furukawa, T., Kohno, H., Tokunaga, R., & Taketani, S. (1995). Nitric oxide-mediated inactivation of mammalian ferrochelatase in vivo and in vitro : Possible involvement of the iron-sulphur cluster of the enzyme. *Biochemical Journal, 310*(2), 533–538. https://doi.org/10.1042/bj3100533

Gali-Muhtasib, H., Roessner, A., & Schneider-Stock, R. (2006). Thymoquinone: A promising anti-cancer drug from natural sources. *The International Journal of Biochemistry and Cell Biology, 38*(8), 1249–1253. https://doi.org/10.1016/j.biocel.2005.10.009

Gantner, B. N., LaFond, K. M., & Bonini, M. G. (2020). Nitric oxide in cellular adaptation and disease. *Redox Biology, 34*. https://doi.org/10.1016/j.redox.2020.101550

Garg, U. C., Devi, L., Turndorf, H., Goldfrank, L. R., & Bansinath, M. (1992). Effect of nitric oxide on mitogenesis and proliferation of cerebellar glial cells. *Brain Research, 592*(1–2), 208–212. https://doi.org/10.1016/0006-8993(92)91678-8

Garg, U. C., & Hassid, A. (1989). Nitric oxide-generating vasodilators and 8-bromo-cyclic guanosine monophosphate inhibit mitogenesis and proliferation of cultured rat vascular smooth muscle cells. *Journal of Clinical Investigation, 83*(5), 1774–1777. https://doi.org/10.1172/JCI114081

Ghosh, A., Dai, Y., Biswas, P., & Stuehr, D. J. (2019). Myoglobin maturation is driven by the hsp90 chaperone machinery and by soluble guanylyl cyclase. *Federation of American Societies for Experimental Biology Journal, 33*(9), 9885–9896. https://doi.org/10.1096/fj.201802793RR

Ghosh, A., Garee, G., Sweeny, E. A., Nakamura, Y., & Stuehr, D. J. (2018). Hsp90 chaperones hemoglobin maturation in erythroid and nonerythroid cells. *Proceedings of the National Academy of Sciences of the United States of America, 115*(6), E1117–E1126. https://doi.org/10.1073/pnas.1717993115

Ghosh, A., Koziol-White, C. J., Asosingh, K., Cheng, G., Ruple, L., Groneberg, D., Friebe, A., Comhair, S. A., Stasch, J. P., Panettieri, R. A., Aronica, M. A., Erzurum, S. C., & Stuehr, D. J. (2016). Soluble guanylate cyclase as an alternative target for bronchodilator therapy in asthma. *Proceedings of the National Academy of Sciences of the United States of America, 113*(17), E2355–E2362. https://doi.org/10.1073/pnas.1524398113

Ghosh, A., Koziol-White, C. J., Jester, W. F., Erzurum, S. C., Asosingh, K., Panettieri, R. A., & Stuehr, D. J. (2021). An inherent dysfunction in soluble guanylyl cyclase is present in the airway of severe asthmatics and is associated with aberrant redox enzyme expression and compromised NO-cGMP signaling. *Redox Biology, 39*. https://doi.org/10.1016/j.redox.2020.101832

Ghosh, A., Sumi, M. P., Tupta, B., Okamoto, T., Aulak, K., Tsutsui, M., Shimokawa, H., Erzurum, S. C., & Stuehr, D. J. (2022). Low levels of nitric oxide promotes heme maturation into several hemeproteins and is also therapeutic. *Redox Biology, 56*. https://doi.org/10.1016/j.redox.2022.102478

Ghosh, A., Stasch, J. P., Papapetropoulos, A., & Stuehr, D. J. (2014). Nitric oxide and heat shock protein 90 activate soluble guanylate cyclase by driving rapid change in its subunit interactions and heme content. *Journal of Biological Chemistry, 289*(22), 15259–15271. https://doi.org/10.1074/jbc.M114.559393

Ghosh, A., & Stuehr, D. J. (2012). Soluble guanylyl cyclase requires heat shock protein 90 for heme insertion during maturation of the NO-active enzyme. *Proceedings of the National Academy of Sciences of the United States of America, 109*(32), 12998–13003. https://doi.org/10.1073/pnas.1205854109

Ghosh, C., Hossain, M., Solanki, J., Dadas, A., Marchi, N., & Janigro, D. (2016). Pathophysiological implications of neurovascular P450 in brain disorders. *Drug Discovery Today, 21*(10), 1609–1619. https://doi.org/10.1016/j.drudis.2016.06.004

Ghosh, S., & Erzurum, S. C. (2011). Nitric oxide metabolism in asthma pathophysiology. *Biochimica et Biophysica Acta - General Subjects, 1810*(11), 1008–1016. https://doi.org/10.1016/j.bbagen.2011.06.009

Glynos, C., Dupont, L. L., Vassilakopoulos, T., Papapetropoulos, A., Brouckaert, P., Giannis, A., Joos, G. F., Bracke, K. R., & Brusselle, G. G. (2013). The role of soluble Guanylyl Cyclase in chronic obstructive pulmonary disease. *American Journal of Respiratory and Critical Care Medicine, 188*(7), 789–799. https://doi.org/10.1164/rccm.201210-1884OC

Goldberg, N. D., Haddox, M. K., Nicol, S. E., Glass, D. B., Sanford, C. H., Kuehl, F. A., & Estensen, R. (1975). Biologic regulation through opposing influences of cyclic GMP and cyclic AMP: The yin Yang hypothesis. *Advances in Cyclic Nucleotide Research, 5*, 307–330.

Grosse, G. M., Schwedhelm, E., Worthmann, H., & Choe, C. U. (2020). Arginine derivatives in cerebrovascular diseases: Mechanisms and clinical implications. *International Journal of Molecular Sciences, 21*(5). https://doi.org/10.3390/ijms21051798

Guo, F. H., Uetani, K., Haque, S. J., Williams, B. R., Dweik, R. A., Thunnissen, F. B., Calhoun, W., & Erzurum, S. C. (1997). Interferon gamma and interleukin 4 stimulate prolonged expression of inducible nitric oxide synthase in human airway epithelium through synthesis of soluble mediators. *Journal of Clinical Investigation, 100*(4), 829–838. https://doi.org/10.1172/JCI119598

Heinemann, I. U., Jahn, M., & Jahn, D. (2008). The biochemistry of heme biosynthesis. *Archives of Biochemistry and Biophysics, 474*(2), 238–251. https://doi.org/10.1016/j.abb.2008.02.015

Hickok, J. R., & Thomas, D. D. (2010). Nitric oxide and cancer therapy: The emperor has NO clothes. *Current Pharmaceutical Design, 16*(4), 381–391. https://doi.org/10.2174/138161210790232149

Hoeijmakers, J. H. J. (2009). DNA damage, aging, and cancer. *New England Journal of Medicine, 361*(15), 1475–1485. https://doi.org/10.1056/nejmra0804615

Huang, Z., Fu, J., & Zhang, Y. (2017). Nitric oxide donor-based cancer therapy: Advances and prospects. *Journal of Medicinal Chemistry, 60*(18), 7617–7635. https://doi.org/10.1021/acs.jmedchem.6b01672

Iachini Bellisarii, F., Radico, F., Muscente, F., Horowitz, J., & De Caterina, R. (2012). Nitrates and other nitric oxide donors in cardiology - current positioning and perspectives. *Cardiovascular Drugs and Therapy, 26*(1), 55–69. https://doi.org/10.1007/s10557-011-6354-0

Ikuta, T., Sellak, H., Odo, N., Adekile, A. D., Gaensler, K. M. L., & Roberts, D. D. (2016). Nitric oxide-cGMP signaling stimulates erythropoiesis through multiple lineage-specific transcription factors: Clinical implications and a novel target for erythropoiesis. *PLoS One, 11*(1). https://doi.org/10.1371/journal.pone.0144561

Intapad, S., Dimitropoulou, C., Snead, C., Piyachaturawat, P., & Catravas, J. D. (2012). Regulation of asthmatic airway relaxation by estrogen and heat shock protein 90. *Journal of Cellular Physiology, 227*(8), 3036–3043. https://doi.org/10.1002/jcp.23045

Jing, X., Yang, F., Shao, C., Wei, K., Xie, M., Shen, H., & Shu, Y. (2019). Role of hypoxia in cancer therapy by regulating the tumor microenvironment. *Molecular Cancer, 18*(1). https://doi.org/10.1186/s12943-019-1089-9

John, M., Hoernig, S., Doehner, W., Okonko, D. D., Witt, C., & Anker, S. D. (2005). Anemia and inflammation in COPD. *Chest, 127*(3), 825–829. https://doi.org/10.1378/chest.127.3.825

Kapil, V., Khambata, R. S., Jones, D. A., Rathod, K., Primus, C., Massimo, G., Fukuto, J. M., Ahluwalia, A., & Garland, C. J. (2020). The noncanonical pathway for in vivo nitric oxide generation: The nitrate-nitrite-nitric oxide pathway. *Pharmacological Reviews, 72*(3), 692–766. https://doi.org/10.1124/pr.120.019240

Kashfi, K., & Rigas, B. (2005). Molecular targets of nitric-oxide-donating aspirin in cancer. *Biochemical Society Transactions, 33*(4), 701–704. https://doi.org/10.1042/bst0330701

Keller, C. C., Kremsner, P. G., Hittner, J. B., Misukonis, M. A., Weinberg, J. B., & Perkins, D. J. (2004). Elevated nitric oxide production in children with malarial anemia: Hemozoin-induced nitric oxide synthase type 2 transcripts and nitric oxide in blood mononuclear cells. *Infection and Immunity, 72*(8), 4868–4873. https://doi.org/10.1128/IAI.72.8.4868-4873.2004

Kim, Y. M., Bergonia, H. A., Müller, C., Pitt, B. R., Watkins, W. D., & Lancaster, J. R. (1995). Loss and degradation of enzyme-bound heme induced by cellular nitric oxide synthesis. *Journal of Biological Chemistry, 270*(11), 5710–5713. https://doi.org/10.1074/jbc.270.11.5710

Klausner, R. D., & Rouault, T. A. (1993). A double life: Cytosolic aconitase as a regulatory RNA binding protein. *Molecular Biology of the Cell, 4*(1), 1–5.

Klinger, J. R., Abman, S. H., & Gladwin, M. T. (2013). Nitric oxide deficiency and endothelial dysfunction in pulmonary arterial hypertension. *American Journal of Respiratory and Critical Care Medicine, 188*(6), 639–646. https://doi.org/10.1164/rccm.201304-0686PP

Lee, M. R., Lin, C., Lu, C. C., Kuo, S. C., Tsao, J. W., Juan, Y. N., et al. (2017). YC-1 induces G(0)/G(1) phase arrest and mitochondria-dependent apoptosis in cisplatin-resistant human oral cancer CAR cells. *Biomedicine, 7.*

Liu, Y. S., Chuang, M. T., Tsai, Y. S., Tsai, H. M., & Lin, X. Z. (2012). Nitroglycerine use in transcatheter arterial (chemo)embolization in patients with hepatocellular carcinoma and dual-energy CT assessment of lipiodol retention. *European Radiology, 22*(10), 2193–2200. https://doi.org/10.1007/s00330-012-2484-4

Loscalzo, J., & Jin. (2010). Vascular nitric oxide: Formation and function. *Journal of Blood Medicine.* https://doi.org/10.2147/JBM.S7000

Loscalzo, J. (2001). Nitric oxide insufficiency, platelet activation, and arterial thrombosis. *Circulation Research, 88*(8), 756–762. https://doi.org/10.1161/hh0801.089861

Malerba, Radaeli, A., Olivini, A., Damiani, G., Ragnoli, Montuschi, P., et al. (2014). *COPD and related comorbidities.*

Maman, S., Sagi-Assif, O., Yuan, W., Ginat, R., Meshel, T., Zubrilov, I., Keisari, Y., Lu, W., Lu, W., & Witz, I. P. (2017). The beta subunit of hemoglobin (HBB2/HBB) suppresses neuroblastoma growth and metastasis. *Cancer Research, 77*(1), 14–26. https://doi.org/10.1158/0008-5472.can-15-2929

Marletta, M. A. (1994). Nitric oxide synthase: Aspects concerning structure and catalysis. *Cell, 78*(6), 927–930. https://doi.org/10.1016/0092-8674(94)90268-2

Megson, I. L., & Webb, D. J. (2002). Nitric oxide donor drugs: Current status and future trends. *Expert Opinion on Investigational Drugs, 11*(5), 587–601. https://doi.org/10.1517/13543784.11.5.587

Milo, R., Hou, J. H., Springer, M., Brenner, M. P., & Kirschner, M. W. (2007). The relationship between evolutionary and physiological variation in hemoglobin. *Proceedings of the National Academy of Sciences of the United States of America, 104*(43), 16998–17003. https://doi.org/10.1073/pnas.0707673104

Mocellin, S., Bronte, V., & Nitti, D. (2007). Nitric oxide, a double edged sword in cancer biology: Searching for therapeutic opportunities. *Medicinal Research Reviews, 27*(3), 317–352. https://doi.org/10.1002/med.20092

Moras, M., Lefevre, S. D., & Ostuni, M. A. (2017). From erythroblasts to mature red blood cells: Organelle clearance in mammals. *Frontiers in Physiology, 8.* https://doi.org/10.3389/fphys.2017.01076

Morbidelli, L., Pyriochou, A., Filippi, S., Vasileiadis, I., Roussos, C., Zhou, Z., Loutrari, H., Waltenberger, J., Stössel, A., Giannis, A., Ziche, M., & Papapetropoulos, A. (2010). The soluble guanylyl cyclase inhibitor NS-2028 reduces vascular endothelial growth factor-induced angiogenesis and permeability. *American Journal of Physiology - Regulatory, Integrative and Comparative Physiology, 298*(3), R824–R832. https://doi.org/10.1152/ajpregu.00222.2009

Mujoo, K., Sharin, V. G., Martin, E., Choi, B.-K., Sloan, C., Nikonoff, L. E., Kots, A. Y., & Murad, F. (2010). Role of soluble guanylyl cyclase–cyclic GMP signaling in tumor cell proliferation. *Nitric Oxide, 22*(1), 43–50. https://doi.org/10.1016/j.niox.2009.11.007

Murad, F. (2006). Nitric oxide and cyclic GMP in cell signaling and drug development. *New England Journal of Medicine, 355*(19), 2003–2011. https://doi.org/10.1056/NEJMsa063904

Murillo-Carretero, M., Ruano, M. J., Matarredona, E. R., Villalobo, A., & Estrada, C. (2002). Antiproliferative effect of nitric oxide on epidermal growth factor-responsive human neuroblastoma cells. *Journal of Neurochemistry, 83*(1), 119–131. https://doi.org/10.1046/j.1471-4159.2002.01116.x

Napoli, C., Paolisso, G., Casamassimi, A., Al-Omran, M., Barbieri, M., Sommese, L., Infante, T., & Ignarro, L. J. (2013). Effects of nitric oxide on cell proliferation: Novel insights. *Journal of the American College of Cardiology, 62*(2), 89–95. https://doi.org/10.1016/j.jacc.2013.03.070

Nguyen, T., Brunson, D., Crespi, C. L., Penman, B. W., Wishnok, J. S., & Tannenbaum, S. R. (1992). DNA damage and mutation in human cells exposed to nitric oxide in vitro. *Proceedings of the National Academy of Sciences of the United States of America, 89*(7), 3030–3034. https://doi.org/10.1073/pnas.89.7.3030

Niikawa, H., Okamoto, T., Ayyat, K. S., Itoda, Y., Farver, C. F., & McCurry, K. R. (2019). The protective effect of prone lung position on ischemia–reperfusion injury and lung function in an ex vivo porcine lung model. *The Journal of Thoracic and Cardiovascular Surgery, 157*(1), 425–433. https://doi.org/10.1016/j.jtcvs.2018.08.101

Nitti, M., Furfaro, A. L., & Mann, G. E. (2020). Heme oxygenase dependent bilirubin generation in vascular cells: A role in preventing endothelial dysfunction in local tissue microenvironment? *Frontiers in Physiology, 11*. https://doi.org/10.3389/fphys.2020.00023

Okamoto, T., Wheeler, D., Farver, C. F., & McCurry, K. R. (2019). Transplant suitability of rejected human donor lungs with prolonged cold ischemia time in low-flow acellular and high-flow cellular ex vivo lung perfusion systems. *Transplantation, 103*(9), 1799–1808. https://doi.org/10.1097/TP.0000000000002667

Otterbein, L. E., Foresti, R., & Motterlini, R. (2016). Heme oxygenase-1 and carbon monoxide in the heart: The balancing act between danger signaling and pro-survival. *Circulation Research, 118*(12), 1940–1959. https://doi.org/10.1161/CIRCRESAHA.116.306588

Owusu, B. Y., Stapley, R., & Patel, R. P. (2012). Nitric oxide formation versus scavenging: The red blood cell balancing act. *Journal of Physiology, 590*(20), 4993–5000. https://doi.org/10.1113/jphysiol.2012.234906

Palmer, R. M. J., Ferrige, A. G., & Moncada, S. (1987). Nitric oxide release accounts for the biological activity of endothelium-derived relaxing factor. *Nature, 327*(6122), 524–526. https://doi.org/10.1038/327524a0

Pankey, E. A., Bhartiya, M., Badejo, A. M., Haider, U., Stasch, J. P., Murthy, S. N., Nossaman, B. D., & Kadowitz, P. J. (2011). Pulmonary and systemic vasodilator responses to the soluble guanylyl cyclase activator, BAY 60-2770, are not dependent on endogenous nitric oxide or reduced heme. *American Journal of Physiology - Heart and Circulatory Physiology, 300*(3), H792–H802. https://doi.org/10.1152/ajpheart.00953.2010

Paoli, M., Marles-Wright, J., & Smith, A. (2002). Structure–function relationships in heme-proteins. *DNA and Cell Biology, 21*(4), 271–280. https://doi.org/10.1089/104454902753759690

Papapetropoulos, A., García-Cardeña, G., Madri, J. A., & Sessa, W. C. (1997). Nitric oxide production contributes to the angiogenic properties of vascular endothelial growth factor in human endothelial cells. *Journal of Clinical Investigation, 100*(12), 3131–3139. https://doi.org/10.1172/JCI119868

Paul, T., Salazar-Degracia, A., Peinado, V. I., Tura-Ceide, O., Blanco, I., Barreiro, E., Barberà, J. A., & Su, Y. (2018). Soluble guanylate cyclase stimulation reduces oxidative stress in experimental chronic obstructive pulmonary disease. *PLoS One, 13*(1). https://doi.org/10.1371/journal.pone.0190628

Peak, T. C., Richman, A., Gur, S., Yafi, F. A., & Hellstrom, W. J. G. (2016). The role of PDE5 inhibitors and the NO/cGMP pathway in cancer. *Sexual Medicine Reviews, 4*(1), 74–84. https://doi.org/10.1016/j.sxmr.2015.10.004

Peng, T., Gong, J., Jin, Y., Zhou, Y., Tong, R., Wei, X., Bai, L., & Shi, J. (2018). Inhibitors of phosphodiesterase as cancer therapeutics. *European Journal of Medicinal Chemistry, 150*, 742–756. https://doi.org/10.1016/j.ejmech.2018.03.046

Pezzulo, A. A., Tudas, R. A., Stewart, C. G., Vargas Buonfiglio, L. G., Lindsay, B. D., Taft, P. J., Gansemer, N. D., & Zabner, J. (2019). HSP90 inhibitor geldanamycin reverts IL-13–and IL-17–induced airway goblet cell metaplasia. *Journal of Clinical Investigation, 129*(2), 744–758. https://doi.org/10.1172/JCI123524

Pinto, R. V., Fernandes, A. C., Antunes, F., Lin, Z., Rocha, J., Pires, J., & Pinto, M. L. (2019). New generation of nitric oxide-releasing porous materials: Assessment of their potential to regulate biological functions. *Nitric Oxide, 90*, 29–36. https://doi.org/10.1016/j.niox.2019.05.010

Ponka, P. (1999). Cell biology of heme. *The American Journal of the Medical Sciences, 318*(4), 241–256. https://doi.org/10.1016/s0002-9629(15)40628-7

Ponzetti, M., Capulli, M., Angelucci, A., Ventura, L., Monache, S. D., Mercurio, C., Calgani, A., Sanità, P., Teti, A., & Rucci, N. (2017). Non-conventional role of haemoglobin beta in breast malignancy. *British Journal of Cancer, 117*(7), 994–1006. https://doi.org/10.1038/bjc.2017.247

Poulos, T. L. (2005). Structural biology of heme monooxygenases. *Biochemical and Biophysical Research Communications, 338*(1), 337–345. https://doi.org/10.1016/j.bbrc.2005.07.204

Poulos, T. L. (2007). The Janus nature of heme. *Natural Product Reports, 24*(3), 504–510. https://doi.org/10.1039/b604195g

Premont, R. T., Reynolds, J. D., Zhang, R., & Stamler, J. S. (2020). Role of nitric oxide carried by hemoglobin in cardiovascular physiology: Developments on a three-gas respiratory cycle. *Circulation Research, 126*(1), 129–158. https://doi.org/10.1161/CIRCRESAHA.119.315626

Pyriochou, A., Beis, D., Koika, V., Potytarchou, C., Papadimitriou, E., Zhou, Z., & Papapetropoulos, A. (2006). Soluble guanylyl cyclase activation promotes angiogenesis. *Journal of Pharmacology and Experimental Therapeutics, 319*(2), 663–671. https://doi.org/10.1124/jpet.106.108878

Radomski, M. W., Palmer, R. M. J., & Moncada, S. (1987). Comparative pharmacology of endothelium-derived relaxing factor, nitric oxide and prostacyclin in platelets. *British Journal of Pharmacology, 92*(1), 181–187. https://doi.org/10.1111/j.1476-5381.1987.tb11310.x

Rayner, B. S., Hua, S., Sabaretnam, T., & Witting, P. K. (2009). Nitric oxide stimulates myoglobin gene and protein expression in vascular smooth muscle. *Biochemical Journal, 423*(2), 169–177. https://doi.org/10.1042/BJ20090716

Reinmuth, N., Meyer, A., Hartwigsen, D., Schaeper, C., Huebner, G., Skock-Lober, R., Bier, A., Gerecke, U., Held, C.-P., & Reck, M. (2014). Randomized, double-blind phase II study to compare nitroglycerin plus oral vinorelbine plus cisplatin with oral vinorelbine plus cisplatin alone in patients with stage IIIB/IV non-small cell lung cancer (NSCLC). *Lung Cancer, 83*(3), 363–368. https://doi.org/10.1016/j.lungcan.2014.01.001

Richard-Fogal, C. L., Frawley, E. R., Bonner, E. R., Zhu, H., San Francisco, B., & Kranz, R. G. (2009). A conserved haem redox and trafficking pathway for cofactor attachment. *EMBO Journal, 28*(16), 2349–2359. https://doi.org/10.1038/emboj.2009.189

Ryan, W. L., & Heidrick, M. L. (1968). Inhibition of cell growth in vitro by adenosine 3′,5′- monophosphate. *Science, 162*(3861), 1484–1485. https://doi.org/10.1126/science.162.3861.1484

Samama, C. M., Diaby, M., Fellahi, J. L., Mdhafar, A., Eyraud, D., Arock, M., Guillosson, J. J., Coriat, P., & Rouby, J. J. (1995). Inhibition of platelet aggregation by inhaled nitric oxide in patients with acute respiratory distress syndrome. *Anesthesiology, 83*(1), 56–65. https://doi.org/10.1097/00000542-199507000-00007

Sandner, P., Berger, P., & Zenzmaier, C. (2017). The potential of sGC modulators for the treatment of age-related fibrosis: A mini-review. *Gerontology, 63*(3), 216–227. https://doi.org/10.1159/000450946

Sandner, P. (2018). From molecules to patients: Exploring the therapeutic role of soluble guanylate cyclase stimulators. *Biological Chemistry, 399*(7), 679–690. https://doi.org/10.1515/hsz-2018-0155

Sandner, P., Zimmer, D. P., Milne, G. T., Follmann, M., Hobbs, A., & Stasch, J.-P. (2019). *Soluble guanylate cyclase stimulators and activators* (pp. 355–394). Springer Science and Business Media LLC. https://doi.org/10.1007/164_2018_197

Sibinska, Z., Tian, X., Korfei, M., Kojonazarov, B., Kolb, J. S., Klepetko, W., Kosanovic, D., Wygrecka, M., Ghofrani, H. A., Weissmann, N., Grimminger, F., Seeger, W., Guenther, A., & Schermuly, R. T. (2017). Amplified canonical transforming growth factor-β signalling via heat shock protein 90 in pulmonary fibrosis. *European Respiratory Journal, 49*(2). https://doi.org/10.1183/13993003.01941-2015

Simon, D. I., Stamler, J. S., Jaraki, O., Keaney, J. F., Osborne, J. A., Francis, S. A., Singel, D. J., & Loscalzo, J. (1993). Antiplatelet properties of protein S-nitrosothiols derived from nitric oxide and endothelium-derived relaxing factor. *Arteriosclerosis, Thrombosis, and Vascular Biology, 13*(6), 791–799. https://doi.org/10.1161/01.ATV.13.6.791

Singh, V. K., Mehrotra, S., Narayan, P., Pandey, C. M., & Agarwal, S. S. (2000). Modulation of autoimmune diseases by nitric oxide. *Immunologic Research, 22*(1), 1–19. https://doi.org/10.1385/ir:22:1:1

Siow, R. C. M., Sato, H., & Mann, G. E. (1999). Heme oxygenase-carbon monoxide signalling pathway in atherosclerosis: Anti-atherogenic actions of bilirubin and carbon monoxide? *Cardiovascular Research, 41*(2), 385–394. https://doi.org/10.1016/S0008-6363(98)00278-8

Sneddon, J. M., & Vane, J. R. (1988). Endothelium-derived relaxing factor reduces platelet adhesion to bovine endothelial cells. *Proceedings of the National Academy of Sciences, 85*(8), 2800–2804. https://doi.org/10.1073/pnas.85.8.2800

Snyder, A. H., McPherson, M. E., Hunt, J. F., Johnson, M., Stamler, J. S., & Gaston, B. (2002). Acute effects of aerosolized S-nitrosoglutathione in cystic fibrosis. *American Journal of Respiratory and Critical Care Medicine, 165*(7), 922–926. https://doi.org/10.1164/ajrccm.165.7.2105032

Stasch, J.-P., Schmidt, P., Alonso-Alija, C., Apeler, H., Dembowsky, K., Haerter, M., Heil, M., Minuth, T., Perzborn, E., Pleiss, U., Schramm, M., Schroeder, W., Schröder, H., Stahl, E., Steinke, W., & Wunder, F. (2002). NO- and haem-independent activation of soluble guanylyl cyclase: Molecular basis and cardiovascular implications of a new pharmacological principle. *British Journal of Pharmacology, 136*(5), 773–783. https://doi.org/10.1038/sj.bjp.0704778

Stasch, J. P., Pacher, P., & Evgenov, O. V. (2011). Soluble guanylate cyclase as an emerging therapeutic target in cardiopulmonary disease. *Circulation, 123*(20), 2263–2273. https://doi.org/10.1161/CIRCULATIONAHA.110.981738

Straub, A. C., Lohman, A. W., Billaud, M., Johnstone, S. R., Dwyer, S. T., Lee, M. Y., Bortz, P. S., Best, A. K., Columbus, L., Gaston, B., & Isakson, B. E. (2012). Endothelial cell expression of haemoglobin α regulates nitric oxide signalling. *Nature, 491*(7424), 473–477. https://doi.org/10.1038/nature11626

Stuehr, D. J., & Haque, M. M. (2019). Nitric oxide synthase enzymology in the 20 years after the Nobel Prize. *British Journal of Pharmacology, 176*(2), 177–188. https://doi.org/10.1111/bph.14533

Stuehr, D. J. (1999). Mammalian nitric oxide synthases. *Biochimica et Biophysica Acta - Bioenergetics, 1411*(2–3), 217–230. https://doi.org/10.1016/S0005-2728(99)00016-X

Sumi, M. P., Tupta, B., Roychowdhury, S., Comhair, S., Asosingh, K., Stuehr, D. J., Erzurum, S. C., & Ghosh, A. (2023). Hemoglobin resident in the lung epithelium is protective for smooth muscle soluble guanylate cyclase function. *Redox Biology, 63*. https://doi.org/10.1016/j.redox.2023.102717

Sun, F., Wang, Y., Luo, X., Ma, Z., Xu, Y., Zhang, X., Lv, T., Zhang, Y., Wang, M., Huang, Z., & Zhang, J. (2019). Anti-CD24 antibody–nitric oxide conjugate selectively and potently suppresses hepatic carcinoma. *Cancer Research, 79*(13), 3395–3405. https://doi.org/10.1158/0008-5472.can-18-2839

Takemoto, D. J., Dunford, C., Vaughn, D., Kramer, K. J., Smith, A., & Powell, R. G. (1982). Guanylate cyclase activity in human leukemic and normal lymphocytes. Enzyme inhibition and cytotoxicity of plant extracts. *Enzyme, 27*(3), 179–188. https://doi.org/10.1159/000459047

Thomas, D. D., Miranda, K. M., Colton, C. A., Citrin, D., Espey, M. G., & Wink, D. A. (2003). Heme proteins and nitric oxide (NO): The neglected, eloquent chemistry in NO redox signaling and regulation. *Antioxidants and Redox Signaling, 5*(3), 307–317. https://doi.org/10.1089/152308603322110887

Thomas, D. D., Ridnour, L. A., Isenberg, J. S., Flores-Santana, W., Switzer, C. H., Donzelli, S., Hussain, P., Vecoli, C., Paolocci, N., Ambs, S., Colton, C. A., Harris, C. C., Roberts, D. D., & Wink, D. A. (2008). The chemical biology of nitric oxide: Implications in cellular signaling. *Free Radical Biology and Medicine, 45*(1), 18–31. https://doi.org/10.1016/j.freeradbiomed.2008.03.020

Thomas, S., Lowe, J. E., Knowles, R. G., Green, I. C., & Green, M. H. L. (1998). Factors affecting the DNA damaging activity of superoxide and nitric oxide. *Mutation Research - Fundamental and Molecular Mechanisms of Mutagenesis, 402*(1–2), 77–84. https://doi.org/10.1016/S0027-5107(97)00284-4

Toda, N., Ayajiki, K., & Okamura, T. (2005). Nitric oxide and penile erectile function. *Pharmacology and Therapeutics, 106*(2), 233–266. https://doi.org/10.1016/j.pharmthera.2004.11.011

Tsai, E. J., & Kass, D. A. (2009). Cyclic GMP signaling in cardiovascular pathophysiology and therapeutics. *Pharmacology and Therapeutics, 122*(3), 216–238. https://doi.org/10.1016/j.pharmthera.2009.02.009

Tsiftsoglou, A. S., Tsamadou, A. I., & Papadopoulou, L. C. (2006). Heme as key regulator of major mammalian cellular functions: Molecular, cellular, and pharmacological aspects. *Pharmacology and Therapeutics, 111*(2), 327–345. https://doi.org/10.1016/j.pharmthera.2005.10.017

Tsutsui, M., Tanimoto, A., Tamura, M., Mukae, H., Yanagihara, N., Shimokawa, H., & Otsuji, Y. (2015). Significance of nitric oxide synthases: Lessons from triple nitric oxide synthases null mice. *Journal of Pharmacological Sciences, 127*(1), 42–52. https://doi.org/10.1016/j.jphs.2014.10.002

Tupta, B., Stuehr, E., Sumi, M. P., Sweeny, E. A., Smith, B., Stuehr, D. J., & Ghosh, A. (2022). GAPDH is involved in the heme-maturation of myoglobin and hemoglobin. *Federation of American Societies for Experimental Biology Journal, 36*(2). https://doi.org/10.1096/fj.202101237RR

Tuttle, T. R., Takiar, V., Kumar, B., Kumar, P., & Ben-Jonathan, N. (2017). Soluble guanylate cyclase stimulators increase sensitivity to cisplatin in head and neck squamous cell carcinoma cells. *Cancer Letters, 389*, 33–40. https://doi.org/10.1016/j.canlet.2016.12.020

Vahora, H., Khan, M. A., Alalami, U., & Hussain, A. (2016). The potential role of nitric oxide in halting cancer progression through chemoprevention. *Journal of Cancer Prevention, 21*(1), 1–12. https://doi.org/10.15430/jcp.2016.21.1.1

Verbeke, P., Fonager, J., Clark, B. F. C., & Rattan, S. I. S. (2001). Heat shock response and ageing: Mechanisms and applications. *Cell Biology International, 25*(9), 845–857. https://doi.org/10.1006/cbir.2001.0789

Visconti, R., Della Monica, R., & Grieco, D. (2016). Cell cycle checkpoint in cancer: A therapeutically targetable double-edged sword. *Journal of Experimental and Clinical Cancer Research, 35*(1). https://doi.org/10.1186/s13046-016-0433-9

Waheed, S. M., Ghosh, A., Chakravarti, R., Biswas, A., Haque, M. M., Panda, K., & Stuehr, D. J. (2010). Nitric oxide blocks cellular heme insertion into a broad range of heme proteins. *Free Radical Biology and Medicine, 48*(11), 1548–1558. https://doi.org/10.1016/j.freeradbiomed.2010.02.038

Wang, G. K., Li, S. H., Zhao, Z. M., Liu, S. X., Zhang, G. X., Yang, F., Wang, Y., Wu, F., Zhao, X. X., & Xu, Z. Y. (2016). Inhibition of heat shock protein 90 improves pulmonary arteriole remodeling in pulmonary arterial hypertension. *Oncotarget, 7*(34), 54263–54273. https://doi.org/10.18632/oncotarget.10855

Weed, D. T., Vella, J. L., Reis, I. M., De La Fuente, A. C., Gomez, C., Sargi, Z., Nazarian, R., Califano, J., Borrello, I., & Serafini, P. (2015). Tadalafil reduces myeloid-derived suppressor cells and regulatory t cells and promotes tumor immunity in patients with head and neck squamous cell carcinoma. *Clinical Cancer Research, 21*(1), 39–48. https://doi.org/10.1158/1078-0432.CCR-14-1711

Weiss, G., Werner-Felmayer, G., Werner, E. R., Grünewald, K., Wachter, H., & Hentze, M. W. (1994). Iron regulates nitric oxide synthase activity by controlling nuclear transcription. *Journal of Experimental Medicine, 180*(3), 969–976. https://doi.org/10.1084/jem.180.3.969

Wen, H. C., Chuu, C. P., Chen, C. Y., Shiah, S. G., Kung, H. J., King, K. L., Su, L. C., Chang, S. C., & Chang, C. H. (2015). Elevation of soluble guanylate cyclase suppresses proliferation and survival of human breast cancer cells. *PLoS One, 10*(4). https://doi.org/10.1371/journal.pone.0125518

West, A. R., & Oates, P. S. (2008). Mechanisms of heme iron absorption: Current questions and controversies. *World Journal of Gastroenterology, 14*(26), 4101–4110. https://doi.org/10.3748/wjg.14.4101

Wijayanti, N., Katz, N., & Immenschuh, S. (2004). Biology of heme in health and disease. *Current Medicinal Chemistry, 11*(8), 981–986. https://doi.org/10.2174/0929867043455521

Wink, D. A., Hines, H. B., Cheng, R. Y. S., Switzer, C. H., Flores-Santana, W., Vitek, M. P., Ridnour, L. A., & Colton, C. A. (2011). Nitric oxide and redox mechanisms in the immune response. *Journal of Leukocyte Biology, 89*(6), 873–891. https://doi.org/10.1189/jlb.1010550

Yang, W. D., Ando, J., Korenaga, R., Toyooka, T., & Kamiya, A. (1994). Exogenous nitric oxide inhibits proliferation of cultured vascular endothelial cells. *Biochemical and Biophysical Research Communications, 203*(2), 1160–1167. https://doi.org/10.1006/bbrc.1994.2304

Yasuda, H., Yamaya, M., Nakayama, K., Sasaki, T., Ebihara, S., Kanda, A., Asada, M., Inoue, D., Suzuki, T., Okazaki, T., Takahashi, H., Yoshida, M., Kaneta, T., Ishizawa, K., Yamanda, S., Tomita, N., Yamasaki, M., Kikuchi, A., Kubo, H., & Sasaki, H. (2006). Randomized phase II trial comparing nitroglycerin plus vinorelbine and cisplatin with vinorelbine and cisplatin alone in previously untreated stage IIIB/IV non–small-cell lung cancer. *Journal of Clinical Oncology, 24*(4), 688–694. https://doi.org/10.1200/jco.2005.04.0436

Yi, L., Jenkins, P. M., Leichert, L. I., Jakob, U., Martens, J. R., & Ragsdale, S. W. (2009). Heme regulatory motifs in heme oxygenase-2 form a thiol/disulfide redox switch that responds to the cellular redox state. *Journal of Biological Chemistry, 284*(31), 20556–20561. https://doi.org/10.1074/jbc.M109.015651

Zhang, R., Wang, X. J., Zhang, H. D., Sun, X. Q., Zhao, Q. H., Wang, L., He, J., Jiang, X., Liu, J. M., & Jing, Z. C. (2016). Profiling nitric oxide metabolites in patients with idiopathic pulmonary arterial hypertension. *European Respiratory Journal, 48*(5), 1386–1395. https://doi.org/10.1183/13993003.00245-2016

Zheng, Y., Miyamoto, D. T., Wittner, B. S., Sullivan, J. P., Aceto, N., Jordan, et al. (2017). Expression of beta-globin by cancer cells promotes cell survival during blood-borne dissemination. *Nature Communications, 8*.

Zhong, X., Zhang, H., Zhu, Y., Liang, Y., Yuan, Z., Li, J., Li, J., Li, X., Jia, Y., He, T., Zhu, J., Sun, Y., Jiang, W., Zhang, H., Wang, C., & Ke, Z. (2020). Circulating tumor cells in cancer patients: Developments and clinical applications for immunotherapy. *Molecular Cancer, 19*(1). https://doi.org/10.1186/s12943-020-1141-9

Zhu, H., Li, J. T., Zheng, F., Martin, E., Kots, A. Y., Krumenacker, J. S., Choi, B. K., McCutcheon, I. E., Weisbrodt, N., Bögler, O., Murad, F., & Bian, K. (2011). Restoring soluble guanylyl cyclase expression and function blocks the aggressive course of glioma. *Molecular Pharmacology, 80*(6), 1076–1084. https://doi.org/10.1124/mol.111.073585

Zhu, Y., & Silverman, R. B. (2008). Revisiting heme mechanisms. A perspective on the mechanisms of nitric oxide synthase (NOS), heme oxygenase (HO), and cytochrome P450s (CYP450s). *Biochemistry, 47*(8), 2231–2243. https://doi.org/10.1021/bi7023817

PART VI

Educational session

CHAPTER

12

Using artificial intelligence to discover new cancer therapies

Apostolos Zaravinos[1,2]

[1]Department of Life Sciences, School of Sciences, European University Cyprus, Nicosia, Cyprus; [2]Cancer Genetics, Genomics and Systems Biology Laboratory, Basic and Translational Cancer Research Center (BTCRC), Nicosia, Cyprus

BTCRC BASIC & TRANSLATIONAL CANCER RESEARCH CENTER

European University Cyprus

HR EXCELLENCE IN RESEARCH

Therapeutic Applications of Nitric Oxide in Cancer and Inflammatory Disorders
https://doi.org/10.1016/B978-0-443-22354-9.00012-7

"Artificial intelligence (AI), the ability of machines to learn from new input, is a broad term for a range of computing methods"

AI is changing drug discovery

- Why we need AI
- What are AI models
- Advantages
- Present status
- Challenges

**AI/ML in drug development,
Need?**

"Evolution is not a force but a process"

- High attrition rate
- Increased expenditure
- Increase volume of data
- Increase global regulatory requirement
- Challenge in timely processing

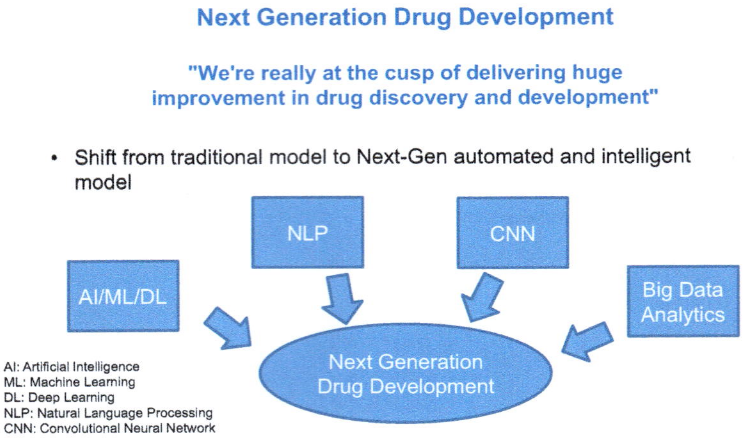
Next Generation Drug Development
"We're really at the cusp of delivering huge improvement in drug discovery and development"
• Shift from traditional model to Next-Gen automated and intelligent model
NLP
CNN
AI/ML/DL
Big Data Analytics
Next Generation Drug Development
AI: Artificial Intelligence
ML: Machine Learning
DL: Deep Learning
NLP: Natural Language Processing
CNN: Convolutional Neural Network

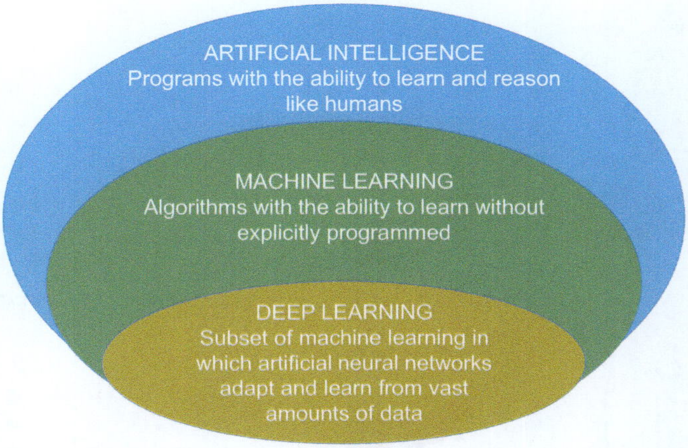
ARTIFICIAL INTELLIGENCE
Programs with the ability to learn and reason like humans
MACHINE LEARNING
Algorithms with the ability to learn without explicitly programmed
DEEP LEARNING
Subset of machine learning in which artificial neural networks adapt and learn from vast amounts of data

MACHINE
LEARNING
HOW
TO

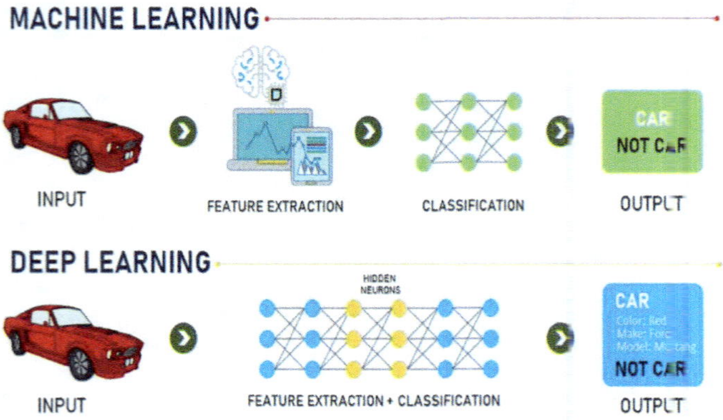
MACHINE LEARNING
INPUT
FEATURE EXTRACTION
CLASSIFICATION
OUTPUT
CAR
NOT CAR
DEEP LEARNING
HIDDEN
NEURONS
INPUT
FEATURE EXTRACTION + CLASSIFICATION
OUTPUT
CAR
NOT CAR

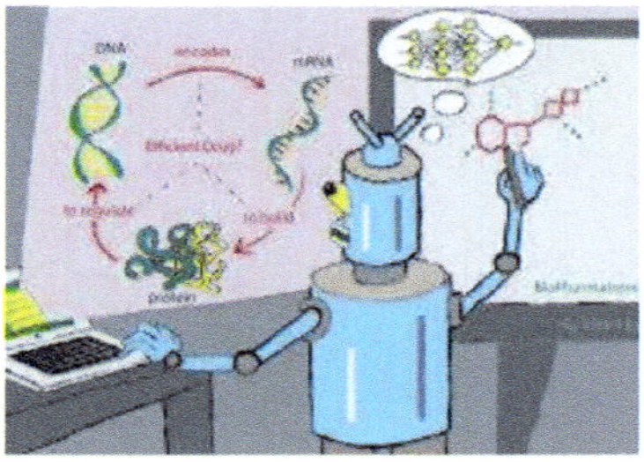
DNA
mRNA

Table 2 Widely used AI techniques in drug discovery

Category	Task	Method	Representative application	Reference
Supervised learning	Regression analysis	MLR	DTI	Talevi et al.[29]
		DT	Adverse drug reactions	Hammann et al.[33]
		LR	Drug-drug interaction	Schober and Vetter[34]
	Classification	SVM	Compound classification	Maltarollo et al.[35]
		CNN	Bioactivity prediction	El-Attar et al.[36]
		RNN	*De novo* drug design	Gupta et al.[37]
		GAN	Molecule discovery	Blanchard et al.[38]
Unsupervised learning	Clustering	*k*-means	Drug candidate selection	Shen et al.[39]
		Hierarchical	Molecular scaffold analysis	Manelfi et al.[40]
	Dimension reduction	PCA	QSAR	Yoo and Shahlaei[41]
		t-SNE	Chemical space mapping	Karlov et al.[42]

CNN, convolution neural network; DT, drug target; GAN, generative adversarial network; LR, logistic regression; MLR, multiple linear regression; PCA, principal-component analysis; RNN, recurrent neural network; SVM, support vector machine; t-SNE, T-distributed stochastic neighbor embedding.

Chen W et al. Mol. Ther. Nucleic Acids (2023)

Role of AI in Drug-Discovery

- Better compounds going into clinical trials
- Better target understanding
- Better conductance of trials
- Better Post-Marketing surveillance

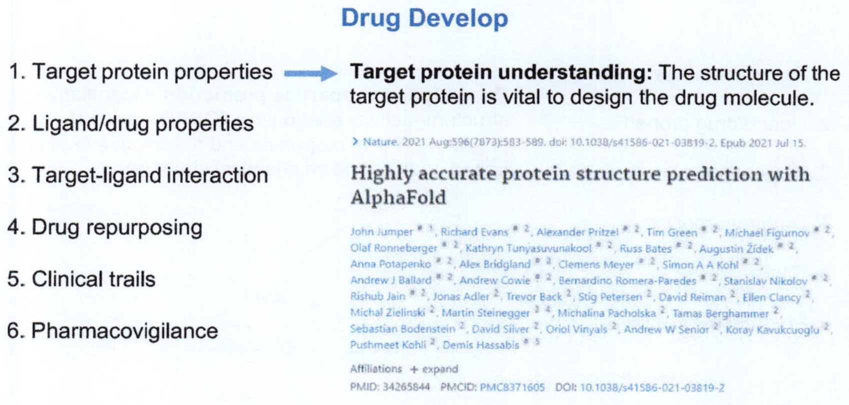
Drug Develop
1. Target protein properties
2. Ligand/drug properties
3. Target-ligand interaction
4. Drug repurposing
5. Clinical trails
6. Pharmacovigilance
Target protein understanding: The structure of the target protein is vital to design the drug molecule.
Nature. 2021 Aug;596(7873):583-589. doi: 10.1038/s41586-021-03819-2. Epub 2021 Jul 15.
Highly accurate protein structure prediction with AlphaFold
John Jumper [# 1], Richard Evans [# 2], Alexander Pritzel [# 2], Tim Green [# 2], Michael Figurnov [# 2], Olaf Ronneberger [# 2], Kathryn Tunyasuvunakool [# 2], Russ Bates [# 2], Augustin Žídek [# 2], Anna Potapenko [# 2], Alex Bridgland [# 2], Clemens Meyer [# 2], Simon A A Kohl [# 2], Andrew J Ballard [# 2], Andrew Cowie [# 2], Bernardino Romera-Paredes [# 2], Stanislav Nikolov [# 2], Rishub Jain [# 2], Jonas Adler [2], Trevor Back [2], Stig Petersen [2], David Reiman [2], Ellen Clancy [2], Michal Zielinski [2], Martin Steinegger [3 4], Michalina Pacholska [2], Tamas Berghammer [2], Sebastian Bodenstein [2], David Silver [2], Oriol Vinyals [2], Andrew W Senior [2], Koray Kavukcuoglu [2], Pushmeet Kohli [2], Demis Hassabis [# 5]
Affiliations + expand
PMID: 34265844 PMCID: PMC8371605 DOI: 10.1038/s41586-021-03819-2

Drug Develop

1. Target protein properties
2. Ligand/drug properties
3. Target-ligand interaction
4. Drug repurposing
5. Clinical trails
6. Pharmacovigilance

Drug/Ligand properties prediction: Quantitative structure–activity relationship (QSAR) - predict physicochemical properties and biological activities based on their known chemical structure

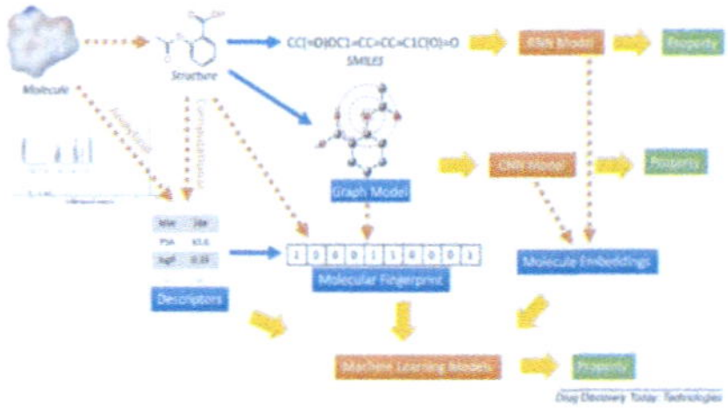

TABLE 1 Targets, selected descriptors, and statistics (classification accuracy, sensitivity, specificity) for selected models

Target	Descriptors	Statistics
Carcinogenicity prediction[a]	H bond acceptors H bond donors Content of H and C	Sensitivity 89.6%, specificity 95.8%, accuracy 92.09%
	H bond acceptor H bond donors Rotatable bonds Polarizability Polar surface area	Sensitivity 35.1%, specificity 88.3%, accuracy 69.3%
Oral acute toxicity	Molecular fingerprint	Accuracy 95.5%
		Accuracy 71%
	Molecular fingerprint (e.g., atom pairs, topological torsion, substructure, hybridization)	Sensitivity 83.9%, specificity 99.6%, accuracy 82%
Cardiotoxicity	Molecular fingerprint and 2D ChemoPy[83] descriptors (e.g., connectivity, topology, Kappa, Burden) or MOE 2D descriptors (surface areas, connectivity, shape indices, atom, and bond counts)	Accuracy 78%
	Molecular fingerprint	Sensitivity 78%, specificity 61%
	Molecular fingerprint and principal component analysis (PCA) (e.g., H bond acceptors, H bond donors, rotatable bond number, number of rings and aromatic rings, molecular fractional polar surface area)	Accuracy 87%
	Molecular descriptor Molecular fingerprint Molecular graph-based features (atom types, number of degrees, number of bound hydrogens, implicit valence, size of ring containing the atom, and	Sensitivity 83.3%, accuracy 77.3%

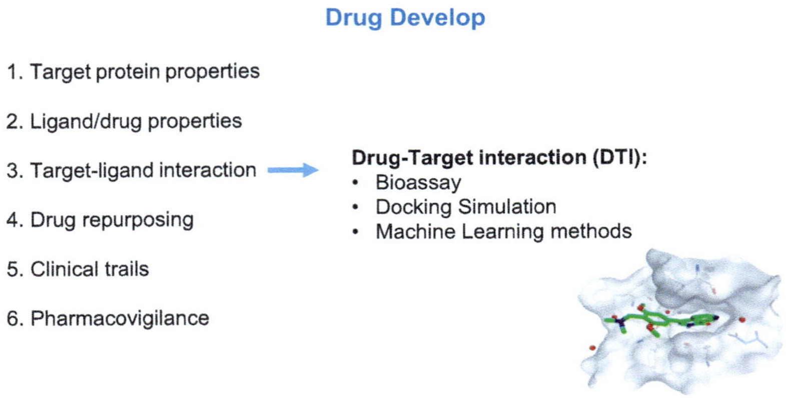
Drug Develop
1. Target protein properties
2. Ligand/drug properties
3. Target-ligand interaction
4. Drug repurposing
5. Clinical trails
6. Pharmacovigilance
Drug-Target interaction (DTI):
• Bioassay
• Docking Simulation
• Machine Learning methods

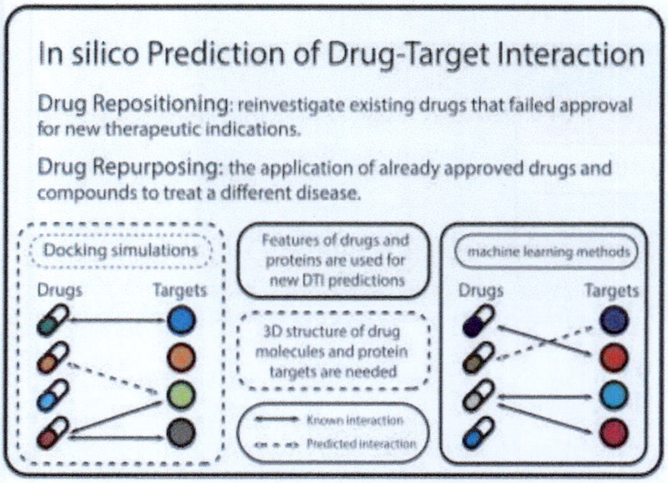

Bagherian M, et al. Briefings in Bioinformatics (2021)

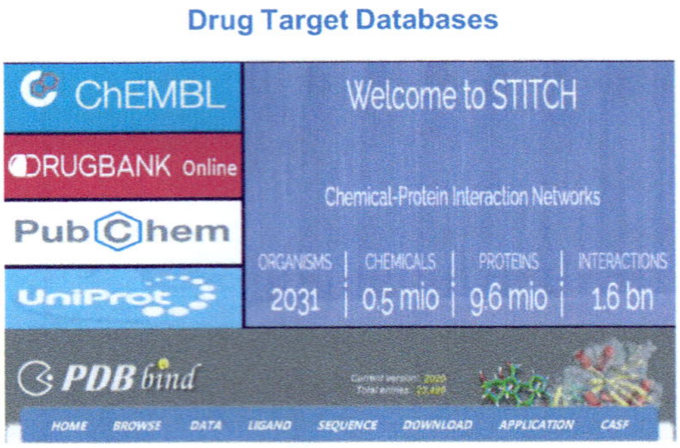
Drug Target Databases
ChEMBL
DRUGBANK Online
PubChem
UniProt
Welcome to STITCH
Chemical-Protein Interaction Networks
ORGANISMS 2031
CHEMICALS 0.5 mio
PROTEINS 9.6 mio
INTERACTIONS 1.6 bn
PDBbind
HOME BROWSE DATA LIGAND SEQUENCE DOWNLOAD APPLICATION CASF

Framework of AI technique application to drug discovery and evaluation

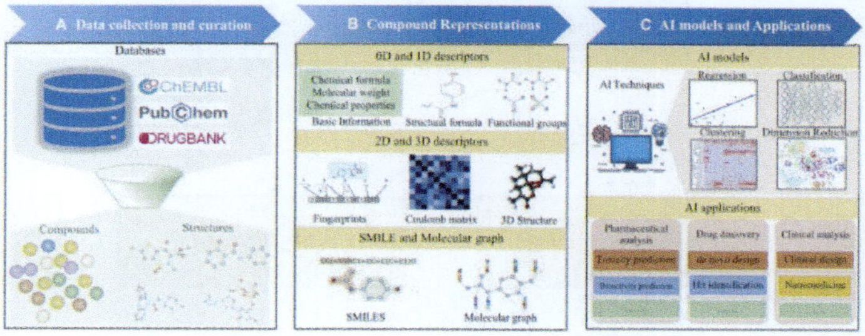

Chen W et al. Mol. Ther. Nucleic Acids (2023)

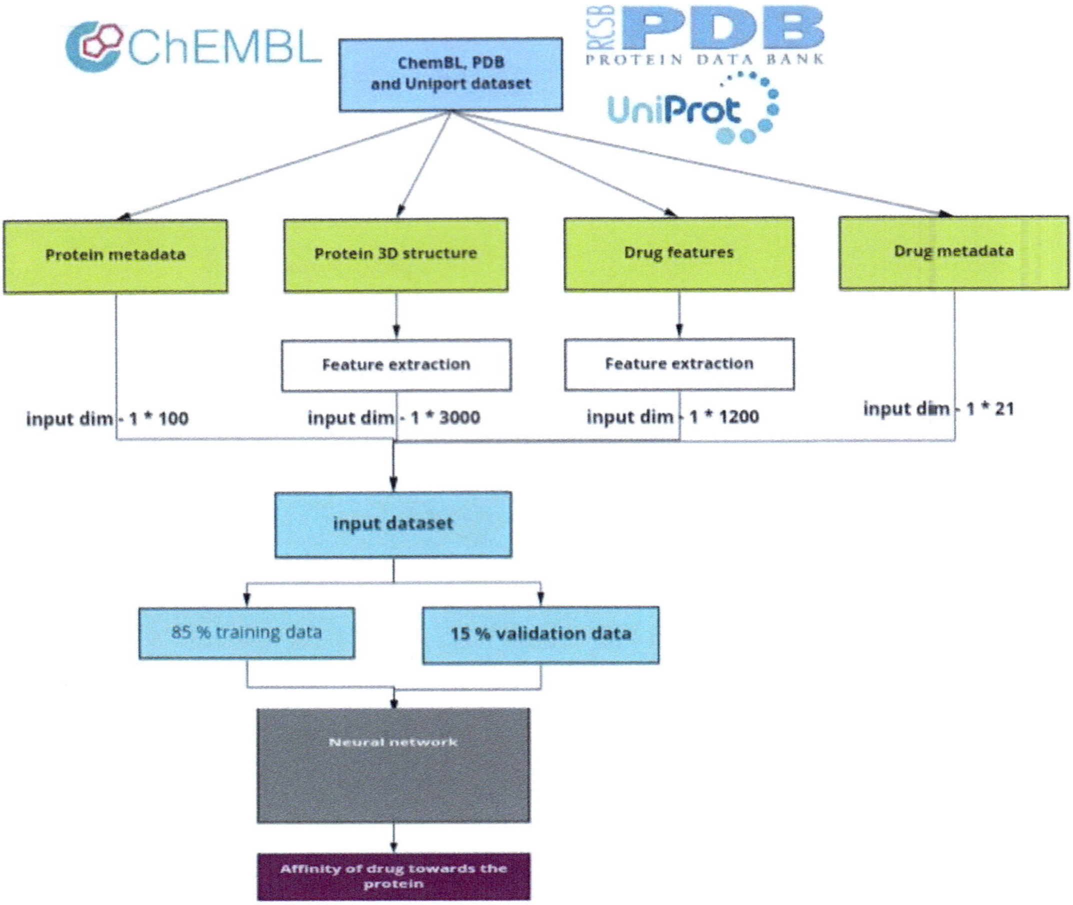
ChEMBL
ChemBL, PDB
and Uniport dataset
RCSB PDB
PROTEIN DATA BANK
UniProt
Protein metadata
Protein 3D structure
Drug features
Drug metadata
Feature extraction
Feature extraction
input dim - 1 * 100
input dim - 1 * 3000
input dim - 1 * 1200
input dim - 1 * 21
input dataset
85 % training data
15 % validation data
Neural network
Affinity of drug towards the protein

Drug Develop

1. Target protein properties
2. Ligand/drug properties
3. Target-ligand interaction
4. Drug repurposing →
   - Drug repositioning, drug retasking, drug reprofiling, drug rescuing, drug recycling, drug redirection, and therapeutic switching
   - Process of identifying new therapeutic use(s) for old/existing/available drugs
   - Highly efficient, time saving, low-cost and minimum risk of failure, increases the success rate
5. Clinical trails
6. Pharmacovigilance

Drug Develop

1. Target protein properties
2. Ligand/drug properties
3. Target-ligand interaction
4. Drug repurposing
5. Clinical trails →
    - Half of time and investment
    - 86% of all trials do not meet enrolment timelines
    - 1/3rd of all Phase III trials fail owing to enrolment problems
    - Patient recruitment takes up 1/3rd of duration
    - A 32% failure rate because of patient recruitment problems
6. Pharmacovigilance

AI models used to enhance patient cohort selection

- By reducing population heterogeneity
- Prognostic enrichment
- Predictive enrichment
- Electronic phenotyping

AI techniques used to automatically analyze Electronic Medical Record (EMR) and clinical trial eligibility databases, find matches between specific patients and recruiting trials, and recommend these matches to doctors and patients

Predict the risk of dropout for a specific patient

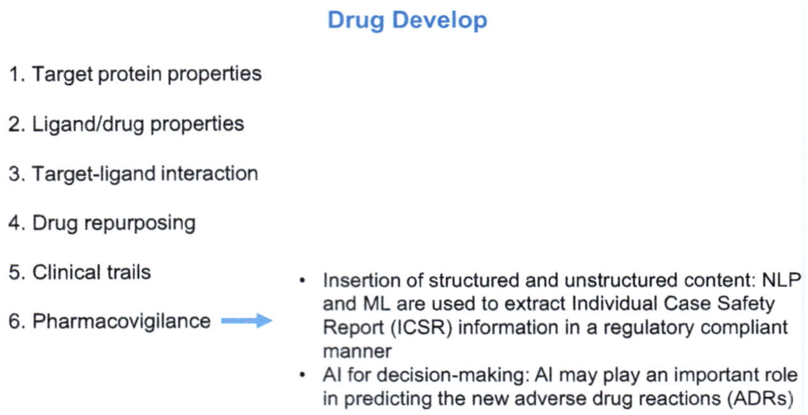
Drug Develop
1. Target protein properties
2. Ligand/drug properties
3. Target-ligand interaction
4. Drug repurposing
5. Clinical trails
6. Pharmacovigilance
• Insertion of structured and unstructured content: NLP and ML are used to extract Individual Case Safety Report (ICSR) information in a regulatory compliant manner
• AI for decision-making: AI may play an important role in predicting the new adverse drug reactions (ADRs)

NLP example 1 - input

Input Text: The 82-year-old Chinese patient was ADM to the hospital last week because she wasn't able to alternate between sitting and standing the whole day. She also had a swelling on knee joint. Her entire left foot was in bad shape due to burn. She experienced short of breath. After examination, COVID-19 ruled out. The test reported HbA1c 8.2%, temperature 100, blood pressure 129/95, respiration 12, pulse 60, wbc count 8.4, AST 349, ALT 186, ALP 155 and lipase 239. Her weight was 240 lbs and height 5 feet 7 inches. The patient indicated using Betaxolol once a while. She was recommended zinc 50 oral capsule twice daily for 2 weeks and ACETAZOLAMIDE 250 mg tab once daily for 10 days . After a few days under medical care she was discharged from hospital this morning at 10 am as she was able to alternate between sitting and standing comfortably.

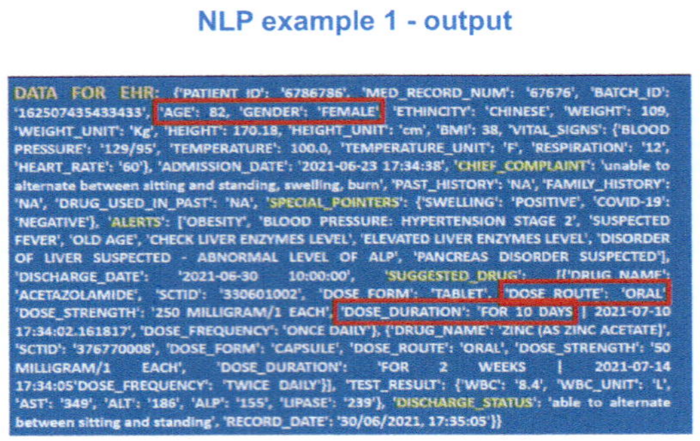
NLP example 1 - output
DATA FOR EHR: {'PATIENT ID': '6786786', 'MED_RECORD_NUM': '67676', 'BATCH_ID': '162507435433433', 'AGE': 82, 'GENDER': 'FEMALE' 'ETHINCITY': 'CHINESE', 'WEIGHT': 109, 'WEIGHT_UNIT': 'Kg', 'HEIGHT': 170.18, 'HEIGHT_UNIT': 'cm', 'BMI': 38, 'VITAL_SIGNS': {'BLOOD PRESSURE': '129/95', 'TEMPERATURE': 100.0, 'TEMPERATURE_UNIT': 'F', 'RESPIRATION': '12', 'HEART_RATE': '60'}, 'ADMISSION_DATE': '2021-06-23 17:34:38', 'CHIEF_COMPLAINT': 'unable to alternate between sitting and standing, swelling, burn', 'PAST_HISTORY': 'NA', 'FAMILY_HISTORY': 'NA', 'DRUG_USED_IN_PAST': 'NA', 'SPECIAL_POINTERS': {'SWELLING': 'POSITIVE', 'COVID-19': 'NEGATIVE'}, 'ALERTS': ['OBESITY', 'BLOOD PRESSURE: HYPERTENSION STAGE 2', 'SUSPECTED FEVER', 'OLD AGE', 'CHECK LIVER ENZYMES LEVEL', 'ELEVATED LIVER ENZYMES LEVEL', 'DISORDER OF LIVER SUSPECTED - ABNORMAL LEVEL OF ALP', 'PANCREAS DISORDER SUSPECTED'], 'DISCHARGE_DATE': '2021-06-30 10:00:00', 'SUGGESTED_DRUG': [{'DRUG_NAME': 'ACETAZOLAMIDE', 'SCTID': '330601002', 'DOSE_FORM': 'TABLET', 'DOSE_ROUTE': 'ORAL' 'DOSE_STRENGTH': '250 MILLIGRAM/1 EACH', 'DOSE_DURATION': 'FOR 10 DAYS | 2021-07-10 17:34:02.161817', 'DOSE_FREQUENCY': 'ONCE DAILY'}, {'DRUG_NAME': 'ZINC (AS ZINC ACETATE)', 'SCTID': '376770008', 'DOSE_FORM': 'CAPSULE', 'DOSE_ROUTE': 'ORAL', 'DOSE_STRENGTH': '50 MILLIGRAM/1 EACH', 'DOSE_DURATION': 'FOR 2 WEEKS | 2021-07-14 17:34:05'DOSE_FREQUENCY': 'TWICE DAILY'}], 'TEST_RESULT': {'WBC': '8.4', 'WBC_UNIT': 'L', 'AST': '349', 'ALT': '186', 'ALP': '155', 'LIPASE': '239'}, 'DISCHARGE_STATUS': 'able to alternate between sitting and standing', 'RECORD_DATE': '30/06/2021, 17:35:05'}}

Advantages

- Improve the quality and accuracy
- Cost-effective
- Timely manner
- Can handle diverse types data formats
- Transparent data sharing with regulators, prescribers
- Prediction of new ADR

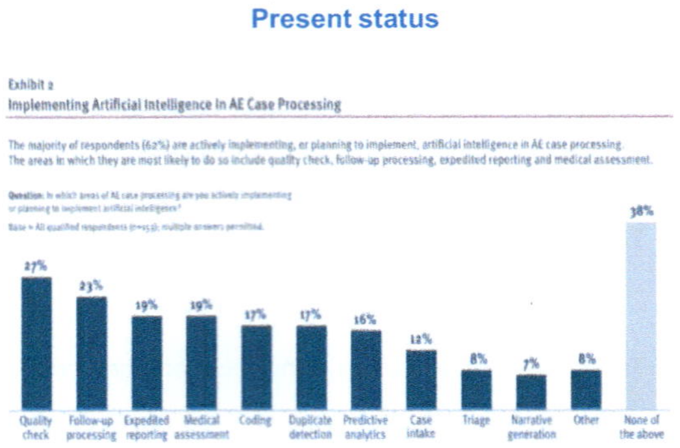
Present status
Exhibit 2
Implementing Artificial Intelligence In AE Case Processing
The majority of respondents (62%) are actively implementing, or planning to implement, artificial intelligence in AE case processing.
The areas in which they are most likely to do so include quality check, follow-up processing, expedited reporting and medical assessment.
27%
23%
19%
19%
17%
17%
16%
12%
8%
7%
8%
38%
Quality check
Follow-up processing
Expedited reporting
Medical assessment
Coding
Duplicate detection
Predictive analytics
Case intake
Triage
Narrative generation
Other
None of the above

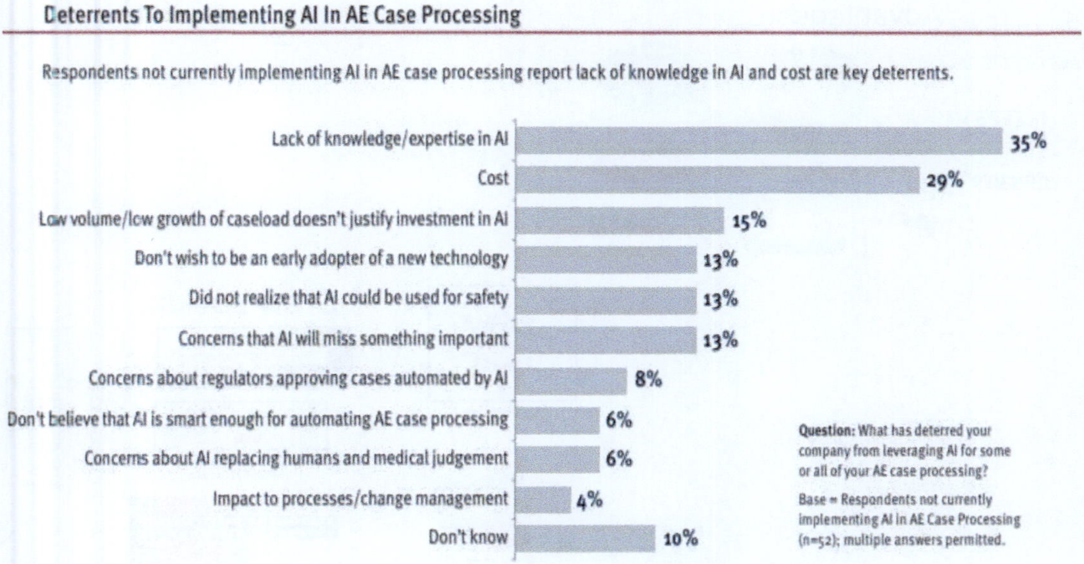
Deterrents To Implementing AI In AE Case Processing
Respondents not currently implementing AI in AE case processing report lack of knowledge in AI and cost are key deterrents.
Lack of knowledge/expertise in AI
35%
Cost
29%
Low volume/low growth of caseload doesn't justify investment in AI
15%
Don't wish to be an early adopter of a new technology
13%
Did not realize that AI could be used for safety
13%
Concerns that AI will miss something important
13%
Concerns about regulators approving cases automated by AI
8%
Don't believe that AI is smart enough for automating AE case processing
6%
Concerns about AI replacing humans and medical judgement
6%
Impact to processes/change management
4%
Don't know
10%
Question: What has deterred your company from leveraging AI for some or all of your AE case processing?
Base = Respondents not currently implementing AI in AE Case Processing (n=52); multiple answers permitted.

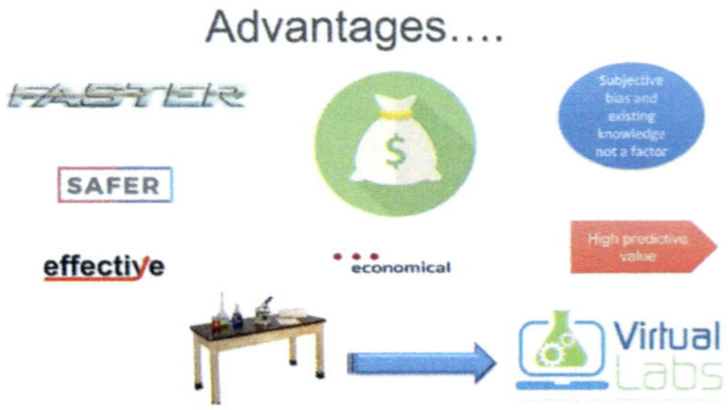
Advantages....
FASTER
SAFER
effective
economical
Subjective bias and existing knowledge not a factor
High predictive value
Virtual Labs

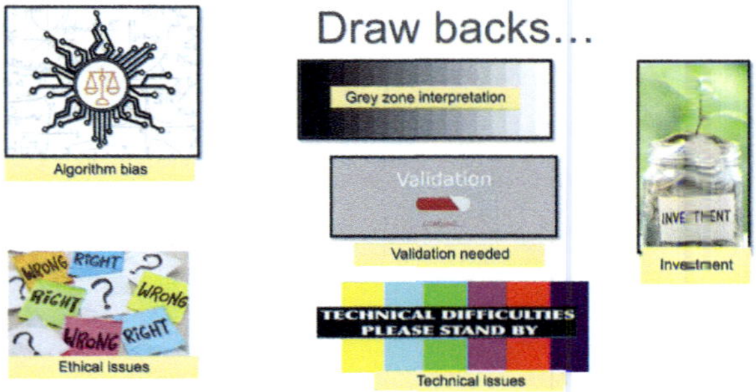
Draw backs...
Algorithm bias
Grey zone interpretation
Validation
Validation needed
INVESTMENT
Investment
WRONG
RIGHT
Ethical issues
TECHNICAL DIFFICULTIES PLEASE STAND BY
Technical issues

Challenges

- Not applicable on monocausal diseases
- Significant data challenges, such as the scale, growth, diversity, and uncertainty of the data.

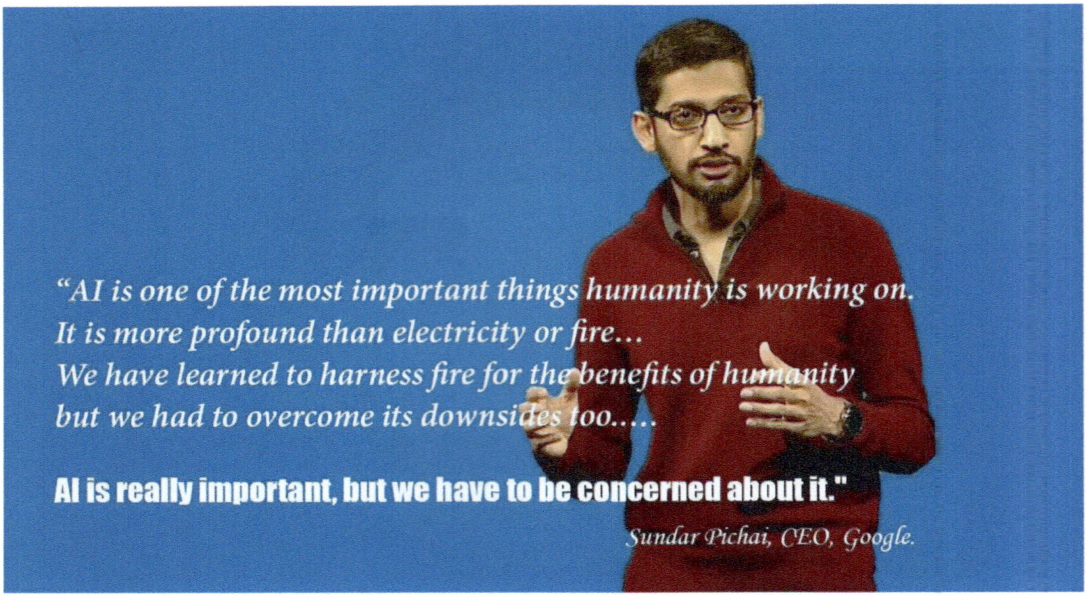
"AI is one of the most important things humanity is working on.
It is more profound than electricity or fire...
We have learned to harness fire for the benefits of humanity
but we had to overcome its downsides too.....
AI is really important, but we have to be concerned about it."
Sundar Pichai, CEO, Google.

Thank you!
Organizers
Valentina Rapozzi
Luigi E. Xodo
Benjamin Bonavida
Co-organizer
Massimo Libra
UDINE 2023
Third International Conference
Therapeutic Applications of Nitric Oxide in Cancer and Inflammatory-Related Disorders
4-5 September 2023
Organizers: Valentina Rapozzi, Luigi Xodo, & Benjamin Bonavida
Co-Organizer: Massimo Libra
UNIVERSITÀ DEGLI STUDI DI UDINE
Università di Catania

Index

Note: "Page numbers followed by *f* indicate figures and *t* indicate tables."

T

U

V

Printed and bound by CPI Group (UK) Ltd, Croydon, CR0 4YY
19/11/2024
01791103-0004